ENVIRONMENTAL TECHNOLOGY AND SUSTAINABILITY

Physical, Chemical and Biological Technologies for Environmental Protection

ENVIRONMENTAL TECHNOLOGY AND SUSTAINABILITY

Physical, Chemical and Biological Technologies for Environmental Protection

Edited by

Tamara Tatrishvili, PhD

Ann Rose Abraham, PhD

A. K. Haghi, PhD

AAP | APPLE ACADEMIC PRESS

First edition published 2024

Apple Academic Press Inc.
1265 Goldenrod Circle, NE,
Palm Bay, FL 32905 USA

760 Laurentian Drive, Unit 19,
Burlington, ON L7N 0A4, CANADA

CRC Press
2385 NW Executive Center Drive,
Suite 320, Boca Raton FL 33431

4 Park Square, Milton Park,
Abingdon, Oxon, OX14 4RN UK

Library and Archives Canada Cataloguing in Publication

Title: Environmental technology and sustainability : physical, chemical and biological technologies for environmental protection / edited by Tamara Tatrishvili, PhD, Ann Rose Abraham, PhD, A. K. Haghi, PhD.
Names: Tatrishvili, Tamara, editor. | Abraham, Ann Rose, editor. | Haghi, A. K., editor.
Description: First edition. | Includes bibliographical references and index.
Identifiers: Canadiana (print) 20230568165 | Canadiana (ebook) 20230568181 | ISBN 9781774914342 (hardcover) | ISBN 9781774914359 (softcover) | ISBN 9781003397960 (ebook)
Subjects: LCSH: Green technology. | LCSH: Environmental protection.
Classification: LCC TD145 .E65 2024 | DDC 628—dc23

ISBN: 978-1-77491-434-2 (hbk)
ISBN: 978-1-77491-435-9 (pbk)
ISBN: 978-1-00339-796-0 (ebk)

About the Editors

Tamara Tatrishvili, PhD
Assistant Professor; Main Specialist, Office of the Academic Process Management, Ivane Javakhishvili Tbilisi State University (TSU); Director of the Institute of Macromolecular Chemistry and Polymeric Materials at TSU, Tbilisi, Georgia

Tamara Tatrishvili, PhD, Assistant Professor, is the main specialist in the Office of Academic Process Management (Faculty of Exact and Natural Sciences) at Ivane Javakhishvili Tbilisi State University, as well as Director of the Institute of Macromolecular Chemistry and Polymeric Materials at TSU, Tbilisi, Georgia. Dr. Tatrishvili is a DAAD alumnus, and a member of Georgian Chemical Society. Her research interests included polymer chemistry, polymeric materials, and chemistry of silicon-organic compounds. She is the Executive Editor of the *Journal of the Georgian Chemical Society*. Dr. Tatrishvili is the author of more than 190 scientific publications, 12 books, and monographs.

Ann Rose Abraham, PhD
Assistant Professor, Sacred Heart College (Autonomous), Thevara, Kochi, Kerala, India

Ann Rose Abraham, PhD, is currently an Assistant Professor at the Department of Physics, Sacred Heart College (Autonomous), Thevara, Kochi, Kerala, India. She has expertise in the field of condensed matter physics, nanomagnetism, multiferroics, and polymeric nanocomposites, etc. She has research experience at various reputed national institutes, including Bose Institute, Kolkata, India; SAHA Institute of Nuclear Physics, Kolkata, India; UGC-DAE CSR Centre, Kolkata, India; and she has collaborated with various international laboratories. She is a recipient of a Young Researcher Award in physics and Best Paper Awards 2020–2021. She served as assistant professor and examiner at the Department of Basic Sciences, Amal Jyothi College of Engineering, under APJ Abdul Kalam Technological University, Kerala, India. Dr. Abraham is a frequent speaker at national and international conferences. She has a good number of publications to her credit in many peer-reviewed high impact journals of international repute. She has authored many book chapters and edited more than 10 books with Taylor and Francis, Elsevier, etc. Dr. Abraham received her MSc, MPhil, and PhD degrees in Physics from the School of Pure and Applied Physics, Mahatma Gandhi University, Kerala, India. Her PhD thesis was on the title "Development of Hybrid Mutliferroic Materials for Tailored Applications."

A. K. Haghi, PhD

Professor Emeritus of Engineering Sciences, Former Editor-in-Chief, International Journal of Chemoinformatics and Chemical Engineering; Member, Canadian Research and Development Center of Sciences and Culture

A. K. Haghi, PhD, has published over 250 academic research-oriented books as well as over 1000 research papers published in various journals and conference proceedings. He has received several grants, consulted for several major corporations, and is a frequent speaker to national and international audiences. He is founder and former Editor-in-Chief of the *International Journal of Chemoinformatics and Chemical Engineering*, published by IGI Global (USA), as well as *Polymers Research Journal*, published by Nova Science Publishers (USA). Professor Haghi has acted as an editorial board member of many international journals. He has served as a member of the Canadian Research and Development Center of Sciences and Cultures (CRDCSC) and the Research Chemistry Centre, Coimbra, Portugal. Dr. Haghi holds a BSc in urban and environmental engineering from the University of North Carolina (USA), an MSc in mechanical engineering from North Carolina A&T State University (USA) and an MSc in applied mechanics, acoustics, and materials from the Université de Technologie de Compiègne (France), and a PhD in engineering sciences from the Université de Franche-Comté (France).

Contents

Contributors

Sh. Akhobadze
Institute of Cybernetics of Georgian Technical University, Tbilisi, Georgia
Department of Polymer Materials, Tbilisi State University, Tbilisi, Georgia

Olga V. Alekseeva
G. A. Krestov Institute of Solution Chemistry, Russian Academy of Sciences, Ivanovo, Russia

K. G. Ambady
Department of Special Education, National Institute for the Empowerment of Persons with Intellectual Disabilities, Telengana, India

Thomas Abraham
Catholicate College, Pathanamthitta, Kerala, India

Zurab Amiridze
Petre Melikishvili Institute of Physical & Organic Chemistry, Tbilisi State University, Tbilisi, Georgia

K. Archvadze
Petre Melikishvili Institute of Physical and Organic Chemistry of Ivane Javakhishvili Tbilisi State University, Tbilisi, Georgia

M. Areshidze
Institute of Cybernetics of Georgian Technical University, Tbilisi, Georgia
Department of Polymer Materials, Tbilisi State University, Tbilisi, Georgia

Koduvarathodi Vadakkethil Arundhathi
School of Chemical Sciences, Mahatma Gandhi University, Kottayam, Kerala, India

Marina Avaliani
R. Agladze Institute of Inorganic Chemistry and Electrochemistry, I. Javakhishvili Tbilisi State University, Georgia

Giorgi Beridze
Al. Janelidze Geological Institute, Department of Petrology, Mineralogy and Lithology, IvaneJavakhishvili Tbilisi State University, Tbilisi, Georgia

Michael Bratychak
Institute of Chemistry and Chemical Technology, Lviv Polytechnic National University, Ukraine

T. Bukia
Institute of Cybernetics of Georgian Technical University, Tbilisi, Georgia
Department of Polymer Materials, Tbilisi State University, Tbilisi, Georgia

Vaja Chagelishvili
R. Agladze Institute of Inorganic Chemistry and Electrochemistry, I. Javakhishvili Tbilisi State University, Georgia

I. Chitrekashvili
Petre Melikishvili Institute of Physical and Organic Chemistry of Ivane Javakhishvili Tbilisi State University, Tbilisi, Georgia

K. Chubinidze
Institute of Cybernetics of Georgian Technical University, Tbilisi, Georgia
Department of Polymer Materials, Tbilisi State University, Tbilisi, Georgia

M. Chubinidze
Institute of Cybernetics of Georgian Technical University, Tbilisi, Georgia
Department of Polymer Materials, Tbilisi State University, Tbilisi, Georgia

Yuriy Demchuk
Institute of Chemistry and Chemical Technology, Lviv Polytechnic National University, Ukraine

L. Devadze
Institute of Cybernetics of Georgian Technical University, Tbilisi, Georgia
Department of Polymer Materials, Tbilisi State University, Tbilisi, Georgia

N. Dokhturishvili
Petre Melikishvili Institute of Physical and Organic Chemistry of Ivane Javakhishvili Tbilisi State University, Tbilisi, Georgia

Nanuli Dolaberidze
Petre Melikishvili Institute of Physical & Organic Chemistry, Tbilisi State University, Tbilisi, Georgia

Maia Dzagania
Petre Melikishvili Institute of Physical and Organic Chemistry, Ivane Javakhishvili Tbilisi State Univeristy, Tbilisi, Georgia

Luba Eprikashvili
Petre Melikishvili Institute of Physical and Organic Chemistry, Ivane Javakhishvili Tbilisi State Univeristy, Tbilisi, Georgia

Ramaz Gakhokidze
Department of Chemistry, Iv. Javakhishvili Tbilisi State University, Tbilisi, Georgia

E. Gavashelidze
Petre Melikishvili Institute of Physical and Organic Chemistry of Ivane Javakhishvili Tbilisi State University, Tbilisi, Georgia

Ioseb Gejadze
Iv. Javakhishvili Tbilisi State University, Al. Tvalchrelidze Caucasian Institute of Mineral Resources; Tbilisi, Georgia

N. Gelashvili
Petre Melikishvili Institute of Physical and Organic Chemistry of Ivane Javakhishvili Tbilisi State University, Tbilisi, Georgia

Nazi Goliadze
Petre Melikishvili Institute of Physical and Organic Chemistry of Ivane Javakhishvili Tbilisi State University, Tbilisi, Georgia

Volodymyr Gunka
Institute of Chemistry and Chemical Technology, Lviv Polytechnic National University, Ukraine

M. Gurgenishvili
Petre Melikishvili Institute of Physical and Organic Chemistry of Ivane Javakhishvili Tbilisi State University, Tbilisi, Georgia

Sabir S. Guseynov
G. A. Krestov Institute of Solution Chemistry, Russian Academy of Sciences, Ivanovo, Russia

Yurii Hrynchuk
Institute of Chemistry and Chemical Technology, Lviv Polytechnic National University, Ukraine

N. Imnaishvili
Institute of Cybernetics of Georgian Technical University, Tbilisi, Georgia
Department of Polymer Materials, Tbilisi State University, Tbilisi, Georgia

Neenamol John
School of Chemical Sciences, Mahatma Gandhi University, Kottayam, Kerala, India

Aruna Joseph
Department of Physics and Centre for Research, St. Teresa's College (Autonomous) Ernakulam, Kerala, India

E. Kalandia
Institute of Cybernetics of Georgian Technical University, Tbilisi, Georgia
Department of Polymer Materials, Tbilisi State University, Tbilisi, Georgia

Evgeni Khuchua
Iv. Javakhishvili Tbilisi State University, Al. Tvalchrelidze Caucasian Institute of Mineral Resources; Tbilisi, Georgia

Bela Khutsishvili
Petre Melikishvili Institute of Physical & Organic Chemistry, Tbilisi State University, Tbilisi, Georgia

V. Kinkladze
Institute of Cybernetics of Georgian Technical University, Tbilisi, Georgia
Department of Polymer Materials, Tbilisi State University, Tbilisi, Georgia

Teimuraz Kordzakhia
Petre Melikishvili Institute of Physical and Organic Chemistry, Ivane Javakhishvili Tbilisi State Univeristy, Tbilisi, Georgia

Beena Mathew
School of Chemical Sciences, Mahatma Gandhi University, Kottayam, Kerala, India

Hiran Mayookh Lal
School of Energy Materials, Mahatma Gandhi University, Kerala, India

Vera Maisuradze
Iv. Javakhishvili Tbilisi State University, Al. Tvalchrelidze Caucasian Institute of Mineral Resources; Tbilisi, Georgia

Tamar Makharadze
R. Agladze Institute of Inorganic Chemistry and Electrochemistry, Ivane Javakhishvili Tbilisi State University, Tbilisi, Georgia

Teona Makharadze
The State Laboratory of Agriculture, Tbilisi, Georgia

Nato Mirdzveli
Petre Melikishvili Institute of Physical & Organic Chemistry, Tbilisi State University, Tbilisi, Georgia

Burkiashvili N.
P. Melikishvili Institute of Physical and Organic Chemistry, Ivane Javakhishvili Tbilisi State Univeristy, Tbilisi, Georgia

Osipova N.
P. Melikishvili Institute of Physical and Organic Chemistry, Ivane Javakhishvili Tbilisi State Univeristy, Tbilisi, Georgia

Marine Nadirashvili
Iv. Javakhishvili Tbilisi State University, Al. Tvalchrelidze Caucasian Institute of Mineral Resources; Tbilisi, Georgia

Anju K. Nair
Department of Physics and Centre for Research, St. Teresa's College (Autonomous) Ernakulam

Manana Nijaradze
Petre Melikishvili Institute of Physical & Organic Chemistry, Tbilisi State University, Tbilisi, Georgia

Andrew V. Noskov
G. A. Krestov Institute of Solution Chemistry, Russian Academy of Sciences, Ivanovo, Russia

G. Papava
Petre Melikishvili Institute of Physical and Organic Chemistry of Ivane Javakhishvili Tbilisi State University, Tbilisi, Georgia

A. Petriashvili
Institute of Cybernetics of Georgian Technical University, Tbilisi, Georgia
Department of Polymer Materials, Tbilisi State University, Tbilisi, Georgia

G. Petriahsvili
Institute of Cybernetics of Georgian Technical University, Tbilisi, Georgia
Department of Polymer Materials, Tbilisi State University, Tbilisi, Georgia

Tamar Petriashvili
Iv. Javakhishvili Tbilisi State University, Al. Tvalchrelidze Caucasian Institute of Mineral Resources; Tbilisi, Georgia

Nino Pirtskhalava
Petre Melikishvili Institute of Physical and Organic Chemistry, Ivane Javakhishvili Tbilisi State Univeristy, Tbilisi, Georgia

Yuriy Prysiazhnyi
Institute of Chemistry and Chemical Technology, Lviv Polytechnic National University, Ukraine

Volodymyr Reutskyy
Institute of Chemistry and Chemical Technology, Lviv Polytechnic National University, Ukraine

Sesili Rogava
Petre Melikishvili Institute of Physical and Organic Chemistry of Ivane Javakhishvili Tbilisi State University, Tbilisi, Georgia

Imeda Rubashvili
Petre Melikishvili Institute of Physical and Organic Chemistry, Ivane Javakhishvili Tbilisi State Univeristy, Tbilisi, Georgia

Muhammed Shah S.
School of Energy Materials, Mahatma Gandhi University, Kerala, India

Levan Samarguliani
Department of Chemistry, Iv. Javakhishvili Tbilisi State University, Tbilisi, Georgia

N. Sepashvili
Institute of Cybernetics of Georgian Technical University, Tbilisi, Georgia
Department of Polymer Materials, Tbilisi State University, Tbilisi, Georgia

Lana Shamanauri
R. Dvali Institute of Machine Mechanics, Tbilisi, Georgia

Elena Shapakidze
Iv. Javakhishvili Tbilisi State University, Al. Tvalchrelidze Caucasian Institute of Mineral Resources; Tbilisi, Georgia
A. Tvalchrelidze Caucasian Institute of Mineral Resources, Tbilisi, Georgia

L. Sharashidze
Institute of Cybernetics of Georgian Technical University, Tbilisi, Georgia
Department of Polymer Materials, Tbilisi State University, Tbilisi, Georgia

Tinatin Sharashenidze
Petre Melikishvili Institute of Physical and Organic Chemistry, Ivane Javakhishvili Tbilisi State University, Tbilisi, Georgia

Natia Shengelia
Sokhumi State University, Tbilisi, Georgia

Neli Sidamonidze
Department of Chemistry, Iv. Javakhishvili Tbilisi State University, Tbilisi, Georgia

Iurii Sidun
Institute of Building and Environmental Engineering, Lviv Polytechnic National University, Ukraine

Rajden Skhvitaridze
Scientific Center "Nano Dugabi," Georgian Technical University, Tbilisi, Georgia

Aiswarya T. Soman
School of Energy Materials, Mahatma Gandhi University, Kerala, India

Kvernadzc T.
P. Melikishvili Institute of Physical and Organic Chemistry, Ivane Javakhishvili Tbilisi State Univeristy, Tbilisi, Georgia

T. Tatrishvili
Institute of Cybernetics of Georgian Technical University, Tbilisi, Georgia
Department of Polymer Materials, Tbilisi State University, Tbilisi, Georgia

Mariyam Thomas
Department of Physics and Centre for Research, St. Teresa's College (Autonomous) Ernakulam

Sabu Thomas
School of Energy Materials, Mahatma Gandhi University, Kerala, India

Giorgi Tsintskaladze
Petre Melikishvili Institute of Physical and Organic Chemistry, Ivane Javakhishvili Tbilisi State University, Tbilisi, Georgia

Vladimer Tsitsishvili
Petre Melikishvili Institute of Physical & Organic Chemistry, Tbilisi State University, Tbilisi, Georgia

Arya Uthaman
School of Energy Materials, Mahatma Gandhi University, Kerala, India

Rusudan Vardiashvili
Department of Chemistry, Iv. Javakhishvili Tbilisi State University, Tbilisi, Georgia

Marine Zautashvili
Petre Melikishvili Institute of Physical and Organic Chemistry, Ivane Javakhishvili Tbilisi State Univeristy, Tbilisi, Georgia

Ts. Zurabishvili
Institute of Cybernetics of Georgian Technical University, Tbilisi, Georgia
Department of Polymer Materials, Tbilisi State University, Tbilisi, Georgia

Abbreviations

AA	adipic acid
AAC	alkali-activated types of cement
BJH	Barrett–Joyner–Halenda
BOD	biological oxygen demand
CA	cellulose acetate
$CaCO_3$	calcium carbonate
CD	cross direction
CFU	colony-forming unit
CMC	carboxymethylcellulose
CNT	carbon nanotubes
COD	chemical oxygen demand
CSTD	Commission for Science and Technology for Development
CuO	copper oxide
CZTS	copper zinc tin sulfide
DAF	dissolved air flotation
DEC	diethyl carbonate
DEGDME	diethyl ethylene glycol dimethyl ether
DFT	density functional theory
DMF	dimethylformamide
DoE	design of experiments
DSC	differential scanning calorimetry
DSSCs	dye-sensitized solar cells
EA	epoxy asphalt concrete
EC	ethylene carbonate
EEA	European Environmental Agency
ERO	rapeseed oil epoxy
ERO-AA	bitumen modifier—a mixture of epoxidized rapeseed oil with AA
ERO-FA	bitumen modifier—a mixture of epoxidized rapeseed oil with FA
ERO-In	bitumen modifier—a mixture of epoxidized rapeseed oil with initiator (AA, MA or PEPA)
ERO-MA	bitumen modifier—a mixture of epoxidized rapeseed oil with MA
ERO-PEPA	bitumen modifier—a mixture of epoxidized rapeseed oil with PEPA
FA	formic acid
FA	fulvic acids
FTIR	Fourier transform infrared
GPM	geopolymer materials
HESs	hydrophobic eutectic solvents

HMA	hot-mix asphalt concrete
HPC	hierarchical porous carbon
IR	infrared
LIB	lithium-ion battery
LiOH	lithium hydroxide
LiPF6	lithium hexafluorophosphate
LiTf	lithium triflate
LiTFSI	lithium bis-trifluoromethanesulfonimide
LOB	lithium-oxygen battery
LOD	limits of detection
LOQ	limits of quantitation
LSD	least significant difference
LTO	lithium titanium oxide
MA	maleic anhydride
MCL	maximum contaminated level
MD	machine direction
MM	molecular mechanics
MP	mobile phase
Mw	molecular weight
NCAP	nanostructured coated absorber plates
NiMH	nickel-metal-hydride
NF	nano-filtration
NMP	n-methyl-2-pyrrolidone
NMs	nanomaterials
OB1	oxidized bitumen
OER	oxygen evolution reaction
OPC	ordinary Portland cement
OPVs	organic photovoltaics
ORR	oxygen reduction reaction
PC	propylene carbonate
PEG	polyethylene glycol
PEO	polyethylene oxide
PEPA	polyethylene polyamine
PP	peak purity
PS	polystyrene
PVA	polyvinyl alcohol
PVDF	polyvinylidene fluoride
PVP	polyvinyl pyrrolidone
QDs	quantum dots
RO	reverse osmosis
RO	rapeseed oil
ROS	reactive oxygen species
RSD	relative standard deviation

RT	relaxation transition
SEM	Society of Mechanical Engineering
SIB	sodium-ion battery
SLMP	stabilized lithium metal powered
SP	spiropyran
SPLCP	spiropyran-doped liquid crystal polymer
SST	system suitability test
TEGDME	tetra ethylene glycol dimethyl ether
TENG	triboelectric nanogenerators
TEOS	tetraethoxysilane
TG	thermogravimetry
TGA	thermos-gravimetric analysis
TS	tensile strength
TSS	total suspended solids
USEPA	United States Environmental Protection Agency
UV	ultraviolet
XRD	X-ray diffraction
XRED	X-ray energy-dispersive

Preface

Our society today is facing grave challenges regarding improvement in the quality of soil, water, air, and the environment to maintain ecological balance. The abnormal growth in industrialization for the generation of new products and advances in chemical processes plays an important role in the deterioration of the ecological balance and the rise of new pollutants.

Sustainable development proposes utilization of natural resources in a sustainable manner without causing changes in our natural world. As a result, modern environmental technologies aim at reducing pollution and have gained widespread interest in the recent years with a broad range of science and engineering sectors.

This new title covers the most recent theoretical and practical advancements in clean and healthy environment, and provides an improved understanding of the research and development of green technologies, which are becoming increasingly important for ensuring sustainability. This research-oriented book provides vital information on advanced materials and green composites from the works of experts in the subject.

It also expounds environmental chemistry for a sustainable world by providing information on different characterization methods as well as new techniques. It provides case studies to make the book more interesting and provide a general idea of the central themes. The volume also considers the most recent developments and applications of clean energy materials.

The book is written for a broad readership, including postgraduate engineering students, scientists, research scholars, and faculty members who are engaged in the area of green technologies and development of sustainable materials used in different sectors.

To summarize, this unique reference book comprehensively reports on the most recent research ideas in green technologies and achievements in various aspects of sustainable materials from a chemistry and engineering point of view.

PART I

Materials and Modern Processes for Environmental Remediation

CHAPTER 1

A Study on the Thermal Behavior of Polymer Composites Based on Polystyrene for Environmental Protection

OLGA V. ALEKSEEVA, ANDREW V. NOSKOV, and SABIR S. GUSEYNOV

G.A. Krestov Institute of Solution Chemistry, Russian Academy of Sciences, Ivanovo, Russia

ABSTRACT

The development of the polymer composites with controllable properties is one of priority areas in the modern chemistry and materials science. A promising method for modifying polymers is in fact associated with inorganic particles along with polymer matrix. The produced composites can acquire novel physical and chemical properties: thermal, electrical, strength, etc.

In the present chapter, we describe the study results of the thermal behavior of polymer composites based on polystyrene (PS). Inorganic particles such as fullerenes, silica, bentonite, and bentonite-magnetite were chosen as fillers. As a matter of fact, a solvent casting of main elements from targeted solutions was considered for the preparation of the PS/filler film composites containing 0–7 wt.% of filler. Pure PS films and PS granules were also researched.

By applying the DSC (differential scanning calorimetry), we can easily inspect the polymeric materials (i.e., the phase transitions from the glassy state to the elastic state). New data connected to the impact of filler (in low weight fraction) on the glass transition temperature, T_g, of composites were found.

Environmental Technology and Sustainability: Physical, Chemical, and Biological Technologies for Environmental Protection. Tamara Tatrishvili, Ann Rose Abraham, & A. K. Haghi (Eds.)

To be able to study the (TS) thermal stability of PS/filler composites at high temperatures, we have applied the reliable Thermogravimetry (TG) method. We have also investigated dependency of the typical thermal destruction temperatures on nature and concentration of the filler in the composite studied. Applying the Van Krevelen method carried out the kinetic analysis of the recorded thermo-gravimetric curves. The main parameters determining the rate constant of the process (activation energy, pre-exponential factor) were found, and their dependences on the composition of the composites were analyzed.

These studies are of great importance for environmental protection, since a substantial amount of dangerous elements can be discharged in the course of thermal degradation of polymers.

1.1 INTRODUCTION

During the last decade there has been steadily increasing interest in the creation of multifunctional hybrid polymer composites, containing the inorganic micro- and nanoparticles. Modification of the polymer with inorganic particles increases its rigidity, impact strength, thermal stability, changes electrical and gas barrier properties, as well as can lead to the appearance of absorption properties and of biological activity.[1–3] Such composites may have applications in many sectors and industries such as:

- Medicine;
- Food;
- Building;
- Electronics.

A lot of publications deal with various methods of composites production and various types of fillers for polymers such as silica,[4,5] fullerene and carbon nanotubes,[6–8] layered silicates,[9–11] and metals.[12]

Among the most commonly used polymers, polystyrene (PS) can be distinguished as matrices for the production of organic–inorganic composites. Polystyrene is a thermoplastic mass marketable polymer with low cost, good film-forming properties, high electrical resistance, ease of machining, and resistance to environmental influences.[13–15] Polystyrene has attracted extensive interest in polymer science due to its use in a range of applications including packaging, appliances, consumer electronics, construction, medical, and various other uses.[4,16]

The expansion of the application fields of such polymers depends on the requirements for their operational and technical characteristics, the possibilities of processing by modern industrial methods. An important problem is the study of ways of stabilizing polymers and inhibiting thermal destruction processes.

The data on thermal stability are extremely necessary for practical purposes; they open up the possibility of obtaining polymer composite materials with the required set of properties. Changes in polymer composites that occur when exposed to high temperatures or oxygen contained in the air lead to a change in the physicochemical characteristics of materials.

One of the most important factor influences in the practical applications of polymer composites is the degradability and durability under various environmental conditions. As being acknowledged, thermogravimetry (TG) is a common method to study the kinetics of polymer degradation. Also, differential scanning calorimetry (DSC) is employed to elucidate the decomposition steps observed in the TG data. Thermal analysis techniques have been widely used to study thermal properties of polymer nanocomposites to evaluate the response of material after being heated or cooled.[17]

An important benchmark used to compare the thermal behavior of composites is the glass transition temperature (T_g), which is associated with RT (relaxation transition). It should be noted that this relaxation transition ranges from a glassy state or crystalline to a very high elastic domain. To be able to detect the glass transition temperature, we should consider the effects of the following parameters:

- Kinetic flexibility of macromolecules and
- Fluctuation network formation.

And it is desired to note that for the 2nd parameter mentioned above, we need to keep in mind the chemical composition and structure of the polymer chain as well. The T_g value can also depend on the sample thickness.

The incorporation of inorganic of filler into a polymer matrix can bring about changes in the thermal characteristics of the resulting composite. Changes in the T_g as a function of filler content have been reported for polymer composites containing a wide variety of filler and polymer materials. Many researchers reported an increase in the T_g as a function of filler content.[18,19] Other authors[5,20] noted that the glass transition temperature of polymer composites decreases or passes through a maximum with the increase in the inorganic fillers content.

Thus, the study of the effect of the filler concentration on the thermal characteristics of composites based on polystyrene is an urgent task. Many aspects of the role of the filler remain unclear. One can hope that the data presented in the current chapter will be of great scientific and practical importance.

1.2 PHASE TRANSITIONS IN THE PS/FILLER COMPOSITES

1.2.1 METHOD AND APPARATUS

To study the relaxation phase transitions in the pure polystyrene and polystyrene/filler composites, we have used a German Differential Scanning Calorimeter (DSC) made by Netzsch company with a typical model of DSC 204 F1. For the preparation of samples (films), the vacuumed samples were dried first for about 4 h using a temperature of 313 K. We have then stacked the films (samples) with a diameter of 4 mm and a weight of 4–10 mg in a special aluminum container. Using two-stage argon flow for the measurements process we have operated Differential Scanning Calorimeter (DSC). The above-mentioned two stages are carried out as follows:

- In the first stage our aim was to remove volatiles from the polymer.
- Heating process carried out from room temperature to 423 K.
- Cooling process carried out to 283 K at a rate of 10 K/min.
- In the second stage, the heating process carried out to 293 K.
- We have considered an isothermal regime at 293 K (for 5 min), and the heating process up to 423 K. The rate was 10 (K/min).
- As a reference sample, an empty aluminum vessel was considered.
- We have then performed the measurements relative to the measurements carried out relative to the standards defined for two empty vessels considering an identical heating program.
- For each sample (film) we have performed three Differential Scanning Calorimeter (DSC).
- For 11 standard samples, the DSC was calibrated. The measurement of the temperatures and heat effects of the phase transitions was carried out in the range of 187–749 K.[21]

For the samples we have done the studies in this book chapter, the content of volatiles obtained on a German TG 209 F1 thermal microbalance identical to heating rate in an argon media. We have expected to achieve a decrease in the weight of the samples by 1×10^{-3} mg (as an error factor).

The study of the data obtained by thermo-gravimetric allows us to determine the weight of substances (minus volatiles weight).

From the glassy to highly elastic state, the phase transition was characterized using the following factors:

- The extrapolated temperature in the phase of transition onset, which was determined by the tangent intersection method. This is shown by $T_{1.}$
- The extrapolated temperature in the phase of transition onset, which was determined by the tangent intersection method. This is shown by $T_{2.}$

Temperature of the phase transition (considered as average temperature): $\bar{T}$

Temperature of inflection in the DSC curves (considered as the glass transition temperature): T_g

Temperature range of the phase transition: $\Delta T = T_2 - T_1$

It should be noted that the values of the "characteristic temperatures" are attained on the base of the second heating data.

In the following sections, we describe the results of our studies of phase transitions in PS-based composites containing inorganic fillers of various natures.

1.2.2 PS/FULLERENE COMPOSITE FILMS

In Figure 1.1 we can see Differential Scanning Calorimeter curves for the sample of polystyrene and composite filled with fullerene. According to this figure, for all tested polymeric samples, a reversible phase transition from the glassy state to the elastic state revealed. This displays heat flow in the endothermic direction.

The average characteristics of the glass transition obtained from tree experiments for each polymer films are presented in Table 1.1. It was found that the $C_{60} + C_{70}$ addition affects all characteristic temperatures and temperature range of phase transition, ΔT, for the composite materials.

From Figure 1.2 we observe the ratio of T_g(comp) or the glass transition temperatures for composite films to T_g(PS), for original polystyrene film versus fullerene content. As can be seen, small additions of fullerene stinkly reduce the T_g of the composite compared to original polymer sample (film). The minimum value of T_g is equal to 0.01 wt.% as observed at the concentration level. Evidently, in such a condition, the plasticization of polystyrene

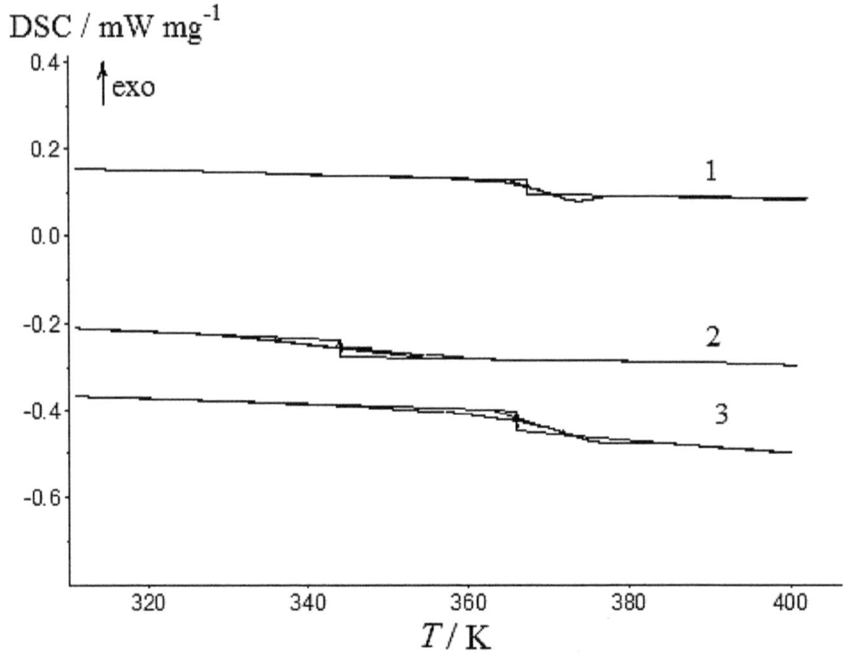

FIGURE 1.1 DSC curves for composite films. This is based on polystyrene at various fullerene contents, wt. %: 0 (*1*), 0.01 (*2*), and 0.1 (*3*).

TABLE 1.1 Parameters of Phase Transition from Glassy State to Elastic for Fullerene–Polystyrene Composites with Varying Weight Fractions of $C_{60} + C_{70}$.[a]

Fullerene content, wt.%	T_1, K	$\bar{T}$, K	Tg, K	T_2, K	ΔT, K
0.0	357.0	361.5	363.8	365.8	8.8
0.010	328.1	343.8	335.8	355.1	27.0
0.020	341.6	348.6	345.0	359.3	17.7
0.035	347.0	358.4	357.4	366.5	19.5
0.100	365.3	369.1	371.6	373.5	8.2

[a]Given values were obtained by averaging the data of three repeated experiments.

with the fullerene is expected.[22,23] Therefore, molecules of $C_{60} + C_{70}$ set among the chains and weaken interchain interactions in the polystyrene. This will cause increment of mobility of polymer chain segments and low value of T_g. The apparent augmentation of ΔT (i.e., the temperature range for phase transition) or the value of $\Delta T = T_2 - T_1$ at low concentration of fullerene (please refre to Table 1.1) shows an evident plasticization effect of these additives on thermal properties of the composite. The value of T_g apparently increases with further augmentation in the ($C_{60} + C_{70}$) content. Any time that the fraction of fullerene is identical to 0.1 wt.%, then it is expected that the glass transition temperature for the composite overpasses that for the original polystyrene.

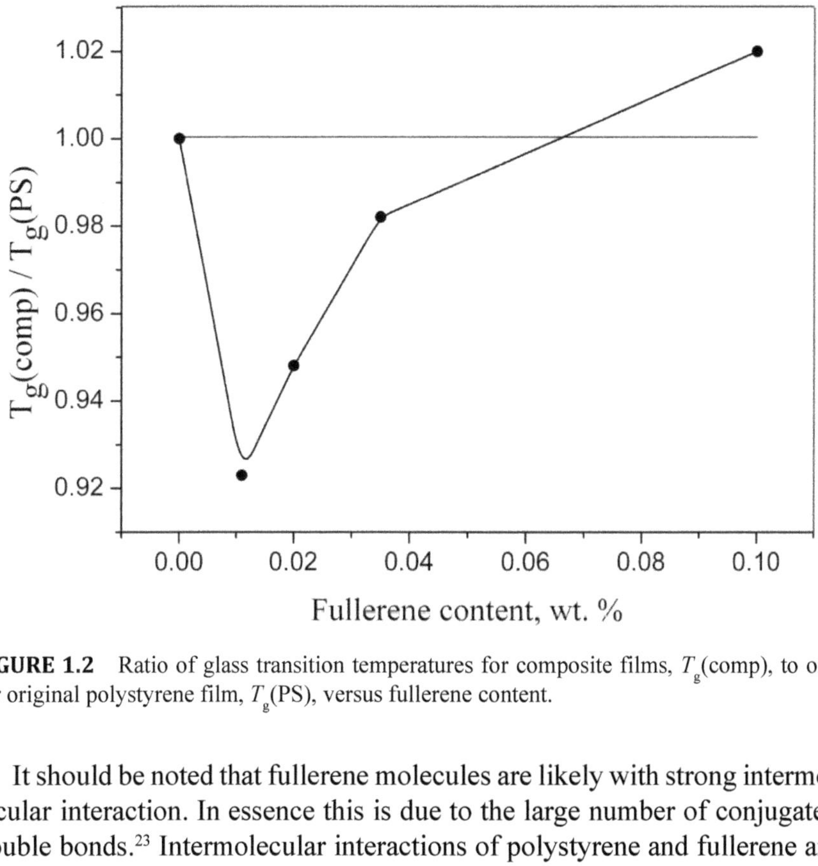

FIGURE 1.2 Ratio of glass transition temperatures for composite films, T_g(comp), to one for original polystyrene film, T_g(PS), versus fullerene content.

It should be noted that fullerene molecules are likely with strong intermolecular interaction. In essence this is due to the large number of conjugated double bonds.[23] Intermolecular interactions of polystyrene and fullerene are appeared on the values of glass transition temperature in the samples (films) with additions of more than 0.01 wt.%. Whenevre fullerenes concentration is equal to 0.1 wt.% or more, the effect of intermolecular interactions is the controlling factor. It is worth mentioning that the probability of interaction between fullerene molecules in the composite is expected to increase as well; this signifies the increasing glass transition temperature (as correlates with the original polystyrene). On the other hand, this can lead to the physical cross-linking of polystyrene chains. Conversely, at exceptionally low concentration (that means for less than 0.01 wt.%), the probability of interaction among fullerene and polystyrene molecules is quite small. This is as a consequence that there is one fullerene molecule per 55–60 polystyrene macromolecules (or one fullerene molecule per 70,000–80,000 monomeric unit) in the composite material. In this mechanism, weakening of interchain interactions in polystyrene is controlling.

In our previous reports,[7] we applied an electron microscopy and the X-ray diffraction to study the influence of the fullerene additives on the structure of the polystyrene films which has been formed by evaporation of the solvent. In this study, we have reported that in general, in films without fullerenes, packing of aligned chains parallel to each other dominated. From this, we have in fact come to a conclusion that a polystyrene molecule, which had the shape of a coil in solution, aligned and stretched itself along the surface of an aggregate when attached to it. Nevertheless, within composite with 0.035 wt.% of fullerene (one fullerene per 7–10 polystyrene macromolecules), fitment to aggregates occurred in the quite similar manner. Meanwhile, at such concentrations the intermolecular interactions between polystyrene and fullerene are substantial. However, under the influence of fullerene molecules, polystyrene molecules aligned with the formation of ordering elements in the arrangement of chains.[7]

According to the results of differential scanning calorimeter studies, we observed specific heat capacities (C_p) for the (samples) films of original polystyrene and polystyrene-doped additive mixture $C_{60} + C_{70}$ (0.01–0.1 wt.%), in the range of 290–420 K. The values of (C_p) for original polystyrene are equal to 1.23 ± 0.03 and 1.95 ± 0.03 J/g/K at 298 and 398 K, respectively. These results are in accordance with our reference data which amount to 1.224 and 1.924 J/g/K, respectively.[21] As we have observed from Figure 1.3, the dependence of (C_p) of the composite on the fullerene content is nonmonotone. The value of (C_p) at 298 K slightly decreases from 1.23 to 1.12 J g^{-1} K^{-1} when the content of $C_{60} + C_{70}$ increases up to 0.02 wt.%. Further increment of fullerene content leads to the significant increase in heat capacity to 1.40 J/g/K (please refer to curve 1).

Now in curve 2, the reliance of the heat capacity on the content of ($C_{60} + C_{70}$) for filled polymer materials in the elastic state at (398 K) passes through a minimum (meanwhile at a concentration of fullerenes equal to 0.01 wt.%). The changes observed in (C_p) affirm the conclusion indicated above. This is more or less about the plasticization of polystyrene by small additions of fullerene. According to the results reported in Ref. [22], the increase in heat capacity at the content of fullerene (with more than 0.02 wt.%) can be

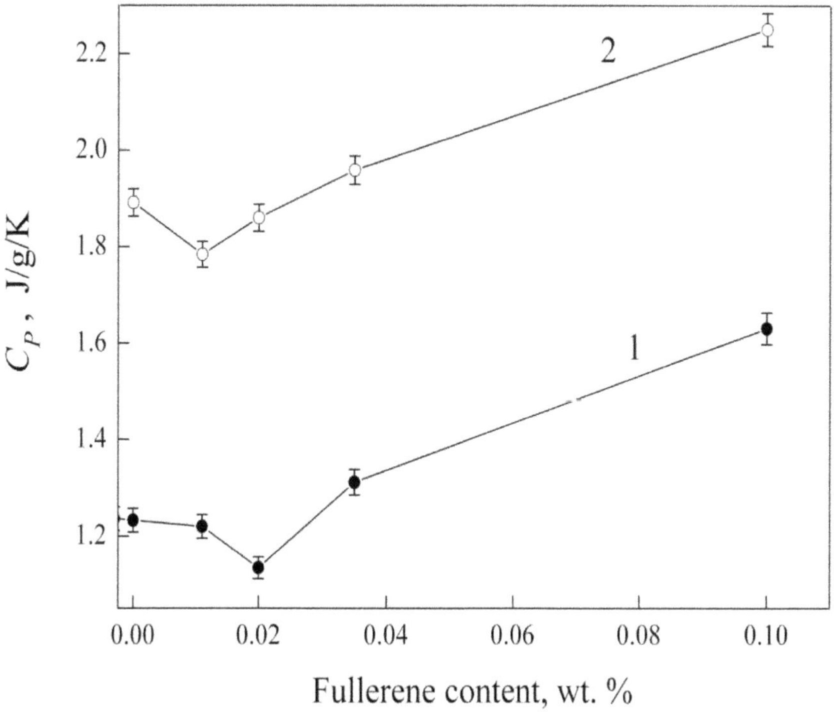

FIGURE 1.3 Dependence of (C_p) or specific heat capacity of composite films on fullerene content: *1*—glassy state at 298 K; *2*—elastic state at 398 K.

correlated with the manifestation of the (polystyrene-fullerene) interactions. And this is in fact consistent with the regularities in change of the glass transition temperatures.[22]

Figure 1.4 indicates the temperature dependence of (C_p) for polystyrene (sample) films for different content of fullerene. As seen, the value of specific heat capacity increases evenly with temperature and increases unexpectedly at the transition phase. The dependence "measured values of specific heat capacity *versus* temperature" is described by the following linear equation:

$$C_p = a + bT \tag{1.1}$$

In Figure 1.5, we can see the temperature-dependent coefficients of the (C_p), (i.e., *b)*, for the composites at different contents of fullerenes. It can be observed that the value of *(b)* for samples in the glassy state in the temperature range of 298 K–328 K escalates strongly (while fullerene content in curve 1 increases up to 0.02 wt.% and decreases with additional increase in the additives concentration). It should be noted that the observed dependence values of (C_p) and (*b*) follow the competing effects of 2 factors (i.e., the plasticizing effect of fullerenes reveals itself with small additions).

In Figure 1.3, we can observe that (C_p) decreases while in Figure 1.5, (*b*) increases. The interactions effect among fullerene and polymer molecules leads to concentrations more than (0.02 wt.%). At this stage, the values of specific heat capacity increased (please refer to Fig. 1.3) and *b* decreased (please refer to Fig. 1.5). In Figure 1.5 (curve 2), it is observed that the concentration dependence of the (*b*) values for the samples in the elastic state for the range of 368–423 K is weak.

1.2.3 PS/SILICA COMPOSITE FILMS

The observed DSC curves for the pure polystyrene and silica-containing polystyrene films are shown in Figure 1.6. It is noted that the reversible phase transition from the glassy to highly elastic state appears as a break in the thermo-gram. This is in fact observed for all studied polymer materials when the temperature is augmented.

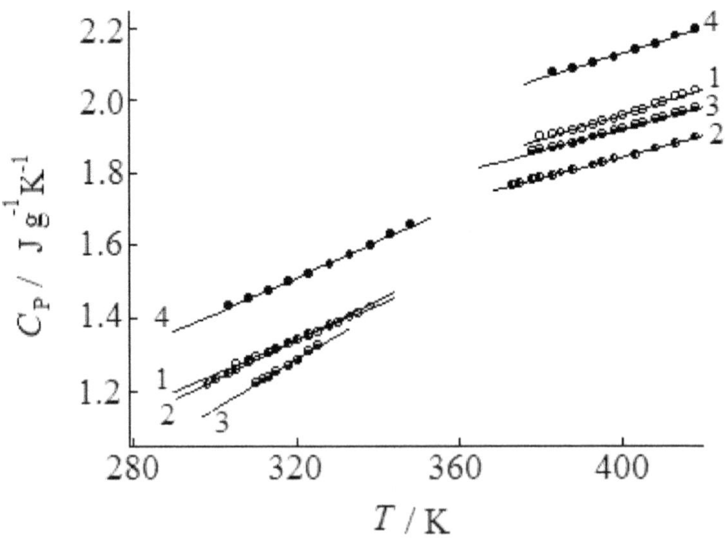

FIGURE 1.4 Temperature dependence of specific heat capacity (C_p) for polystyrene films with various content of fullerene: *1*—0.0, *2*—0.010, *3*—0.020, *4*—0.035 wt.%.

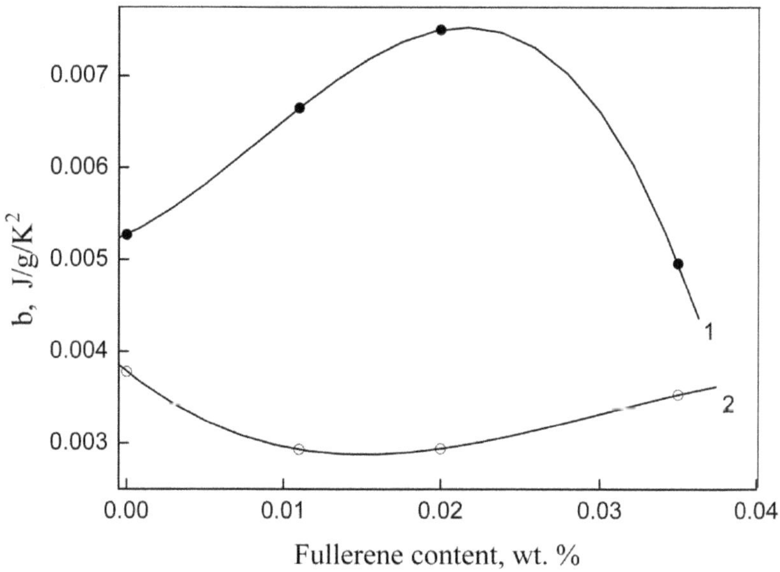

FIGURE 1.5 Coefficient of temperature dependence of specific heat capacity (C_p) for polystyrene films. This is for different content of fullerene: *1*—glassy state in temperature range of 298–328 K; *2*—elastic state in range of 368–423 K.

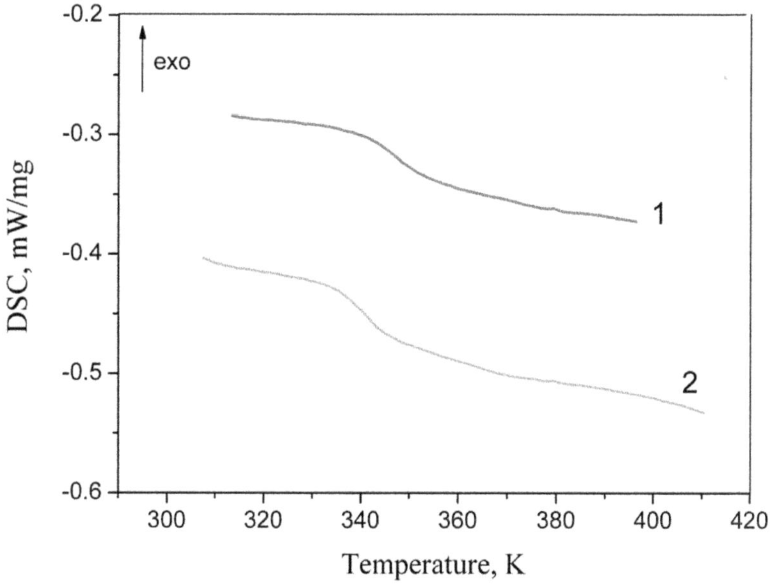

FIGURE 1.6 DSC curves for the polystyrene-based composite films at silica concentrations, wt. %: 0 (*1*) and 0.5 (*2*).

TABLE 1.2 Parameters of the Phase Transition from the Glassy to Highly Elastic State for PS Composites with Various Silica Concentrations.[a]

SiO_2 content, wt.%	T_1, K	$\bar{T}$, K	Tg, K	T_2, K	ΔT, K
0.0	341.3	346.4	347.9	353.7	12.4
0.03	339.6	347.5	347.8	355.5	15.9
0.1	338.0	344.9	344.4	352.3	14.3
0.5	334.6	339.8	342.3	345.9	11.3
1	332.9	341.3	338.0	349.4	16.5
3	334.8	347.1	341.8	360.2	15.4
5	350.6	362.9	366.6	373.1	22.5

[a]Given values were obtained by averaging the data of three repeated experiments.

According to the work presented in Ref. [24], reliance of this type is characteristic of flexible-chain polymers. Meanwhile in Table 1.2 we can see the characteristic parameters of the glass transition for the polystyrene film and silica-containing composites provided by the analysis of the thermo-grams.

Now let us analyze the data in Table 1.2. We simply observe that the SiO_2 concentration in the film (sample) influences the characteristic temperatures of the phase transition. In Figure 1.7, we notice that by the application of insignificant amount of silica, the glass transition temperature of the composite decreases compared to the original polymer film.

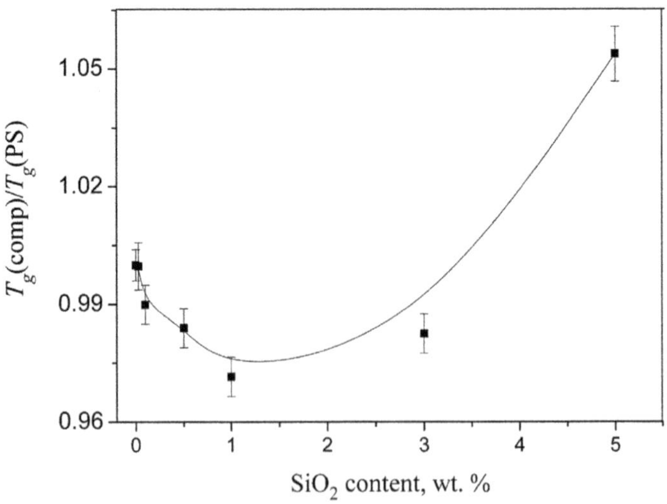

FIGURE 1.7 Ratio of the glass transition temperatures of the composite T_g(comp) to polystyrene T_g(PS) at different silica concentrations.

The minimum value of glass transition temperatures can be reached at the SiO_2 concentration in the composite with (1 wt.%). This of course shows the plasticizing effect for small additives of silica. Most likely we can conclude that SiO_2 particles are incorporated among the polymer chains and then weaken the inter-chain interaction, causing an increase in the segmental mobility of the macromolecular chains and a decrease in glass transition temperatures (T_g).

In Figure 1.7 we can clearly observe that further increase in the silica concentration in the composite increases the (T_g). This might be due to a higher probability of the formation of (polymer–filler) bonds maintained by a decrease in the mobility of the polystyrene chains segments and an increase in glass transition temperatures.

According to Table 1.2, and based on our studies in this chapter, we can easily make a discussion on different characteristic parameters of the phase transition such as (T_1, T_2, $\bar{T}$, and ΔT), as well as the glass transition temperature (T_g).

In this chapter, we have demonstrated that a complicated pathway of changing the (T_g) of the polystyrene/silica composite material occurs with an increase in the SiO_2 content. In general, this is due to nonuniform distribution of the filling agent molecules in the polymer matrix.

1.2.4 PS/BENTONITE COMPOSITE FILMS

Relaxation transitions were studied for the PS/bentonite composites containing 0–5 wt.% of filler. The DSC curves for the unmodified and bentonite containing polystyrene films recorded upon a second heating are shown in Figure 1.8.

From our studies, we observed that, for all the sample (film) materials, an increase in temperature leads to a transition from the glassy state to a highly elastic state (please see the kink in the thermogram). Such reliance is correlated with a change in the heat capacity of the samples during the relaxation transition.

From Table 1.3 we can observe the following:

- the values of the characteristic temperatures of the relaxation transition for a polystyrene film (sample).
- the bentonite-containing composites (found by processing thermograms).
- the clay concentration in the film which affects the above parameters.

It should be noted (from Table 1.3) that for unmodified polystyrene, T_g is 70.9°C, whereas higher values of glass transition temperature are observed in composites. It is worth mentioning that by the introduction of bentonite (5 wt.%) into polystyrene we do expect an increase in the T_g around 15°C. This effect is somehow related with a decrease in the mobility of segments of polymer chains as a result of their interaction with clay. We can observe that such a decrease in the mobility of the segments might be local in nature and takes place near clay particles.

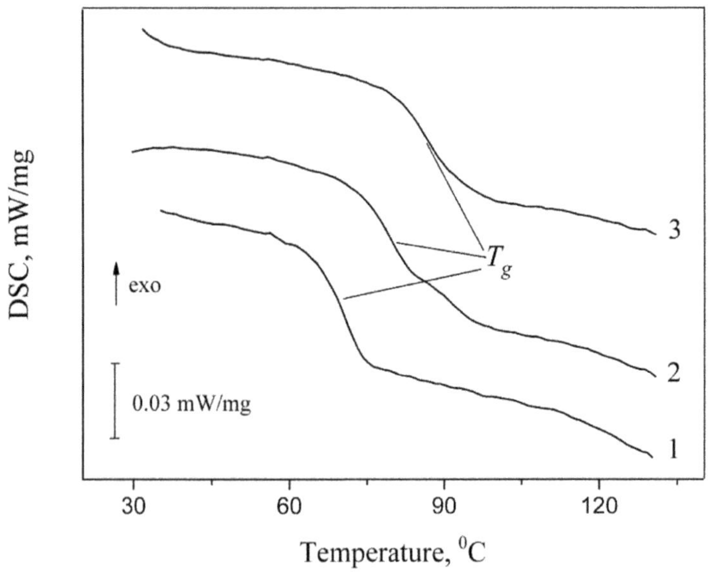

FIGURE 1.8 DSC curves of polystyrene/bentonite film composites with various filler concentrations, wt %: 0 (*1*), 1 (*2*), and 5 (*3*).

TABLE 1.3 Parameters of the Relaxation Transition from a Glassy to Highly Elastic State for PS/Bentonite Composites with Various Filler Concentrations.[a]

Bentonite content, wt.%	T_1, °C	$\bar{T}$, °C	*Tg*, °C	T_2, °C
0.0	64.9	69.8	70.9	74.6
0.1	68.3	77.3	73.9	87.0
0.5	71.1	76.1	77.7	81.0
1.0	72.5	79.9	79.1	87.5
3.0	79.5	84.4	84.8	89.2
5.0	80.2	87.0	85.7	93.5

[a]Given values were obtained by averaging the data of three repeated experiments.

In Figure 1.9 we can see the experimental results. The dependence of the (T_g) of composite on its composition can be observed clearly in this Figure 1.9. It can be seen the difference in the glass transition temperature values between the composite and the starting polymer. This means: T_g(comp)–T_g(PS), equally increased with increasing in the filler concentration. Our analysis of results showed that this dependence can be well described by an equation in the form of:

$$y = \frac{x}{a + bx} \quad (1.2)$$

It should be pointed out that for parametric values $a = 0.055 \pm 0.010$ and $b = 0.053 \pm 0.004$, coefficient of determination R^2 is 0.981. It should be pointed out that a similar analytical expression was reported in Ref. [4] while studying the glass transition processes of polystyrene/silica composites in the range of filler concentrations up to 30 wt.%.

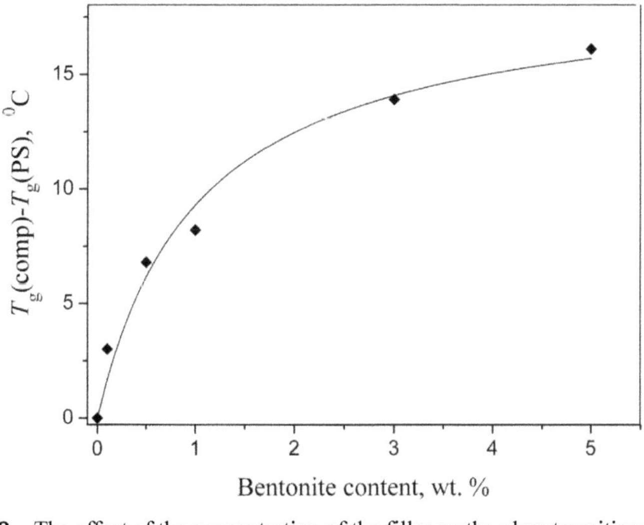

FIGURE 1.9 The effect of the concentration of the filler on the glass transition temperature of the PS/bentonite composite: ♦ experimental data; – fitting curve.

Therefore, our studies show that:

- The introduction of bentonite into polystyrene leads to an increase in the (T_g) of the composite compared to unmodified polymer.
- These characteristics of thermal behavior must be considered while developing and improving technologies for the production of polymer materials with controlled properties.

1.2.5 PS/BENTONITE/MAGNETITE COMPOSITE FILMS

In this paragraph, we describe the research results of relaxation transitions in PS-based composites containing two-component filler such as bentonite/magnetite. Figure 1.10 shows the DSC curves for pure PS films recorded upon the first and second heating. As can be seen, in case of the first heating, an endothermic peak is observed on the thermogram in the temperature range of 65–90°C (trace *1*), apparently associated with the removal of o-xylene residues from the film. This peak is absent upon repeated heating (trace *2*), which indicates the complete removal of the solvent from the film during the first heating. The revealed difference in thermograms for the first and second heating takes place for all studied samples of the PS/bentonite/magnetite composite films, regardless of their composition.

The DSC curves for the PS/bentonite/magnetite composites with various filler concentrations recorded upon the second heating are presented in Figure 1.11. It can be seen, in this case, for all the materials studied, a kink is observed in the thermogram, which can be associated with a relaxation transition from the glassy state to a highly elastic state and is accompanied by a change in the heat capacity of the material during such a transition.

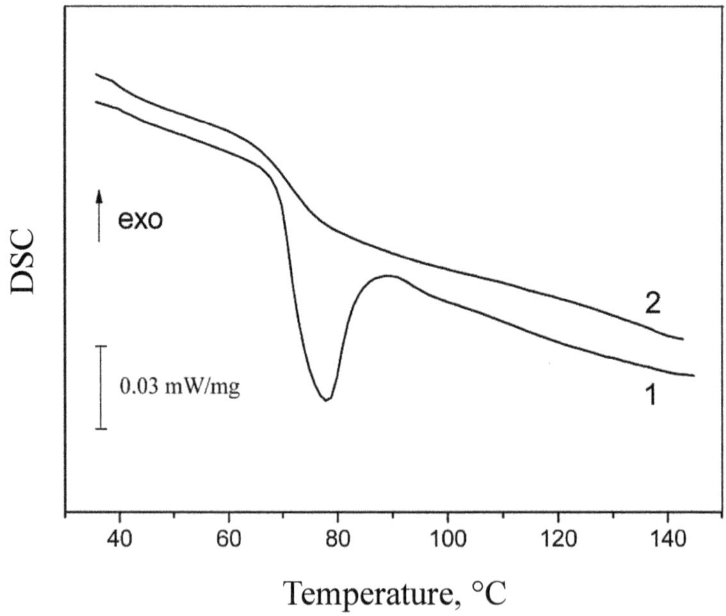

FIGURE 1.10 DSC curves for PS films recorded upon the first (*1*) and second (*2*) heating.

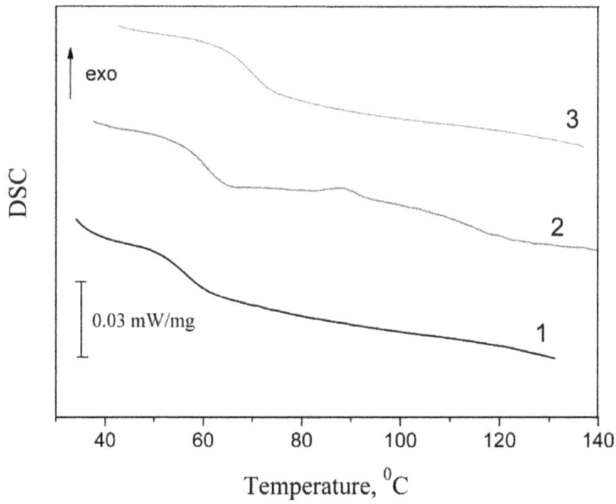

FIGURE 1.11 DSC curves of polystyrene/bentonite/magnetite film composites with various filler concentrations, wt %: 0.1 (*1*); 1 (*2*); 5 (*3*).

The values of the characteristic temperatures found for composites under study are given in Table 1.4 which demonstrates that the concentration of magnetized clay in the film affects the listed parameters. For unmodified polystyrene, the glass transition temperature is 70.4°C. With the introduction of the filler in an amount of 0.1 wt.%, the T_g value decreases to 57.0°C, which indicates the dominant effect of plasticization of polystyrene with small additions of magnetized clay. However, with a further increase in the concentration of bentonite/magnetite in the composite, the glass transition temperature increases, which is apparently associated with a decrease in the mobility of polymer chain segments as a result of their interaction with the filler. It should be noted that the revealed nonmonotonic dependence on the composition of the composite is retained for other characteristic temperatures of the relaxation transition (Table 1.4).

TABLE 1.4 Parameters of the Relaxation Transition from a Glassy to Highly Elastic State for PS/Bentonite/Magnetite Composites with Various Filler Concentrations.[a]

Filler concentration, wt.%	T_1, °C	$\bar{T}$, °C	Tg, °C	T_2, °C	ΔT, °C
0.0	66.0	72.0	70.4	77.7	11.7
0.1	51.3	56.7	57.0	61.8	10.5
1.0	55.4	59.9	60.3	63.9	12.5
5.0	64.3	70.3	69.8	76.0	12.7

[a]Given values were obtained by averaging the data of three repeated experiments.

To explain the results obtained, the thermodynamic model developed in Refs. [25,26] based on the configurational entropy model and the Flory-Huggins theory can be used. The proposed model assumes that the change in the glass transition temperature of the composite (T_g^{comp}) in comparison with the glass transition temperature of the matrix polymer (T_g^{PS}) is determined by the change in entropy:

$$\ln\left[\frac{T_g^{comp}}{T_g^{PS}}\right] \sim S(\varphi) - S(0) \tag{1.3}$$

where φ is the volume fraction of filler particles in composite; $S(\varphi)$ and $S(0)$ are the entropy of the composite and entropy of the polymer in the absence of filler, respectively.

As follows from eq 1.3, with the introduction of filler nanoparticles into the polymer matrix, the glass transition temperature can either increase or decrease depending on the sign of the change in the total entropy of the system, which, in turn, depends on the contributions of the total entropy components. In particular, a decrease in T_g can take place when the contribution of the entropy of the specific polymer-nanoparticle interaction is small compared to the contributions of the mixing entropy and entropy of nanoparticles under conditions of spatial confinement.

The model developed in Refs. [25,26] predicts that this condition can most likely be met at low concentrations of nanoparticles and ceases to be met with an increase in the volume fraction of filler. This means that the theoretical dependence of the glass transition temperature of the composite on its composition has a minimum. As seen from the experimental results presented in Figure 1.12, this minimum predicted by the model is observed for the PS/bentonite/magnetite composites researched in the current study.

It should be noted that we found similar non-monotonic dependences of the glass transition temperature on the filler concentration for the polystyrene/fullerene composite films (Section 1.2.2 in the current chapter), polystyrene/silica (Section 1.2.3 in the current chapter), as well as for polymethyl methacrylate/fullerene composite.[27]

1.3 THERMAL STABILITY OF THE PS/FILLER COMPOSITES

1.3.1 METHOD AND APPARATUS

Thermogravimetric analysis for the pure polystyrene and polystyrene/filler composites was performed on a TG 209 F1 thermo-microbalance (Netzsch

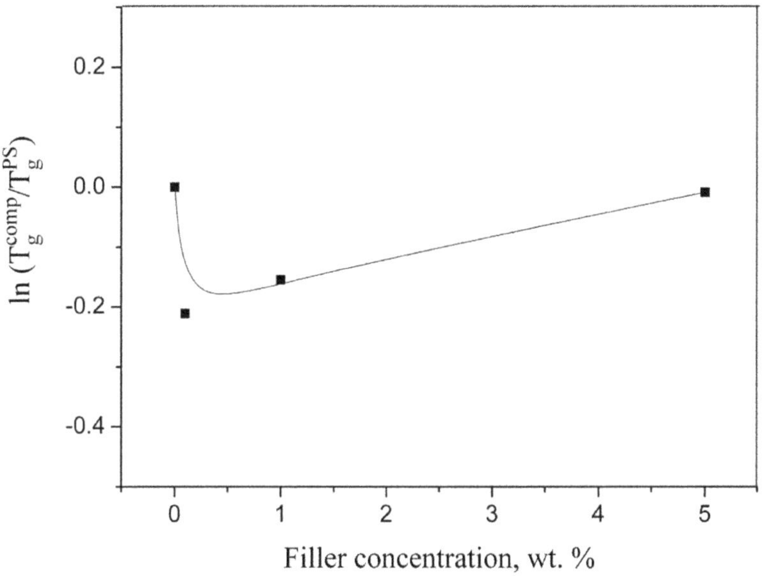

FIGURE 1.12 Effect of the filler concentration on the glass transition temperature for the PS/bentonite/magnetite composites.

Gerätebau GmbH, Germany) using platinum crucibles in an argon flow (30 mL/min). The samples of PS/filler composites with a mass of 3–5 mg were heated at a heating rate of 10 K/min, and the weight loss was measured. The accuracy of the sample mass measurement was 1×10^{-7} g, and the accuracy in temperature measurement was 0.1 K. TG measurements for the PS granule were also performed. For all data reported, three samples were tested.

1.3.2 MATHEMATICAL MODEL OF THE THERMAL DEGRADATION KINETICS

To better analyze the kinetics of thermal degradation of PS/filler composite films, data of TG measurements were used: for instance,

- The values of the main parameters determining the rate constant of the process (such as the activation energy and the pre-exponential factor) were found and

- Analyses of their dependences on the composition of the composite were considered.

To analyze thermogravimetric data, kinetic eq 1.4 is usually used:

$$\frac{da}{dt} = k(T)f(a), \tag{1.4}$$

whereas

α = the mass fraction of the substance that decayed during the process,
t = time,
$f(\alpha)$ = the reaction model, and
$k(T)$ = the rate constant of the destruction depending on temperature according to the Arrhenius law:

$$k(T) = A\exp\left(-\frac{E_a}{RT}\right), \tag{1.5}$$

whereas

A = the pre-exponential factor, E_a is the activation energy of thermal destruction, and;

R = the gas constant.

However, the reaction model $f(\alpha)$ in the case of "Thermal Degradation" of polymers is power function of the order n

$$f(\alpha) = (1 - \alpha)^n \tag{1.6}$$

After coupling eq 1.4–1.6, we obtain the following equation:

$$\frac{d\alpha}{dt} = A\exp\left(-\frac{E_a}{RT}\right)(1-\alpha)^n, \tag{1.7}$$

This can be integrated with the initial conditions

$$\alpha = 0 \text{ at } T = T_0. \tag{1.8}$$

And the final expression can appear in the following form[28]:

$$F(\alpha) = \int_0^\alpha \frac{d\alpha}{(1-\alpha)^n} = \frac{A}{\beta}\int_{T_0}^{T} \exp\left(-\frac{E_a}{RT}\right)dT, \tag{1.9}$$

where $\beta = \frac{dT}{dt}$ is the heating rate of the sample.

To be able to analyze the data of thermo-gravimetric measurements and for the determination of the kinetic parameters of the decomposition process, we applied eq 1.9, van Krevelen method.[28,29] It is worth mentioning that this method is based on an assumption that the temperature range stays between $0.9T_m<T<1.1T_m$ so the following relationship is satisfied:

$$\ln\left[F(\alpha)\right]=\ln\left[\frac{A}{\beta}\left(\frac{0.368}{T_m}\right)^{\frac{E_a}{RT_m}}\left(\frac{E_a}{RT_m}+1\right)^{-1}\right]+\left(\frac{E_a}{RT_m}+1\right)\ln T. \qquad (1.10)$$

From eq 1.10 it can be concluded that with the slope of the dependence $\ln[F(\alpha)]$ vs $\ln T$ we can easily determine the activation energy (E_a). Meanwhile, the pre-exponential factor (A) can be obtained using the value of the segment cutoff by this line on the ordinate axis.

In the following paragraphs, we describe the results of our research on the effect of filler on the thermal stability of composites on the basis of polystyrene.

1.3.3 PS/SILICA COMPOSITE FILMS

In Figures 1.13a and 1.13b, we have shown our own observations and results obtained from the thermo-gravimetric curves (i.e., TG and DTG) for unmodified polystyrene and PS/SiO_2 film composites containing 1 wt.% of silica. This is in fact a typical thermal behavior for all types of film compositions that contained silica in our experimental studies.

It can be seen that there are two steps in the TG curves. The first step is observed in the temperature range from 370 K to 500 K. This mass loss (5–10 wt.%) is caused by solvent evaporation out of the film because it is not observed in the TG curve for the PS granule (Fig. 1.13c). Table 1.5 shows the values of the mass loss in the first stage and nonvolatile residue. It can be seen that the value of nonvolatile residue determined by thermogravimetry is not always agreed with the weight percent of silica incorporated into the polystyrene matrix. We found no nonvolatile residue in composites where the silica concentration does not exceed 1 wt.%. Bera et al.[4] reported on this discrepancy, too. The difference between the nonvolatile residue value and the weight percent of silica was both positive and negative.

TABLE 1.5 Numerical Data on the First Stage of Thermal Decomposition for the PS/Silica Composites.[a]

SiO_2 content (wt.%)	Mass loss (%)	Nonvolatile residue (%)
0	8.8	–
0.5	9.3	–
1	9.5	–
3	6.1	1.1
5	4.5	5.0

[a]Given values were obtained by averaging the data of three repeated experiments.

In the second stage of the TG curve (the actual decomposition of the substance), four characteristic temperatures can be identified:

T_{onset} is the extrapolated temperature of the thermal destruction onset that was determined by the tangent intersection method;

T_m is the temperature corresponding to the peak on the DTG curve or the inflection on the TG curve;

$T_{50\%}$ is the temperature at which the mass of the sample is halved;

T_{end} is the extrapolated temperature of the degradation end that was determined by the tangent intersection method.

Table 1.6 shows the mean values of T_{onset} and T_m for unmodified polystyrene and PS/SiO_2 film composites. The onset of degradation, that is, where the weight loss begins, plays an important role in studying the degradation behavior of the composites. The polymer degradation starts with chain scission. It is followed by depolymerization and formation of the main evolved products (styrene monomer, dimer, and trimer).[30] Figures 13a and 13b show that the degradation rate increases beyond 660 K. This implies almost all polymer chains are degraded abruptly between 660 and 710 K. The temperature corresponding to the DTG peak, T_m, is also an important indicator. At this mass loss, the material does not retain its initial properties. It can be seen for the composite film containing 5 wt.% of SiO_2, the T_{onset} and T_m values are shifted toward higher temperatures in comparison with pure PS (Table 1.6). This indicates an improvement in the thermal stability of the polymer owing to silica loading.

The values of T_{onset} for the PS/silica composites found in this work (Fig. 1.14, curve 1) agree with those found by Bera et al.[4] However, the values of T_m presented in Ref. [4] are higher by 10–12 K than those found by us. This difference may arise from the higher molecular weight (240,000) of PS studied by Bera et al.[4] When the molecular weight is greater, the temperature, T_m, at which the majority of the polymer chains are degraded, is higher.

a

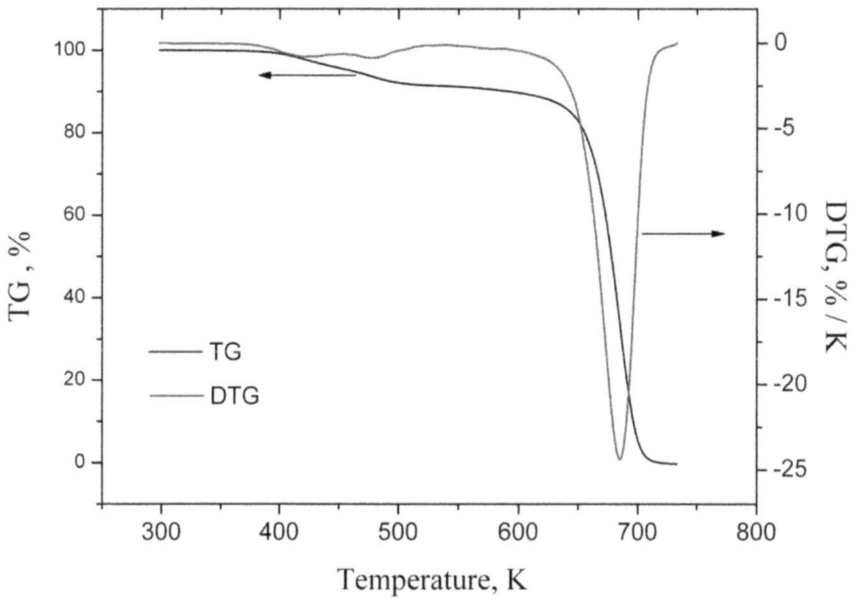

b

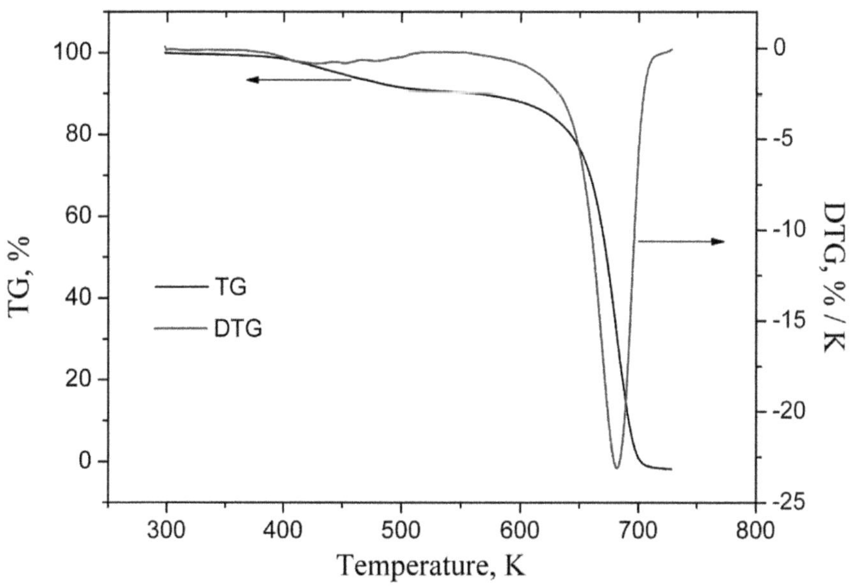

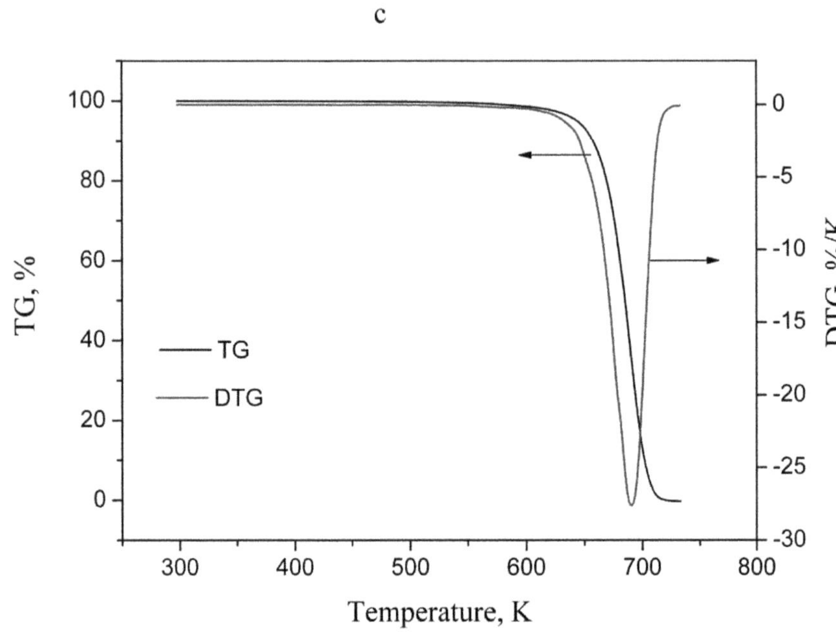

FIGURE 1.13 Thermogravimetric curves (TG and DTG): (a) polystyrene; (b) polystyrene/silica composite (1 wt.% of SiO_2); (c) polystyrene granule.

Data of TG measurements of the PS/silica composites were used to estimate the main parameters determining the rate constant of the process, like the "activation energy" and also the other important factor named "pre-exponential factor." For this purpose, the above-mentioned van Krevelen method was used.

Figure 1.15 shows the dependences ln[$F(\alpha)$] vs ln$T$, constructed at n = 1 for thermal destruction of PS/silica film composites with different filler concentrations. In Figure 1.15 the linearity of the given dependences (R^2 = 0.994–0.998) justifies the viability of our assumption (i.e., n = 1). It is worth mentioning that in a report already published (please refer to Ref. [31]), the possibility of considering the first-order reaction equation for describing the kinetics of thermal destruction of polymers has been explained in detail. Nevertheless, this idea agrees completely with the results of published research work in Ref. [32]. In this Ref. [32] the researchers considered the 15 reliable models. They have then analyzed the experimental data on the thermal destruction of polystyrene (in the form of granules) and at the

TABLE 1.6 Characteristic Temperatures of Thermal Degradation (Second Stage) for the PS/Silica Composites.[a]

SiO_2 content (wt.%)	*Tonset* (K)	T_m (K)
0	662.0	685.1
0.5	666.1	686.9
1	657.7	681.8
3	663.8	686.4
5	670.3	691.5
PS granule	666.2	690.6

[a]Given values were obtained by averaging the data of three repeated experiments.

end, they have concluded that the kinetic model of the first order is much preferable.[32]

Table 1.7 shows the kinetic parameters of thermal decomposition of the studied PS/SiO_2 film composites, obtained by the van Krevelen method for various filler concentrations. When comparing these data with those presented in other publications, we can note a certain difference. For example, according to Ref. [33], in the case of unmodified polystyrene, the activation energy takes values from 154 to 260 kJ/mol depending on the molecular weight. It is likely that the value of E_a may also depend on other characteristics (polydispersity, fluidity of the melt of the polymer material, spatial configuration of macromolecules, etc.). Therefore, the experimental data obtained by different authors (including our results) do not always agree with each other.

TABLE 1.7 Kinetic Parameters of Thermal Degradation of the Polystyrene/Silica Composites Determined by the van Krevelen Method.

SiO_2 content, wt.%	E_a, kJ/mol	$A \cdot 10^{-19}$, min^{-1}
0	272.0	5.1
0.5	274.2	5.9
1	276.8	17.7
3	280.6	21.5
5	284.5	24.5

From Table 1.7 these results can be concluded:

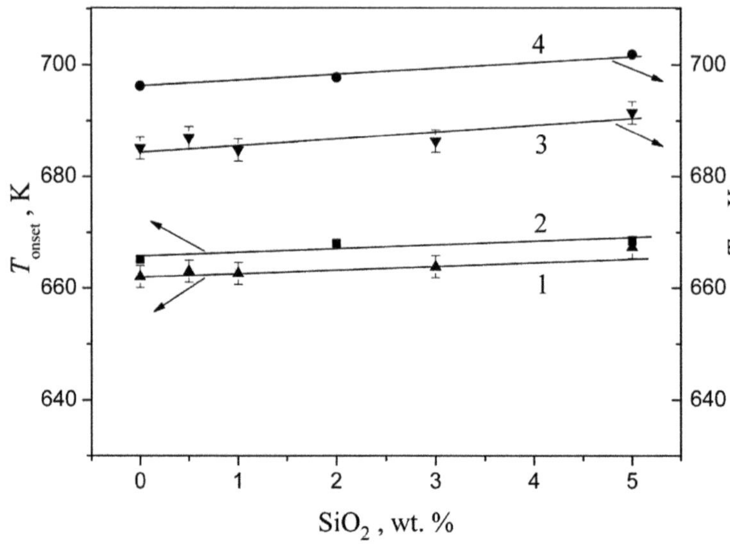

FIGURE 1.14 Concentration dependences of the characteristic temperatures: T_{onset} (graphs 1 and 2) and T_m (graphs 3 and 4). Graphs 1 and 3 were recorded in the current study. Graphs 2 and 4 were reported by Bera et al.[4]

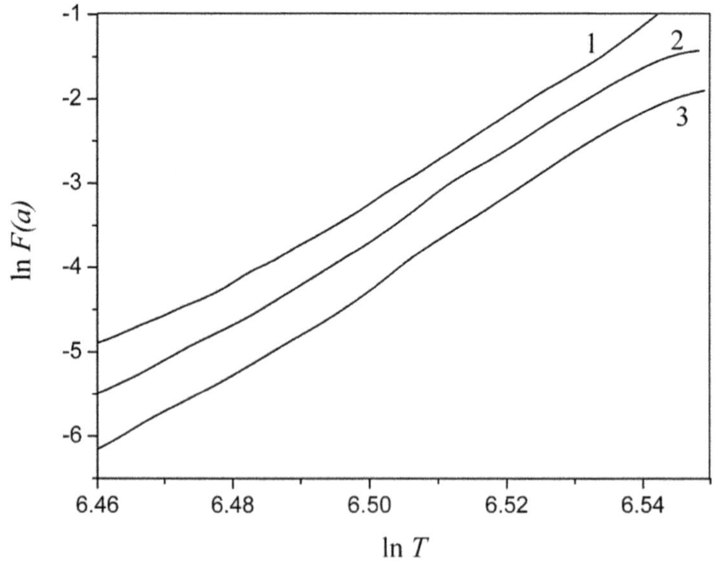

FIGURE 1.15 Thermogravimetry data for PS/SiO_2 composites in van Krevelen coordinates. Concentration of SiO_2, wt.%: 1 (*1*), 3 (*2*), and 5 (*3*).

- An augmentation in concentration of silicon dioxide (in the composite), caused the activation energy and pre-exponential factor to augment as well.
- *A* and E_a are the leading factors that calculate the rate constant (i.e., characterize the thermal stability of the targeted material).
- If the rate constant is reduced with augmentation of activation energy, then an augmentation in the pre-exponential coefficient could lead to the opposite results.

These parameters are therefore very vital to understand the mechanism of thermal decomposition (for investigating the effects of nature and concentration of filler on the values).

As mentioned in the earlier report published in Ref. [34], the introduction of Pd particles into polystyrene avoids the destruction of the polymer. In his published research work,[34] it is emphasized that modification with silicon dioxide could augment the thermal stability of PS. Meanwhile, in such cases, evidently the mechanism of inhibition of destruction seems to be different. Namely, the main inhibiting factor could augment the activation energy upon administration of the filler. Meanwhile, this augmentation is not making up by a change in pre-exponential coefficient (*A*). That means, the introduction of silica into polystyrene could possibly augment the mobility of polymer chains, but it might lead to the augmentation of the "energy barrier," that is needed for their destruction. Nevertheless, the second factor is leading one.

1.3.4 PS/BENTONITE COMPOSITE FILMS

Figure 1.16 shows the TG curves for unmodified polystyrene and PS/bentonite film composites containing 0.1–5 wt. % of filler.

It can be seen that, as for composites with silica, there are two regions in the TG curves. The first region is observed in the temperature range from 110 to 190°C and caused by solvent evaporation out of the film.

The second region of the TG curve is associated with the actual degradation of the substance. Table 1.8 shows the mean values of characteristic temperatures above mentioned such as T_{onset}, T_m, $T_{50\%}$, and T_{end}. The onset of degradation, that is, where the weight loss begins (in the range of 388–404°C depending on the concentration of bentonite), is caused by the processes of polymer chains rupture. In the temperature range of 404–422°C (depending on the concentration of bentonite), a two-fold decrease in the sample masses is observed. Upon reaching 445°C, almost all polymer chains are destroyed.

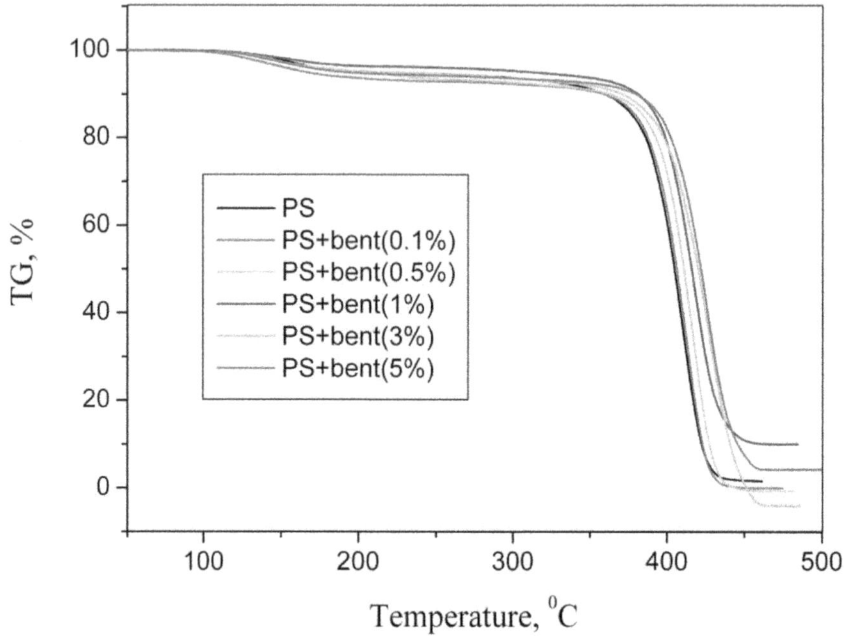

FIGURE 1.16 TG curves for unmodified polystyrene and PS/bentonite film composites.

As can be seen from Figure 1.16, with an increase in the concentration of bentonite in the composite, the TG curves shift toward higher temperatures. Information on concentration dependences of characteristic temperatures (Table 1.8) confirms this trend, which indicates an improvement in the thermal stability of the polymer owing to clay loading.

TABLE 1.8 Characteristic Temperatures of Thermal Degradation (Second Stage) for the PS/Bentonite Composites.[a]

Bentonite content in composite, wt.%	T_{onset}, °C	T_m, °C	$T_{50\%}$, °C	T_{end}, °C
0	388.3	410.4	404.7	423.7
0.1	391.0	412.6	406.6	425.5
0.5	396.1	416.8	411.0	429.8
1	396.5	417.7	415.8	433.1
3	400.2	426.9	420.0	445.6
5	403.1	426.0	421.9	442.2

[a]Given values were obtained by averaging the data of three repeated experiments.

Using the data of TG measurements for the PS/bentonite composites, we found the main quantitative parameters determining the process rate. Namely, we used the van Krevelen method and calculated the activation energy along with pre-exponential factor that is included in eq 1.10.

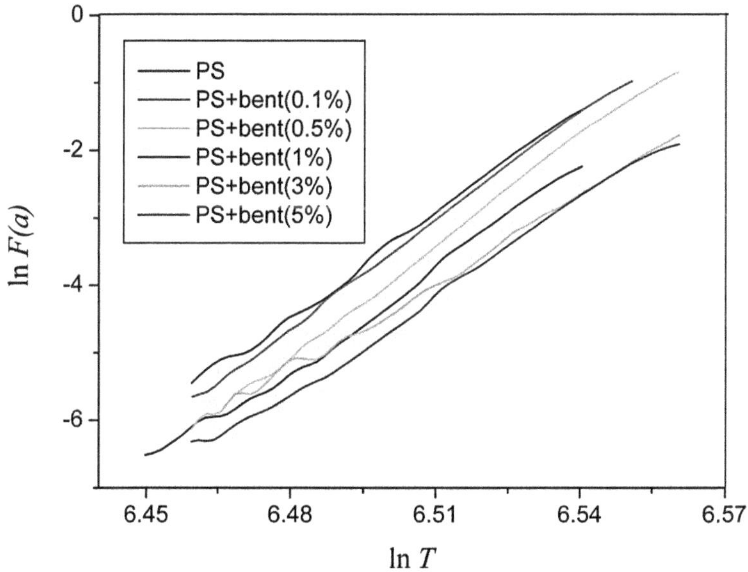

FIGURE 1.17 Thermogravimetry data for PS/bentonite composites in van Krevelen coordinates.

TABLE 1.9 Kinetic Parameters of Thermal Degradation of the PS/Bentonite Composites Determined by the van Krevelen Method.

Bentonite content, wt.%	E_a, kJ/mol	$A \cdot 10^{-19}$, min^{-1}
0	286.7	80.316
0.1	298.1	577.76
0.5	305.7	1524.9
1	279.4	9.19
3	243.8	0.01
5	270.6	1.16

However, the listed parameters change nonmonotonically with an increase in the clay concentration in the composite (Table 1.9). This is especially true for the value of the pre-exponential factor. Apparently, it should be concluded that the van Krevelen method is inapplicable for assessing the main kinetic parameters of thermal decomposition of the PS/bentonite composites.

1.3.5 PS/BENTONITE/MAGNETITE COMPOSITE FILMS

Thermogravimetric curves (TG and DTG) for unmodified polystyrene and PS/bentonite/magnetite film composites containing 0.1, 1, and 5 wt.% of filler are shown in Figures 1.18a and 1.18b. As can be seen, the TG curves are identified by two steps (i.e., thermal decomposition occurs in two stages).

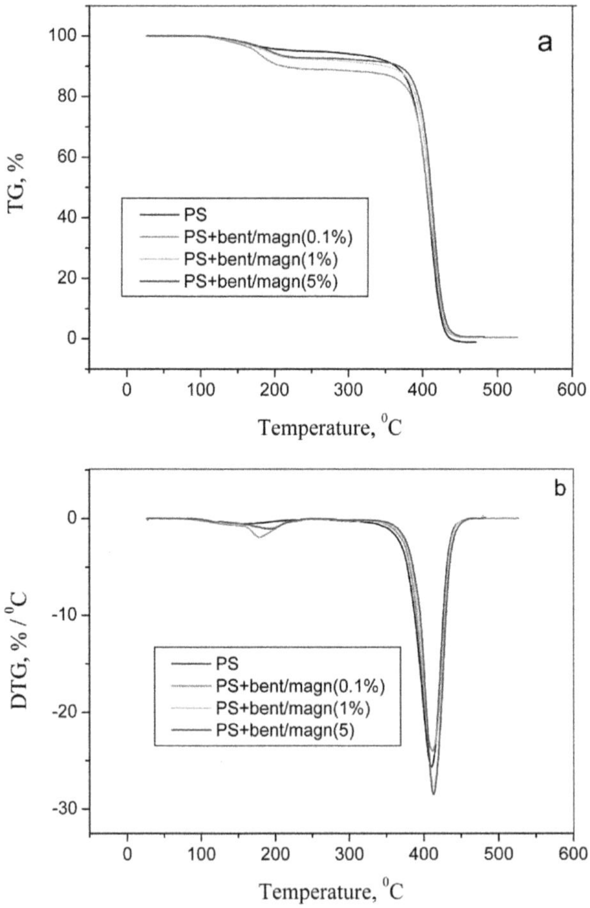

FIGURE 1.18 Thermogravimetric curves (*a*—TG; *b*—DTG) for PS/bentonite/magnetite film composites with various filler concentrations.

At the first stage (for the temperature range of 110 to 220°C), the weight loss (5–11%) is associated with the removal of solvent residues from the sample (film).

At the second stage, the actual decomposition of the film substance takes place.

As can be seen from Table 1.10, for unmodified polystyrene, the temperature of the onset of thermal degradation is 388.4°C and almost does not differ from that given in[35] (388.07°C). With an increase in the concentration of magnetic clay from 0 to 5 wt.%, the T_{onset} value increases by almost 8°C, which indicates a positive effect of the filler on the thermal stability of the polymer. At temperature close to T_{onset}, the destruction of PS polymer chains starts. For instance, depolymerization with the formation of styrene monomers, dimers, trimers, etc. The revealed higher initial temperature of thermal degradation of the composites compared to the original polystyrene may be due to the fact that the addition of filler reduces the mobility of the chains and, therefore, slows down the decomposition reaction.

TABLE 1.10 Characteristic Temperatures of Thermal Destruction of PS/Bentonite/Magnetite Composites with Various Filler Concentrations.[a]

Filler content, wt.%	T_{onset}, °C	T_{50}, °C	T_d, °C	T_{end}, °C
0	388.4	405.0	410.5	425.1
0.1	389.7	405.0	411.7	426.6
1	393.6	408.7	413.7	426.8
5	396.2	410.6	413.1	427.6

[a]Given values were obtained by averaging the data of three repeated experiments.

In the temperature range of 370–430°C (Fig. 1.18a), most of the polymer chains are destroyed. At $T = T_d$, the rate of thermal degradation is maximum (Fig. 1.18b). As can be seen from Table 1.10, compared to the T_{onset} value, the other characteristic temperatures of thermal degradation change less.

1.4 CONCLUSION

Nowadays, there has been a great interest in the development of multifunctional composites based on polymers (such as polystyrene), containing inorganic fillers. This is caused by the desire to obtain materials with low cost and a set of properties that exceed the properties of pristine polymers. Many researches concluded that an improvement in physical and chemical properties of composites compared to ones of original polymers is related to structural changes under the effect of introduced filler. However, the mechanism of these improvements is still far from clear. Therefore, the effect of doping on relaxation transitions from the glassy state to the elastic one, thermal stability etc requires complex analysis. The prolongation of these studies is very important, since the use of the polystyrene/filler composites will allow expansion of the applications borders of polystyrene traditionally used in construction industry, home appliances, electronics, etc., and will allow them to be replaced with composite materials.

KEYWORDS

- **thermal behavior**
- **polystyrene**
- **fullerenes**
- **silica**
- **bentonite**
- **DSC**
- **thermogravimetry**

REFERENCES

1. Bunekar, N.; Tsai, T. Y.; Huang, J. Y.; Chen, S. J. Investigation of Thermal, Mechanical, and Gas Barrier Properties of Polypropylene-Modified Clay Nanocomposites by Micro-Compounding Process. *J. Taiwan Inst. Chem. Eng.* **2018,** *88*, 252–260. doi.org/10.1016/j.jtice.2018.04.016
2. Lopes, A. C.; Martins, P.; Lanceros-Mendez, S. Aluminosilicate and Aluminosilicate based Polymer Composites: Present Status, Applications and Future Trends. *Prog. Surf. Sci.* **2014,** 89, 239–277. doi.org/10.1016/j.progsurf.2014.08.002
3. Badamshina, E.; Gafurova, M. Polymeric Nanocomposites Containing Non-covalently Bonded Fullerene C_{60}: Properties and Applications. *J. Mater. Chem.* 2012, *22*, 9427–9438. DOI: 10.1039/C2JM15472B
4. Bera, O.; Pilic, B.; Pavlicevic, J.; Jovicic, M.; Holló, B.; Mészáros Szécsényi, K.; Spirkova, M. Preparation and Thermal Properties of Polystyrene/Silica Nanocomposites. *Thermochim. Acta* **2011,** *515*, 1–5. doi.org/10.1016/j.tca.2010.12.006
5. Alekseeva, O. V.; Noskov, A. V.; Guseynov, S. S.; Agafonov, A. V. Thermal Behaviour of Polystyrene/Silica Composites. *Phil. Mag. Lett.* **2018,** *98*, 107–117. DOI: 10.1080/09500839.2018.1486046
6. Alekseeva, O. V.; Bagrovskaya, N. A.; Noskov, A. V. Effect of C_{60} Filling on Structure and Properties of Composite Films based on Polystyrene. *Arab. J. Chem.* **2018,** *11*, 1160–1164. Doi: 10.1016/j.arabjc.2014.09.008
7. Alekseeva, O. V.; Rudin, V. N.; Melikhov, I. V.; Bagrovskaya, N. A.; Kuzmin, S. M.; Noskov, A. V. Kinetics of Formation of Hierarchical Nanostructures in Polystyrene

Films Containing Fullerene. *Dokl. Phys. Chem.* **2008,** *422*, 275–278. DOI: 10.1134/S0012501608100084

8. Chafidz, A.; Latief, F. H.; Al-Fatesh, A. S.; Kaavessina, M. Crystallization and Thermal Stability of Polypropylene/Multi-Wall Carbon Nanotube Nanocomposites. *Phil. Mag. Lett.* **2016,** *96*, 367–374. doi.org/10.1080/09500839.2016.1213441
9. Alekseeva, O. V.; Noskov, A. V.; Guseynov, S. S. Thermal Behavior of Polystyrene-based Composite Materials. *Prot. Met. Phys. Chem. Surf.* **2020,** *56*, 469–472. DOI: 10.1134/S2070205120030041
10. George, J.; Ishida, H. A Review on the Very High Nanofiller-Content Nanocomposites: Their Preparation Methods and Properties with High Aspect Ratio Fillers. *Prog. Polym. Sci.* **2018,** *86*, 1–39. doi.org/10.1016/j.progpolymsci.2018.07.006
11. Alekseeva, O. V.; Rodionova, A. N.; Bagrovskaya, N. A.; Agafonov, A. V.; Noskov, A. V. Effect of the Organo-bentonite Filler on Structure and Properties of Composites based on Hydroxyethyl Cellulose. *J. Chem.* **2017,** *2017*, 1603937. DOI: 10.1155/2017/1603937
12. Xu, Q.; Li, X.; Zhang, Z. Preparation of Cu Nanoparticles-Modified PA6 Composites using CuO as Filler. *J. Iran. Chem. Soc.* **2014,** *11*, 1717–1721. DOI: 10.1007/s13738-014-0444-4.
13. Lazar, S. T.; Howell, B. A.; Daniel, Y. G.; Li, K. J. Thermal Decomposition of Poly(styrene) in the Presence of a Hydrogen Atom Transfer Agent. *J. Therm. Anal. Calorim.* **2017,** *127*, 969–974. DOI: 10.1007/s10973-016-5777-z
14. Chigwada, G.; Kandare, E.; Wang, D.; Majoni, S.; Mlambo, D.; Wilkie, C. A.; Hossenlopp, J. M. Thermal Stability and Degradation Kinetics of Polystyrene/Organically-Modified Montmorillonite Nanocomposites. *J. Nanosci. Nanotechno.* **2008,** 8, 1927–1936. DOI: 10.1166/jnn.2008.027
15. Alekseeva, O. V.; Bagrovskaya, N. A.; Kuzmin, S. M.; Noskov, A. V. Study of Structural Properties of Polystyrene-Fullerene Composites. *J. Balkan Tribol. Assoc.* **2010,** *16*, 558–563.
16. Dixit, S.; Yadav, V. L. Comparative Study of Polystyrene/Chemically Modified Wheat Straw Composite for Green Packaging Application. *Polym. Bull.* **2020,** *77*, 1307–1326. DOI: 10.1007/s00289-019-02804-0
17. Keshavarz, M.; Zebarjad, S. M.; Daneshmanesh, H.; Moghim, M. H. On the Role of TiO_2 Nanoparticles on Thermal Behavior of Flexible Polyurethane Foam Sandwich Panels. *J. Therm. Anal. Calorim.* **2017,** *127*, 2037–2048. DOI: 10.1007/s10973-016-5700-7
18. Katiyar, R.; Bag, D. S.; Nigam, I. Thermal Properties of Fullerene (C_{60}) Containing Poly(alkyl methacrylate). *Thermochim. Acta* **2013,** *557*, 55–60. doi.org/10.1016/j.tca.2013.01.027
19. Barabanova, A. I.; Pryakhina, T. A.; Bychko, K. A.; Kazantseva, V. V.; Zavin, B. G.; Vygodskii, Y. S.; Askadskii, A. A.; Shevnin, P. L.; Philippova, O. E.; Khokhlov, A. R. Nanocomposites based on Epoxy Resin and Silicon Dioxide Particles. *Polym. Sci. Ser. A* **2008,** *50*, 808–819. DOI: 10.1134/S0965545X08070110
20. Sanz, A.; Wong, H. C.; Nedoma, A. J.; Douglas, J. F.; Cabral, J. T. Influence of C_{60} Fullerenes on the Glass Formation of Polystyrene. *Polymer* **2015,** *68*, 47–56. DOI: 10.1016/j.polymer.2015.05.001
21. Sabban, J. R.; Xu-wu, A.; Chichos, J. S.; Planas Leitão, M. L.; Roux, M. V.; Torres, L. A. Reference Materials for Calorimetry and Differential Thermal Analysis. *Thermochim. Acta* **1999,** *331*, 93–204.

22. Weng, D.; Lee, H. K.; Levon, K.; Mao, J.; Scrivens, W. A.; Stephens, E. B.; Tour, J. M. The Influence of Buckminsterfullerenes and their Derivatives on Polymer Properties. *Eur. Polym. J.* **1999,** *35*, 867–78. DOI: 10.1016/S0014-3057(98)00055-X
23. Gladchenko, S. V.; Polotskaya, G. A.; Gribanov, A. V.; Zgonnik, V. N. The Study of Polystyrene-Fullerene Solid-Phase Composites. *Techn. Phys.* **2002,** *47*, 102–106. DOI: 10.1134/1.1435898 .
24. Tugov, I. I.; Kostrykina, G. I. *Khimiya i fizika polimerov* (Chemistry and Physics of Polimers); Khimiya: Moscow, 1989.
25. Lee, K. J.; Lee, D. K.; Kim, Y. W.; Choe, W. S.; Kim, J. H. Theoretical Consideration on the Glass Transition Behavior of Polymer Nanocomposites. *J. Polym. Sci. Pol. Phys.* **2007,** *45*, 2232–2238. DOI: 10.1002/polb.21178
26. Roldugin, V. I.; Serenko, O. A.; Getmanova, E. V.; Karmishina, N. A.; Chvalun, S. N.; Muzafarov, A. M. Thermodynamic Analysis of the Glass Transition Temperatures of the Polymer–Hybrid Nanoparticles Systems. *Dokl. Phys. Chem.* **2013,** *449*, 83–87. DOI: 10.1134/S0012501613040088
27. Alekseeva, O. V.; Noskov, A. V.; Guseynov, S. S. Thermal Behaviour of Poly(Methyl Methacrylate)/Fullerene Composite Films. *Phil. Mag. Lett.* **2018,** *98*, 330–340. DOI: 10.1080/09500839.2018.1548786
28. Sinfronio, F. S. M.; Santos, J. C. O.; Pereira, L. G.; Souza, A. G.; Conceicao, M. M.; Fonseca, V. M. Kinetic of Thermal Degradation of Low-Density and High-Density Polyethylene by Non-isothermal Thermogravimetry. *J. Therm. Anal. Calorim.* **2005,** *79*, 393–399. DOI: 10.1007/s10973-005-0072-4
29. Santos, J. C. O.; Santos, I. M. G.; Souza, A. G. Thermal Stability and Kinetic Study on Thermal Decomposition of Commercial Edible Oils by Thermogravimetry. *J. Food Sci.* **2002,** *67*, 1393–1398. DOI: 10.1111/j.1365-2621.2002.tb10296.x
30. De Amorim, M. T. S. P.; Bouster, C.; Vermande, P.; Veron, J. Study of the Pyrolysis of Polystyrenes: II. Study of Transfer Reactions by Identification of the Most Important by-Products. *J. Anal. Appl. Pyrolysis* **1981,** 3, 19–34. DOI: 10.1016/0165-2370(81)80023-X
31. Van Krevelen, D. W. In *Properties of Polymers: Correlations with Chemical Structure*; Elsevier Pub. Co.: Amsterdam, 1972; p 427.
32. Cheng, J.; Pan, Y.; Yao, Y.; Wang, X.; Pan, F.; Jiang, J. Mechanisms and Kinetics Studies on the Thermal Decomposition of Micron poly (Methyl Methacrylate) and Polystyrene. *J. Loss Prevent. Process Ind.* **2016,** *40*, 139–146. DOI: 10.1016/j.jlp.2015.12.017
33. Yildirim, Y.; Doğan, B.; Muğlali, S.; Saltan, F.; Özkan, M.; Akat, H. Synthesis, Characterization, and Thermal Degradation Kinetic of Polystyrene-*g*-Polycaprolactone. *J. Appl. Polym. Sci.* **2012,** *126*, 1236–1246. DOI: 10.1002/app.36888
34. Lee, J. Y.; Liao, Y.; Nagahata, R.; Horiuchi, S. Effect of Metal Nanoparticles on Thermal Stabilization of Polymer/Metal Nanocomposites Prepared by a One-Step Dry Process. *Polymer* **2006,** *47*, 7970–7979. DOI: 10.1016/j.polymer.2006.09.034
35. Alshabanat, M.; Al-Arrash, A.; Mekhamer, W. Polystyrene/Montmorillonite Nanocomposites: Study of the Morphology and Effects of Sonication Time on Thermal Stability. *J. Nanomat.* **2013,** *2013*, 650725. DOI: 10.1155/2013/650725

CHAPTER 2

Spiropyran-Doped Liquid Crystal Film for the Monitoring of Human Ecology upon Exposure to UV Radiation: An Environmentally Friendly Approach

G. PETRIAHSVILI, L. DEVADZE, TS. ZURABISHVILI, N. SEPASHVILI, K. CHUBINIDZE, T. TATRISHVILI, T. BUKIA, E. KALANDIA, M. ARESHIDZE, L. SHARASHIDZE, SH. AKHOBADZE, A. PETRIASHVILI, M. CHUBINIDZE, V. KINKLADZE, and N. IMNAISHVILI

Institute of Cybernetics of Georgian Technical University, Tbilisi, Georgia

Department of Polymer Materials, Tbilisi State University, Tbilisi, Georgia

ABSTRACT

(UV) or ultra-violet radiation is a kind of electromagnetic radiation that originated from the sun and artificial sources. UV radiation can harm the deoxyribonucleic acid or DNA (genes) in cells, which consecutively may lead to cancer. Almost all skin cancers are a result of exposure to ultra-violet radiation. We have developed a novel UV radiation dosimeter that allows a real-time estimation of the dose of harmful ultraviolet UV-A and UV-B radiations discharged by the sun and man-made sources. When exposed to ultra-violet radiation, our UV dosimeter visibly converts color from the starting yellow to blue to dark violet. The converted color corresponds to levels of accumulated UV radiation.

Environmental Technology and Sustainability: Physical, Chemical, and Biological Technologies for Environmental Protection. Tamara Tatrishvili, Ann Rose Abraham, & A. K. Haghi (Eds.)

2.1 INTRODUCTION

Tracking down and monitoring of ultraviolet (UV) radiation are essential in many areas such as:

- ultra-violet radiation measuring,
- monitoring of pollution,
- purification of water,
- medicine,
- furnace control,
- biological/chemical battlefield reagent detectors,
- space-to-space communications, f
- detection of flames,
- ultra-violet astronomy, and
- early missile threat warning.

The ultraviolet (UV) region covers the wavelength range of 100–400 nm. This is divided into three bands as follows:

Band 1: ultraviolet (C), UVC (100–280 nm),

Band 2: ultraviolet (B), UVB (280–320 nm), and

Band 3: ultraviolet (A), UVA (320–400 nm).

As sunlight proceeds through the atmosphere, all ultraviolet (C), and approximately 90% of ultraviolet (B), radiation are absorbed by "ozone," "water vapor," "oxygen," and "carbon dioxide." Ultraviolet (A) radiation is much less influenced by the atmosphere. Therefore, the ultraviolet radiation reaching the Earth's surface is to a great extent composed of ultraviolet (A), with a minor ultraviolet (B), component. It should be noted that exposure to ultraviolet radiation is involved in both unfavorable and favorable health outcomes. Nevertheless, constant human exposure to solar ultraviolet radiation could result in drastic and long-term health effects on the skin, eye, and immune system. Sunburn and tanning are the foremost critical effects of uncontrolled ultraviolet radiation exposure; in the long duration, ultraviolet radiation-associated degenerative changes in cells, fibrous tissue, and blood vessels lead to early skin aging. Ultraviolet radiation can also cause provocative reactions in the eye, such as photokeratitis. But according to published works indicated in Refs. [1,2], the most critical chronic effects include two major public health problems: skin cancers and cataracts. To be able to conduct research on such important environmental and health field, the accurate measurement and monitoring of ultraviolet radiation exposure dose are extremely important. Different measurement systems are available to measure personal exposure to ultraviolet radiation doses and there are different types of ultraviolet radiation sensors in the market. All of these different measurement systems have their own features that have to be taken into account when choosing one of them for personal dosimetry and research studies. Electronic dosimeters can precisely measure ultraviolet radiation doses, can be obtained in watch-like designs, or can be provided like data loggers. These are obviously attached to certain parts of the body such as the chest or the limbs. The other type is biological dosimeter, which in fact uses the germicidal effects of ultraviolet radiation on spores to be able to quantify ultraviolet exposure. These types of dosimeters only measure the progressive amount of ultraviolet exposure during the measurement period. It should be pointed out that these do not allow temporal resolution of measurement data. Therefore, a validity check of the captured measurement data, for example, in terms of perfect attachment to measurement sites, is not possible. Meanwhile, they are not sensitive to radiation with a wavelength greater than 340 nm. This is in fact due to their material properties. And this means that the ultraviolet (A) radiation portion of the spectrum has to be deduced during a calibration process in contrast to the sun spectrum without awareness of the real exposure and the actinic fraction of the ultraviolet (A) radiation. In these systems, typical battery-powered electronics could easily promote wireless operation and digital data collection using photo-detectors and memory modules. This will help to reduce collective costs. Moreover, the limited endurance of the batteries along with their need for recharging and their sensitivity to water, heat, and other environmental conditions set-back practicability and this could end up to throw away the device.[3–5]

The climate erythemal dose can be considered the "incident erythemally weighted irradiance," or it can de expressed as "the dose rate, on a horizontal surface over a specified period." This is shown in Joules per square, that is, meter, J/m^2, or mJ/cm^2. The spectral sensitivity of this effect is homogenous to certain action spectra. As an example, we can mention the followings:

- The "SER" or the erythemal weighting function;
- The "SED" or analogous to the standard erythema dose; and
- The "MED" or minimum erythema dose that are in common use for erythema.

The erythemal weighting function resembles the ultraviolet radiation erythema reaction of the skin and it is then applied to explain the spectral dependence of this effect. It should be noted that radiometers and spectro-radiometers could be used to monitor ambient solar ultraviolet radiation. It should be pointed out that this equipment is not appropriate for monitoring the ultraviolet radiation of individuals, particularly at multiple sites over the body. This is due to their bulk and high cost. The most personal ultraviolet dosimeter extensively used over the past 45 years is:

- The plastic films based on polysulphone,
- The diazo systems,
- The plastic films incorporating photosensitizing drugs,
- The photosensitive papers,
- The thermo-luminescent materials.

However, they have some drawbacks and limitations, to name just a few:

- Has limited sensitivity by covering just the ultraviolet (A), UVA (320–400 nm) and also the blue regions of the spectrum,
- Are designed for single use only,
- Will provide a semiquantitative estimate of exposure.

In this work, we introduce a spiropyran (SP) doped liquid crystal polymer (SPLCP) film-based UVA + UVB reusable dosimeter, designed as a test card that allows a real-time estimation of the dose of harmful ultraviolet UV-A and UV-B irradiation. The operational principle of the dosimeter is based on the change of optical absorbance of the SPLCP film by exposure to UV radiation. In particular, upon exposure to UVA + UVB radiation, the SPLCP film-based dosimeter visibly changes color from the starting yellow to blue to dark violet.

One of the best and remarkable examples of molecular switches is spiropyrans (or SPs). This can be well observed in Figure 2.1. Their closed ring

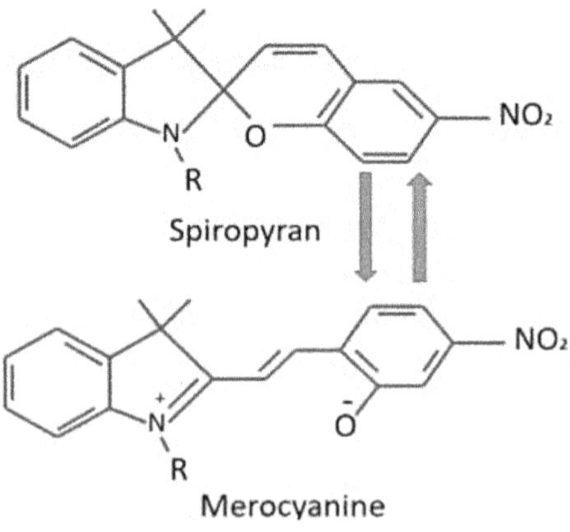

FIGURE 2.1 Light-driven spiropyran to merocyanine conversion.

along with the hydrophobic isomer transforms into a highly polar and the open-ring (MC) or the merocyanine form upon exposure to ultraviolet light, and also the reverse reaction which can be induced by visible light or by heat. As mentioned in Ref. [6], spiropyrans have been extensively used due to their potential applications in equipment such as light-sensitive eyewear for information recording and processing including “3D” recording devices, optical memories, and spiropyran-based dynamic devices.

2.2 SPLCP FILM PREPARATION

To fabricate an SPLCP film, commercially available and certified compounds were used: nematic matrix-BL-036, (from Merck), photochromic material-1’, 3′, 3′-Trimethyl-6-nitro-1’, 3′-dihydrospiro [chromene-2, 2’-indole] (SP), and macro monomer Poly (vinyl alcohol) – PVA with an average molecular weight –Mw 85,000–124,000, 99 + % hydrolyzed (both from Sigma-Aldrich). An SP-doped LC mixture was prepared to mix the nematic LC and SP with the following concentration ratio in weight: 98% BL-036 + 2% SP. The thin film was prepared as follows: 1 g of the SP-doped CLC mixture was added to 10 mL of a 15-weight % PVA solution in distilled water. As the suitable emulsifying agent, 0.5 mL of icy acetic acid was doped into the mixture. The prepared composite was poured into a glass vial and stirred at 600 rpm for 30 min at 75°C. After this procedure, a creamy white emulsion was obtained. By controlling all stages of microencapsulation, such as stirring speed, temperature, and concentration ratio of the constituents, we have obtained the microcapsules with the predictable sizes. To prepare a uniform film with the desired thickness a drop casting-spreading method was utilized. First, the prepared emulsion was let to degas for 0.5 h, and then it was dropped on the glass substrate treated with deionized water and spread homogeneously using a metal stripe. The prepared sample was stored for 48 h at room temperature. After the water evaporation, a film was detached carefully from the glass substrate. As a result, we obtained process-perfect, uniform elastic film, which meets the requirements for practical applications. Prepared SPCLC film contains phase-separated composites aggregated in the form of microcapsules with desired packages and spatial distributions. The microcapsules produced in such a manner tend to be uniform in size in a range from 10 to 15 μm. The photo-switching behavior and absorption spectra of the mixtures were investigated with a UV/VIS Spectrometer (AvaSpec 2048, Avantes) at room temperature.[7–9]

2.3 EXPERIMENTAL

The number of recording-erasing cycles, that a spiropyran (SP)-doped liquid crystal polymer (SPLCP) film can undergo, is a critical experimental parameter. For this reason we are obligated to consider the photostability over time while dealing with spiropyran (SP) photochromic dyes. To be able to carry out a study on the feasibility of repeated cycles, experiments on the fatigue resistance of spiropyran (SP)-doped liquid crystal polymer (SPLCP) films upon ultraviolet and visible light irradiation needed to be performed. At this stage, for a light-induced generation of MC, films (samples) were irradiated using a (100 W) mercury lamp HG 100 AS (Jelosil), with a (2800–400 nm) band-pass filter. Afterward, the reverse switching to the "non-colored" form was attained using the same light source with a (540–630 nm) band pass filter. Now, in a cycle, a system is transformed from the spiropyran configuration to the MC one and, then, back to the spiropyran form. If the degree of degradation in a cycle is x, there is no degraded fraction y after n cycles, and then we have $y = (1-x)^n$. Now, this expression or very small x and very large n can be approximated as $y \approx 1 - nx$. The average measured reduction in absorption was found to be 0.08 (yield = 92%) after fifty cycles on spiropyran (SP)-doped liquid crystal polymer (SPLCP) film. This approximated value can be considered for the determination of the absorption and also to compare experimental results during 300 cycles.

Next step is to optical measurements. Such measurements should be performed to compare the experimental results with the calculated absorbance as a function of the number of cycles. We have then set the exposure time for ultraviolet irradiations to 20 s. Meanwhile, the time is set on 20 s for the exposure time of the reverse switching. Instantly, we have performed the measurements after the irradiation. It is very important to consider the fact that the time interval between two cycles must be fixed to 2 min only. After the first 10 cycles, the efficiency can be found to be practically constant with no apparent hysteresis. Considering the fact that the linear estimation for the absorption approximates a more complex behavior. Now refer to Figure 2.2. As you can see from Figure 2.2, even after 300 erasing-recording cycles, the absorption efficiency of the MC form is high. As a matter of fact, it is high enough to observe images with acceptable optical contrast.

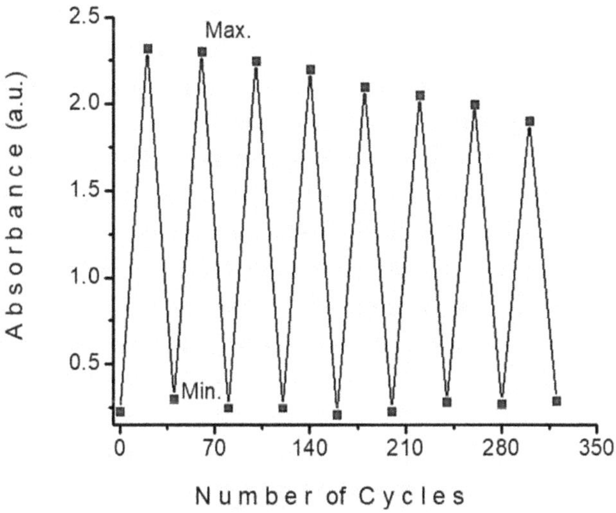

FIGURE 2.2 Absorbance of SPLCP film under recording-erasing cycles.

The calibration of ultraviolate dose was taken into account by exposing a series of UVA + UVB irradiation. This is in fact due to differing exposures on a horizontal plane of SPLCP film. It is worth mentioning simultaneous measurements of the exposures observed with a Solar Light's UV Minder Model 3D UV intensity radiometer. This device has been demonstrated in Figure 2.3. With this device we can accurately measure both ultraviolet (A), UVA (320–400 nm) and ultraviolet (B), UVB (280–320 nm) spectra (Fig. 2.3). This device is capable of displaying the intensity of ultraviolet (B), UVB (280–320 nm). This device is also called Sun-burning ultraviolate "SUV" in minimal erythemal doses per hour (MED/h). This device will give the acceptable measurement ranges for sunburn potential. The ultraviolet (A), UVA (320–400 nm) detector measures ultraviolet intensity in the range between 320 and 400 nm, displaying ultraviolet (A), UVA (320–400 nm) irradiance in mW/cm^2. The ultraviolet (A), UVA (320–400 nm) and SUV sensors are conveniently mounted on the top of the meter, making it easy to point the sensors at the target. The other important factor that should be considered is that the analog output can be provided at the bottom of the meter. This will easily allow the user to continuously monitor and record the intensity irradiated on the device active sensor. The 3D ultraviolet (B), UVB (280–320 nm), (SUV) sensor's spectral response is 280–320 nm, while the ultraviolet (A), UVA (320–400 nm) sensor's spectral response is 320–400 nm. Typical applications include:

- SPF testing,
- Environmental monitoring along with research and development,
- Agricultural studies,
- Dermatology, and
- Cosmetics research and development.

FIGURE 2.3 UV minder model 3D.

Nowadays, the basic consideration of dosimetry in skin photobiology lies in the fact that the potentiality of ultraviolet radiation to obtain erythema in human skin depends strongly on wavelength, encompassing a range of four orders of magnitude. This range is set between 280 nm and 400 nm. The erythema response to UV radiation exposure may be given or expressed as a function of either biological dose (MED) or physical dose (J/m^2). When data are expressed as mJ/cm^2 there is a clear difference between the skin types. However, when erythema is plotted as a function of MED, the apparent skin type difference disappears. Because the intensity of an ambient UV radiation varies by latitude and altitude, depends on season and time of day, and with cloud cover, ambient doses of UVA and UVB were recorded in Tbilisi (Georgia), with the next coordinates: Latitude-41.7151° N, Longitude-44.8271° E. Time of measurement, from 10:00 a.m. to 4:00 p.m. The weather—under the clear sky conditions. Measurements were carried out on 19, 20, 21, 22, and 23 of June 2021.

The SPLCP film, 3 × 1 cm in size, was exposed for intervals of 5, 10, 15, and 20 s. For each of these experiments, the absorbance was measured pre-exposure and immediately post-exposure. Curve 1, in Figure 2.4a, corresponds to the not-irradiated sample. From this figure we can observe a very small absorption at "585 nm," which is related to a quite small quantity of photoactive molecules in the MC form before irradiation. When the film is exposed to UVA + UVB light, the absorption peak at 585 nm grows, indicating the light-induced conversion of SP molecules into MC form. The curves in Figure 2.4b demonstrate the time depending reversible color changing of SPLCP film, upon exposure to solar radiation.

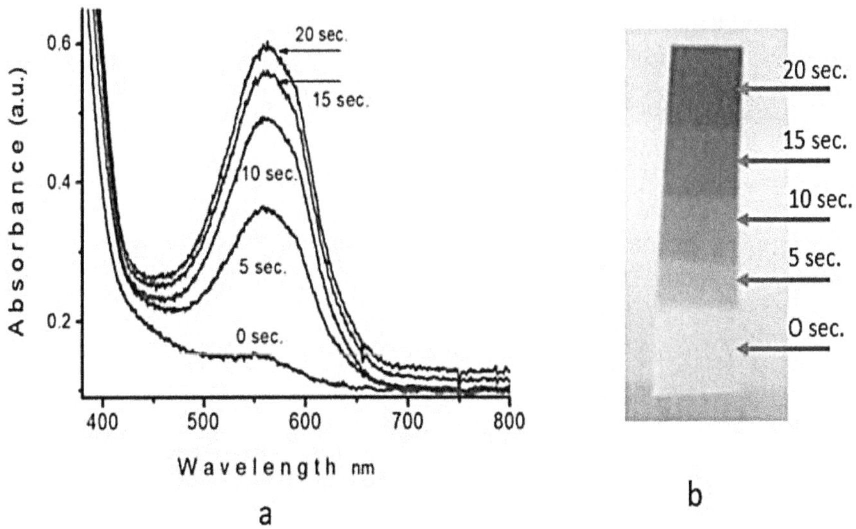

FIGURE 2.4 (a) Absorption dependence on the exposure time for an SPLCP film upon solar irradiation. Curve 1 corresponds to the absorption before UVA + UVB irradiation. Curves show the absorption peak after 5, 10, 15, and 20 s of exposure time, respectively. (b) Time depending reversible color changing of SPLCP film upon exposure to solar radiation.

The solar radiation wavelength-dependent UVB intensity can be calculated as

$$I_{UVB}(\lambda) = \int_{\lambda_m}^{\lambda_n} I(d\lambda) = \int_{\lambda_{280}}^{\lambda_{320}} I(d\lambda) \tag{2.1}$$

where λ_m and λ_n are the starting and ending spectral points of the UVB spectrum. Integrating the solar UVB radiation over the time gives the cumulative exposed intensity on the SPLCP film,

$$I_{UVB}(t,\lambda) = \int_{t_0}^{t_k} I(dt) \int_{\lambda_m}^{\lambda_n} I(d\lambda) = \int_{t_0}^{t_k} I(dt) \int_{\lambda_{280}}^{\lambda_{320}} I(d\lambda) \quad (2.2)$$

where t_0 corresponds to the SPLCP film conditions before the solar exposure, and t_k is the maximum solar expose time. Similarly to the UVB spectrum, the solar radiation wavelength- dependent UVA intensity can be calculated as

$$I_{UVA}(\lambda) = \int_{\lambda_i}^{\lambda_j} I(d\lambda) = \int_{\lambda_{320}}^{\lambda_{400}} I(d\lambda) \quad (2.3)$$

where λ_i and λ_j are the starting and ending spectral points of the UVA spectrum. As in the case of UVB exposure, integrating the solar UVA radiation over the time gives the cumulative exposed intensity on the SPLCP film,

$$I_{UVA}(t,\lambda) = \int_{t_0}^{t_k} I(dt) \int_{\lambda i}^{\lambda j} I(d\lambda) = \int_{t_0}^{t_k} I(dt) \int_{\lambda_{320}}^{\lambda_{400}} I(d\lambda) \quad (2.4)$$

Summing the intensity of UVB and UVA radiations, we get the total solar exposure on the SPLCP film,

$$I_{UVA}(t,\lambda) + I_{UVB}(t,\lambda) = \int_{t_0}^{t_k} I(dt) \int_{\lambda_{320}}^{\lambda_{400}} I(d\lambda) + \int_{t_0}^{t_k} I(dt) \int_{\lambda_{280}}^{\lambda_{320}} I(d\lambda) \quad (2.5)$$

Using eq 2.5, we have calculated the total energy absorbed by SPLCP film, expressed in mJ/cm^2 units upon 5, 10, 15, and 20 s of solar exposure times.

To enhance the precision of the obtained results, we averaged the measurements based on day-to-day experiments, using eq 2.5,

$$\sum_{N}^{M} I_{UVA,B}(t,\lambda) = \sum_{N} I_{UVA}(t,\lambda) + I_{UVB}(t,\lambda)/N + \sum_{M} I_{UVA}(t,\lambda) + I_{UVB}(t,\lambda)/M \quad (2.6)$$

where N is the number of days and M is the number of experiments per day. The images of the SPLCP film captured with a digital camera were quantitatively analyzed to determine the change in color tones and, with appropriate calibration, the exposure dose. Table 2.1 shows the variation of color tonality immediately after UVA + UVB exposure for 5, 10, 15, and 20 s at intensities of 25, 50, 75, and 100 mJ/cm². The increase of the UVA + UVB intensity results in increases in the SPLCP film coloration. Return to a colorless state occurs spontaneously after the removal of solar UVA + UVB light.

TABLE 2.1 The Variation of Color Tonality Immediately after UVA + UVB Exposure for 5, 10, 15, and 20 s.

Exposure time, s	Exposed energy mJ/cm2	Color tonality	Exposure category
0	0 mJ/cm²		Low
5	25 mJ/cm²		Moderate
10	50 mJ/cm²		High
15	75 mJ/cm²		Very high
20	100 mJ/cm²		Extreme

Finally, we have fabricated the UVA + UVB dosimeter, designed as the laminated card, which allows users to estimate the harmful UV radiation by visually monitoring of color tonality variation of the UV-sensitive SPLCP film glued onto the card (Figure 2.5).

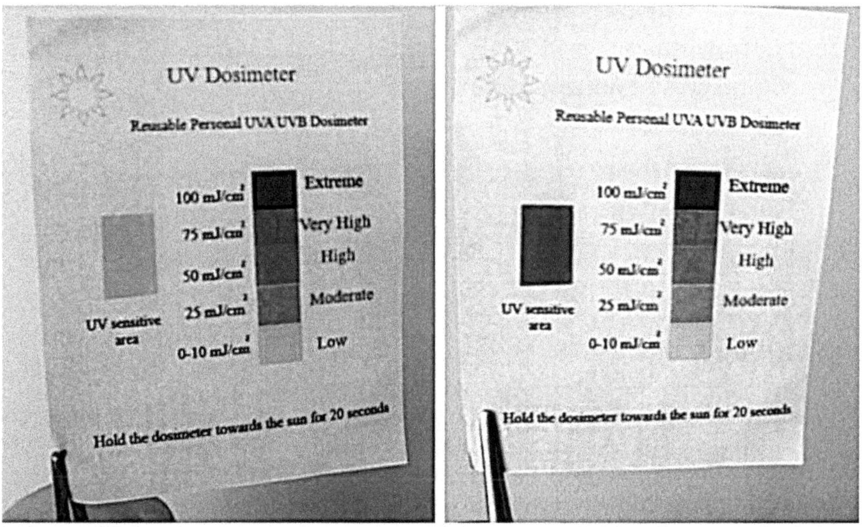

a b

FIGURE 2.5 Reusable personal UV dosimeter before exposure to solar UVA + UVB radiation (a) and after 20 s of exposure to solar UVA + UVB radiation.

2.4 CONCLUSION

In summary, we have developed a low-cost, wireless, reusable UVA + UVB dosimeter based on spiropyran-doped liquid crystal polymer film. The proposed dosimeter enables users to control a skin sunburn level caused by UVA and UVB lights from the sun and artificial sources and visually monitors how much harmful radiation has been delivered to a surface. The fabricated UVA UVB dosimeter is environmentally safe, exhibits excellent photochromic properties, improved photosensitivity, and has a high spatial resolution. It can be efficiently reused over 300 cycles without compromising its readability. A comparative coloring scale enables the user to integrate the measured UV dose in mJ/cm². In addition, the proposed material holds promise for the realization of mechanically flexible, environment-friendly, lightweight, large-area photo-chromic devices that can be fabricated by room-temperature solution processing.

KEYWORDS

- **human ecology**
- **spiropyran**
- **liquid crystal polymer film**
- **UV dosimeter**
- **erythemal dose**
- **color variation**

REFERENCES

1. Bian, X.; Jin, H.; Wang, X.; Dong, S.; Chen, G.; Luo, J. K.; Deen, M. J.; Qi, B. UV Sensing using Film Bulk Acoustic Resonators based on Au/n-ZnO/piezoelectric-ZnO/Al Structure. *Sci. Rep.* **2015,** *5*, 9123. DOI: 10.1038/srep09123, 2014
2. King, L.; Fan, X.; Swaminathan, A.; Lucas, R. M. Measuring Sun Exposure in Epidemiological Studies: Matching the Method to the Research Question. *J. Photochem. Photobiol. B Biol.* **2015,** *153*, 373–379.

3. Diffey, B. The Early Days of Personal Solar Ultraviolet Dosimetry. *Atmosphere* **2020,** *11*, 125.
4. Strehl, C.; Heepenstrick, T.; Knuschke, P.; Wittlich, M. Bringing Light into Darkness—Comparison of Different Personal Dosimeters for Assessment of Solar Ultraviolet Exposure. *Int. J. Environ. Res. Public Health* **2021,** *18*, 9071. https://doi.org/10.3390/ijerph18179071
5. Heo, S. Y.; et al. Wireless, Battery-Free, Flexible, Miniaturized Dosimeters Monitor Exposure to Solar Radiation and to Light for Phototherapy. *Sci. Transl. Med.* **2018,** *10*, eaau 1643.
6. Petriashvili, G.; De Santo, M. P.; Devadze, L.; Zurabishvili, T.; Sepashvili, N.; Gary, R.; Barberi, R. Light Controlled Drug Delivery Containers based on Spiropyran Doped Liquid Crystal Micro Spheres. *Biomed. Opt. Express* **2016,** *7* (2), 442–447.
7. Petriashvili, G.; Devadze, L,; Chanishvili, A.; Zurabishvili, C.; Sepashvili, N.; Ponjavidze, N.; De Santo, M. P.; Barberi, R. Spiropyran Doped Rewritable Cholesteric Liquid Crystal Polymer Film for the Generation of Quick Response Codes. *Opt. Mater. Express* **2018,** *8* (12), 3708–3716.
8. Petriashvili, G.; De Santo, R. H. M. P.; Gary, R.; Barberi, R. Light-Controllable Linear Dichroism in Nematics. *Appl. Opt.* **2015,** *54* (28), 8293–8297.
9. Petriashvili, G.; De Santo, M. P.; Devadze, L.; Zurabishvili, T.; Sepashvili, N.; Gary, R.; Barberi, R. Rewritable Optical Storage with a Spiropyran Doped Liquid Crystal Polymer Film. *Macromol. Rapid Commun.* **2016,** *37*, 500–505.

CHAPTER 3

High-Performance Polymer Composites Based on the Residual Polyethylene and Modified Minerals for Ecological Protection of the Environment

LANA SHAMANAURI[1] and NATIA SHENGELIA[2]

[1]*R. Dvali Institute of Machine Mechanics, Tbilisi, Georgia*

[2]*Sokhumi State University, Tbilisi, Georgia*

ABSTRACT

The solid industrial and household polymer wastes create an extremely hazardous ecological situation in the environment. Such polymer wastes do not degrade for many years, and once getting into nature, they create big environmental problems. Therefore, their utilization is very important.

Based on the scientific and practical studies, it has been established that household and industrial polymer wastes can be successfully used to obtain a targeted, competitive, and inexpensive household product.

This chapter proposes the technology of making polymer composites based on secondary polyethylene and the minerals scattering in Georgia (andesite from Bakuriani, Sachkhere quartz sand, and Okami slag).

Several important and remarkable mechanical and physical properties are studied in this chapter. We have studied absorption of water along with thermal stability as well. It is observed that the amount of optimum strength and the value of the thermal stability of the composites depend on the type of the filler and the concentration of the filler. In this research, the fillers used were modified by tetraethoxysilane into the composites that lead to

Environmental Technology and Sustainability: Physical, Chemical, and Biological Technologies for Environmental Protection. Tamara Tatrishvili, Ann Rose Abraham, & A. K. Haghi (Eds.)

enhancing of the physical properties of the materials due to creation of the buffer zone between polymer matrix and filler particles.

The polymer composites on the basis of secondary polyethylene and the minerals spread in Georgia (andesite from Bakuriani, Sachkhere quartz sand, and Okami slag) have been obtained.

It was concluded that composites consisting of binary fillers, diatomite and andesite at precise ratios, are characterized by the so-called synergistic effect acquiring the maximal physical or chemical properties of composites at definite proportions of the ingredients.

3.1 INTRODUCTION

Ecological protection of the environment and the usage of industrial wastes are considered as very important and actual problems today. The solid industrial and household polymer wastes create an extremely hazardous ecological situation in the environment. Such polymer wastes do not degrade for many years, and once getting into nature, they create big environmental problems. Therefore, their utilization is very important.

From the scientific and published research works, it is concluded that if the quality improvement of the composites could be based on secondary thermoplastic materials, about 40% of the primary ores can be spared. In such process, different types of fillers can be handled, including natural or dispersive along with the artificial fiber fillers. In the secondary polymer composites, we can consider application of the industrial technological wastes. Such wastes can be obtained from industrial sectors, from paring to the injecting process in the molding heads. It is quite important to consider the fact that the percent of such wastes can vary from 10 to 60%. According to most recent research works, polyethylene could be considered as the most widely spread polymer that has desirable engineering properties and low cost.[1–4] In this chapter, we have used the polyethylene industrial wastes (high-pressure) as binders.

3.2 EXPERIMENTAL METHOD

Fine dispersive powders were obtained from polyethylene domestics bags. These bags were made of high-pressure polyethylene with low density. The fillers used for the composites were of three different types of minerals with various ranges of concentrations (viz., Bakuriani andesite, Sachkhere quartz sand, and Okami slag).

We have used the fine dispersive powders obtained as a result of the grinding of different polyethylene bags for domestic use. Most of them are made from polyethylene of high pressure (with low density). The three types of minerals, Bakuriani andesite, Sachkhere quartz sand, and Okami slag, were used as composite fillers with a wide range of concentrations.

3.3 EXPERIMENTAL PROCEDURE

1. We determined dependence of material properties on the type and concentration of fillers.
2. Mineral powders with concentrations of 20–60 wt% were selected as composite fillers.
3. Composite ingredients were mixed in the propeller mill for about 2–3 min.
4. Mixing process of the polymers and different fillers continued until we could obtain a very uniform powder for our experimental work.
5. The drying process was completed at 50–70°C, and the pressing approached using rectangular as well as cylindrical press forms (in standard conditions).
6. The samples were prepared after pressing at 8–10 MPa considering temperature of 140–150°C for 10–15 min.
7. Compressive strength along with mechanical properties of samples was measured.
8. Water absorption was determined separately.
9. Thermal stability of samples was measured as well.

3.4 RESULTS AND DISCUSSION

The different physical and mechanical properties of the composites based on PE are presented in Figures 3.1–3.3.

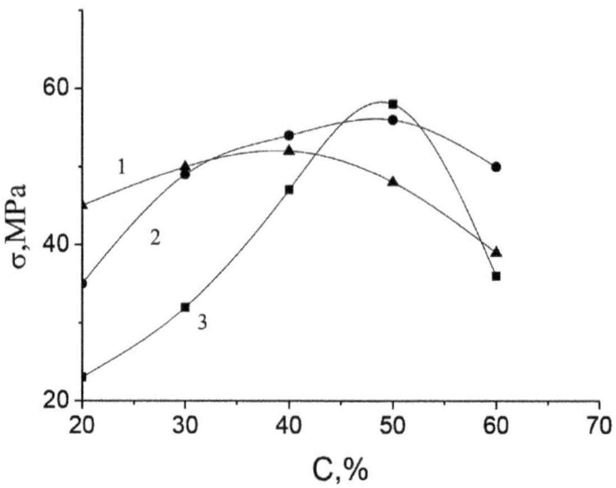

FIGURE 3.1 Dependence of the ultimate compressive strength of the PE on the filler concentration (refer to 1,2,3).

1: Quartz sand
2: Slam
3: Andesite

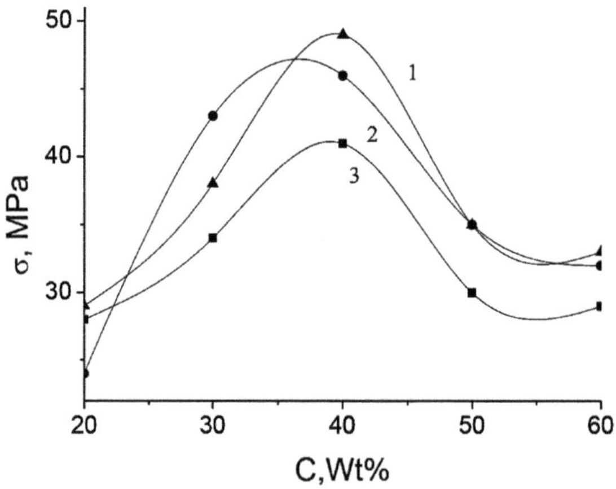

FIGURE 3.2 Dependence of the optimum bending strength of composites on the filler concentration with

1: Andesite
2: Slam
3: Quartz sand.

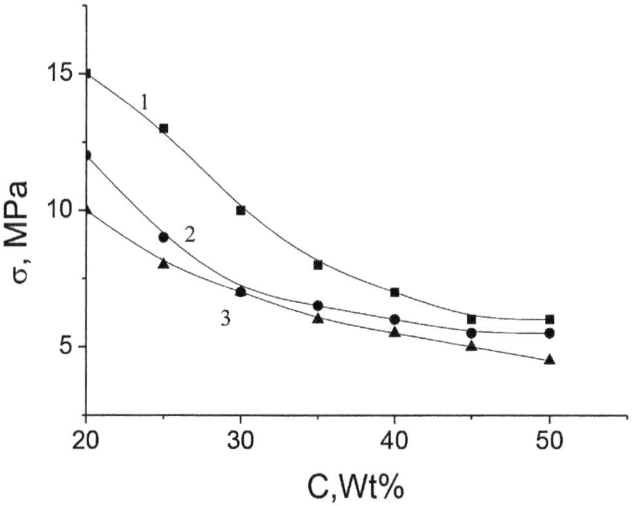

FIGURE 3.3 Dependence of impact viscosity for targeted composites containing andesite (1), slam (2), and quartz sand (3) on the filler concentration.

We have now shown a clear variation in the dependence of the composite's mechanical strength on the filler concentration at:

- Compression stage (please refer to Fig. 3.1),
- Bending stage (please refer to Fig. 3.2), and
- Impact viscosity (please refer to Fig. 3.3).

If we look attentively at the dependences shown in the above three figures, we can better understand the differences among the corresponding curves. These differences can be realized by different levels of interaction intensity between polymer matrices and the filler particles.

We have now noticed that the most important effect on the mechanical characteristics has to be due to the microstructure of the filler particles. In this regard, "the higher surface filler particles result in higher interphase square." Consequently, we do expect to have the adhesive forces between polymer matrix and filler particles. Meanwhile, the actual active chemical groups on the surface of filler particles obviously can have positive effects on the composite due to formation of covalent bonds between the phases. Therefore, it can be concluded that the mineral andesite particle surfaces could be characterized by relatively nonhomogeneous surface structure and roughness on the surface of which some active chemical groups are displaced.

It should be noted that the difference between curves is because of different types of filler. It is important to consider the fact that this difference should be expressed based on the character of particles surface profile (i.e., partially surface smoothness). Meanwhile, the particles with deep irregularities contribute to the penetration of the polymer segments into the micropores of filler particles and the formation of engagements. In such a manner, physical bonds are formed (i.e., formation of Van der Waals bonds) that will cause mechanical strength increment for the composite. However, we should also consider the possibility of formation of chemical bonds between active chemical groups on the filler surface and macromolecules. This will cause incremental strength in the composite.

In Figure 3.4, the dependence character also appears in the properties of thermal stability (Fig. 3.4), namely, the lower the significance of the maximal deepening of the indentor in the sample (measured by the Vicate method), the higher the thermal stability of the composite.

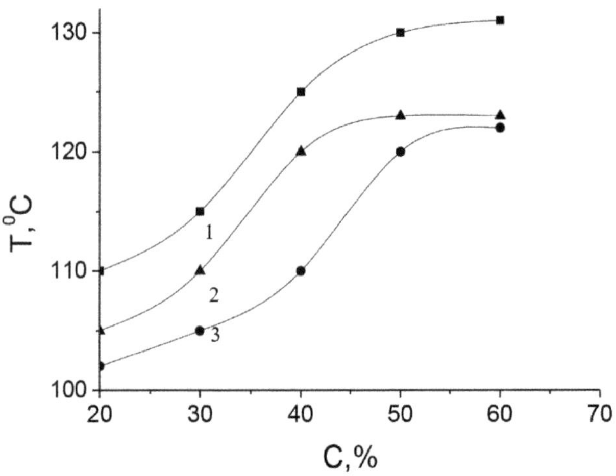

FIGURE 3.4 Composite softening dependence (measured by Vicate method) of the PE composites containing quartz sand (1), slam (2), and andesite (3) on the filler concentration.

From Figure 3.4, we can observe that the thermal stabilities of investigated materials fully correspond to their mechanical properties, and therefore the character of this parameter may be expressed in the same terms as in case of mechanical properties.

It is known that in some cases, the physical properties of composites may be increased by the use of binary fillers.[5] With this objective, we have prepared the composites based on PE and the abovementioned binary fillers, whose concentrations were varied in two groups of sum concentrations: 40 wt% and 50 wt%.

Based on the experimental data on determination of the dependence of ultimate strengthening of composites with various proportions of the fillers, the obtained curves are characterized by maximums.

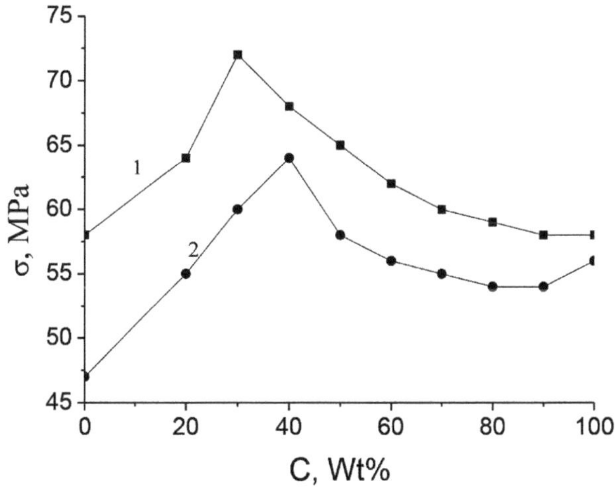

FIGURE 3.5 Dependence of the composite ultimate strength (at compression) of the PE composites on the binary filler (quartz sand + slam) at the filler sum concentrations 50 wt% (1) and 40 wt% (2). On the x-axis, the concentration of the filler sum.

From Figure 3.5, we can observe that the maximum ultimate strength appears for composites containing the fillers slam and quartz sand with a proportion of 30/70 (please see curve 1), when the sum of the fillers is the same as 40/60, and when the sum of the fillers is 40 wt%. The maximums on the curves correspond to the so-called synergistic effect, a nonadditive improvement of the material properties at a definite proportion of ingredients in the binary filler.[6,7]

TABLE 3.1 Composites Properties Based on PP and Binary Fillers.

Filler (proportion of the fillers in composite)	Impact viscosity, kJ/M2	Ultimate strength at binding, MPa	Ultimate strength at compression, MPa	Thermal stability, °C (by Vicate)
Slam	7.0	41.5	62.3	160
Andesite	6.0	46.8	65.1	165
Quartz sand	5.1	45.5	60.4	180
Slam/andesite (2/3)	8.2	50.5	70.0	185
Slam/quartz sand (1/1)	7.7	56.6	69.5	201
Quartz sand/andesite (2/3)	7.8	54.2	80.1	196

In Table 3.1, the numerical data of some physical–mechanical properties of the composites with binary fillers are presented. A comparison of the numerical data of dependences determined for composites containing binary fillers and composites with one filler is done. It is observed that by combination of the fillers, it is possible to augment the mechanical properties of the composites.

In order to improve the targeted properties of our composites, a structural modifier of type tetraethoxysilane (TEOS) was used (it must be noted that early this substance gave essential effect in the composites based on epoxy resin[8]). This modifier was introduced into the composite with amount 5 wt%. It was then added in the toluene to obtain a uniform solution that was evaporated after a definite time.

TABLE 3.2 Technical Characteristics and Properties of the Composites.[a]

#	Filler	Specific weight, kg/m³	Impact viscosity, kJ/M²	Ultimate strength at binding, MPa	Ultimate strength at compression, MPa	Thermal stability, °C (by Vicate)
1	Slam	1082 (1174)	7.5(7.0)	63.9 (62.3)	43.8 (41.5)	170 (160)
2	Quartz sand	1115 (1242)	5.0 (5.0)	58.0 (60.4)	51.3 (45.5)	200 (180)
3	Andesite	1183 (1180)	6.0 (6.0)	53.7 (65.1)	30.5 (46.8)	160 (150)

[a]Modified by TEOS fillers (in brackets corresponding characteristics for analogues with unmodified fillers shown).

Table 3.2 shows engineering properties of the composites studied in this chapter. These composites are modified and unmodified by tetraethoxysilane fillers. From Table 3.2, we can observe that the effect of a modifier on mechanical characteristics of the samples is relatively weak. Meanwhile, the effect of a modifier on the thermal stability of composites should be noted as well. This parameter increases by 10–20°C.

Thus, the protecting role of silicone organic compounds is quite clear. According to the results obtained on hydrophobic properties of the composites, it is concluded that the water absorption by composites is rather low (i.e., no more than 1.5%). This means the microstructure of the composites contains a rather low amount of structural defects (such as partial voids, cracks, etc.).

3.5 CONCLUSIONS

1. Polymer composites can be prepared based on industrial and domestic wastes of PE.

2. Experimental results show that the physical, mechanical, and thermal properties depend on the type and concentration of the fillers.
3. In case of the composites containing binary fillers, the ideal proportion of the fillers in the blend is found. This ensures that some physical properties will improve for composites containing one type of filler from the binary blend.
4. The composites of PE containing tetraethoxysilane-modified fillers differ from the analogues with unmodified fillers by improved technical characteristics.
5. Water absorption by composites is rather low (no more than 1.5%).

ACKNOWLEDGMENTS

This work was supported by Shota Rustaveli National Science Foundation of Georgia (SRNSFG) under GENIE-2021 Young Scientists Research Grant Funding Competition

Project Code YS -21-687.

KEYWORDS

- **polyethylene**
- **minerals**
- **composite**
- **binary filler**
- **synergistic effect**

REFERENCES

1. Achilias, D. S.; Roupakias, C.; Megalokonomos, P.; Lappas, A; Antonakou, E. V. Chemical Recycling of Plastic Wastes made from Polyethylene (LDPE and HDPE) and Polypropylene (PP). *J. Hazard. Mater.* **2007,** *149*, 536–542.
2. Jassim, A. K. Recycling of Polyethylene Waste to Produce Plastic Cement. *Procedia Manuf.* **2017,** *8*, 635–642.
3. Araujo, R. S.; Rezende, C. C.; Marques, M. F.; Gentile, G. *J. Appl. Polym. Sci.* **2017,** *28*, 134–142.

4. Kucukdogan, N.; Ozturk, S.; Sutcu M. Bio-composites based on Polypropylene Filled with Waste of Camellia Sinensis. *Arch. Mater. Sci. Eng.* **2016,** *79* (1), 12–18.
5. Maksimov, R. D.; Plume, E. N. *Mech. Compos. Mater.* **2014,** *50* (5), 613.
6. Asare, E.; Basir, A.; Tu, W.; Porwal, H.; Zhang, H.; Liu, Y. Effect of Mixed Fillers on Positive Temperature Coefficient of Conductive Polymer Composites. *Nanocomposites* **2016,** *2* (2), 58–64.
7. Pedrazzoli, D.; Pegoretti, A.; Kalaitzidou, K. *J. Appl. Polym. Sci.* **2015,** *132*, 4168.
8. Aneli, J.; Shamanauri, L.; Mukbaniani, O.; Markarashvili, E. *Probl. Mech.* **2010,** *N1*, 82–87.

CHAPTER 4

Removal of Some Frequently Used Antibiotics from Wastewaters Using the Natural Zeolite Adsorption Method Combined with Analytical HPLC Procedures: Pollutant Removal from Wastewater

IMEDA RUBASHVILI, LUBA EPRIKASHVILI, MARINE ZAUTASHVILI, TEIMURAZ KORDZAKHIA, NINO PIRTSKHALAVA, and MAIA DZAGANIA

Petre Melikishvili Institute of Physical and Organic Chemistry, Ivane Javakhishvili Tbilisi State University, Tbilisi, Georgia

ABSTRACT

Antibiotics are indispensable and being used by a large group of humans, animals, and pharmaceutical sectors for the treatment of different bacterial diseases. Due to the daily increase of residual antibiotics occurrence in our living environment, many scientists are doing research studies for the removal of such pollutants from wastewaters. Nowadays, for the human safety and to live in a cleaner environmental, we need to pay a special attention to minimize the water pollution due to the extensive use of antibiotics. Antibiotics residues in the wastewater, even in very small amounts, can easily cause resistance in bacterial populations. This will of course automatically reduce their therapeutic effectiveness against infectious diseases. It is necessary to point out the fact that around 30–90% of antibiotic dose can remain undegradable in the human or animal body which greatly discharged as an active compound. In our living environment, the elimination of imperfectly

Environmental Technology and Sustainability: Physical, Chemical, and Biological Technologies for Environmental Protection. Tamara Tatrishvili, Ann Rose Abraham, & A. K. Haghi (Eds.)

metabolized antibiotics by humans and animals should be considered as the primary source of antibiotic residues. Other origins are the transfer of unwanted or unused antibiotics from pharmaceutical industrial sectors. It is worth mentioning that the detection of antibiotics traces in drinking water, surface and ground water, and hospital wastewater, etc. has been reported in many research studies. For this reason, it is vital to increase global research studies into efficient and effective processes for removing antibiotics from the our living environment.

Nowadays, one of the most popular techniques are "adsorption." This popular technique is one of the most utilized methods which has great advantages as well. To highlight these advantages we can consider simple design with very low cost, the unique properties of flexibility, smooth operation, effectiveness, superior performance, relatively simple regeneration, and the robustness for consecutive cycles. Adsorbents aimed at reducing environmental risk, reducing energy intensity of processes, refusal of special sorbents, etc. should be attributed to compliance with the principles of "green economy." This approach implies a high degree of minimization of expenses during water purification from toxic and harmful components. The economic aspects of this process play a very relevant role, allowing the purification and additional treatment of wastewater, thus decreasing the anthropogenic load on water bodies.

This book chapter is the first case in our studies where the adsorptive removal of the most commonly used and the poorly metabolized antibiotics by humans and animals fluoroquinolone antibiotic namely moxifloxacin HCl, norfloxacinfrom, and β-lactam broad-spectrum cephalosporin antibiotic namely ceftriaxone from aqueous solutions as a model of wastewaters by natural zeolites – clinoptilolite, mordenite from the local region, Georgia and their acid-modified H-forms has been studied under static and dynamic conditions. We have used our particular designed laboratory for considering the adsorption under dynamic conditions. These devices are designed as dynamic-type equipment with fixed bed adsorption glass column packed by zeolite adsorbent and chromatography pump. To measure adsorption process and control the above-mentioned pharmaceutical pollutants in the influent and effluent test solutions, new, rapid, and selective HPLC analytical procedures were developed and validated with respect to the validation parameters: robustness, system suitability test (SST), specificity, linearity-range, accuracy, precision, and sensitivity. The upgraded curves for the adsorption of each antibiotic on the selected zeolite adsorbents in terms of the [(effluent)/(influen)] concentrations ratio were investigated by carrying out a set of fixed bed experiments at constant temperature. Considering "Langmuir and Freundlich" isotherm models, the adsorptive removal of antibiotics mechanism and the interaction of zeolite adsorbent with antibiotic was described (by the adsorption isotherms). The results of the adsorption study in static conditions show that the Langmuir isotherm model is a powerful tool to interpret the adsorption process. It should be noted that the adsorption is driven by the formation of antibiotic monolayer on the adsorbent surfaces as well. So as a result, in such a case, the adsorption mechanism is mainly composed of electrostatic interaction between the adsorbent surface and the adsorbate.

Our experimental results in this work show that the selected natural zeolites— clinoptilolite, mordenite, and their acid-modified H-forms with hydrophilic pores have a higher affinity for adsorbing antibiotics in aqueous solution. From the experimental data it is observed that the following three factors, namely:

- high value of the volumetric flow rate,
- the inlet concentration of the antibiotic solution, and
- the pH value accelerate breakthrough of adsorbates and correspondingly reduce the adsorption capacities.

We need to point out that the "capacities of moxifloxacin HCl," "norfloxacin," and "ceftriaxone adsorption" are highly dependent on the volumetric flow rate and the inlet concentration of the antibiotic solution. The value of pH does not have a significant influence on the adsorption process. The removal efficiency increased with a decrease in the concentration of each antibiotic solution. The amount of antibiotic adsorbed per unit mass of zeolite is a function of the contact time of each adsorbate with adsorbent, the structure, the molecular weight, and the concentration of antibiotic solution. The greatest static adsorption capacities were obtained at low initial concentration of influent solution for all the adsorbents (the highest dynamic adsorption capacities at low flow rate of each influent solution). According to the current adsorption research studies and method validation the laboratory adsorption method combined with analytical HPLC procedures was developed and standardized to remove the mentioned antibiotic residues from wastewaters and control routinely the concentration of each antibiotic pollutant in influent/effluent streams.

Thus, this book chapter indicates that natural zeolites—clinoptilolite and mordenite from Georgia are efficient, eco-friendly, alternative, and competitive adsorbents in terms of low cost, shape selectivity along with the adsorption effectiveness for the removal of the commonly used fluoroquinolone and cephalosporin antibiotics from wastewaters.

4.1 INTRODUCTION

Water resources are of great importance for ensuring the living conditions of the population, normal functioning of the economy, and the preservation of the environment. Providing water to the population, industry, energy, and agriculture is a priority for all countries, including Georgia.[1] In addition, surface water pollution issues are part of Georgia's National Environmental Action Program.

Georgia's European policy commitments include: developing basic legislation and basic procedures, as well as improving the cross-border approach and active participation.[2]

Recently, Governments and scientists have expressed concern about the effects of chemicals in the environment (in water) on humans and animals. Antibiotics are problematic in this regard. Due to the extensive use of antibiotics,[3] the development of residual antibiotics in the environment is enhanced on daily basis. They can be disposed into our living environment in several different ways (please refer to Fig. 4.1).[4]

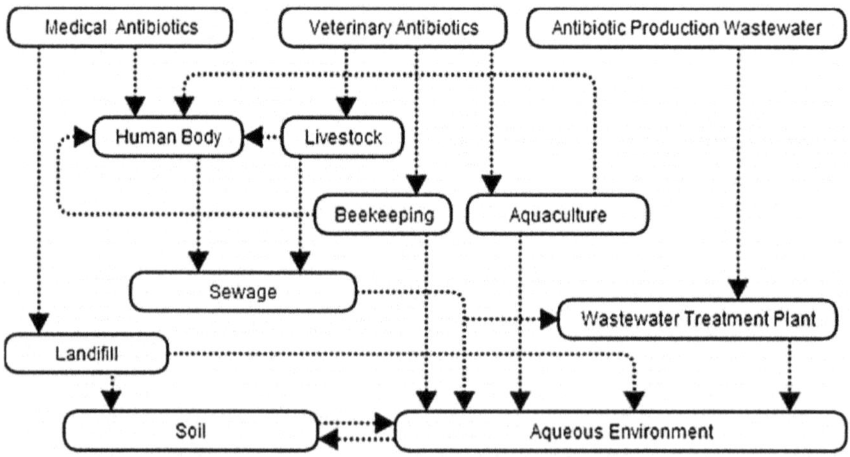

FIGURE 4.1 The existence of antibiotics in our living environment.

About 30–90% of the given antibiotic dose according to the results reported in Ref. [5] could remain undegradable in the animal and human body. This is highly discharged as an active compound.[5] The excretion of poorly metabolized antibiotics is the main source of antibiotics in the environment. Other origins are the disposal of unutilized or undesired antibiotics from pharmaceutical industrial processes.[6,7] Many research studies observed that antibiotics are noticed in our drinking water, ground water, surface water, and hospital wastewater (please refer to Ref. [3] and [8]). In water systems, the recognized antibiotic concentrations were not at low levels (at μg/L level) in hospital effluent and wastewater. The detected antibiotic concentrations were low over a wide range from ng/L to μg/L in various surface and ground waters. The detection of antibiotics in sediments was low to medium at μg/kg level (as reported in Ref. [9]). As a matter of fact, antibiotics in hospital effluent and wastewater treatment plants impact are not relatively low compared with different locations. Nowadays, antibiotics have been classified as a critical risk group. This is mainly due to their high toxicity to algae and bacteria at low concentrations and their capability to cause resistance throughout the natural bacterial population.[8] As indicated in Ref. [4], it is very important to emphasize that the antibiotics in (surface water) show the potential to disrupt bacteria cycles and processes critical to agriculture (soil fertility) and animal production (rudimentary processes) or to water ecology (nitrification and denitrification).

It should be noted that the quinolone antibiotics are synthetic antibacterial drugs with a 4-quinolone basic structure. This in fact slows down bacterial DNA gyrase and could give rise to cells to stop dividing. And it will cause irreversible damage to bacteria. According to Ref. [10] the utmost frequently used quinolones comprehend a new series of quinolones called FQ or (fluoroquinolone) antibiotics. Meanwhile, Moxifloxacin HCl and norfloxacin related to the fluoroquinolone antibiotics family. The presence of fluoroquinolone residues in effluents from pharmaceutical sectors and households along with hospitals is an important source of critical and chronic toxicity, besides the importance of resistant bacteria. Accordingly, removal of fluoroquinolone residues from the environment is a critical problem.[4,11] "Ceftriaxone" as ceftriaxone sodium is a broad-spectrum third-generation semisynthetic cephalosporin antibiotic. According to Refs. [12] and [13] ceftriaxone has a lengthy half-life in comparison to other cephalosporins.

In Table 4.1 we have shown the structure and main physicochemical properties, along with the concentrations of moxifloxacin HCl, norfloxacin, and ceftriaxon in different environmental compartments.[4,14,15–20]

For the treatment of residual antibiotics different methods have been developed. This includes adsorption by activated carbons, ultraviolet irradiation,

TABLE 4.1 The Structure and Main Physicochemical Properties of Antibiotics and Their Content in Surface Water.

Compound	Moxifloxacin	Norfloxacin	Ceftriaxone
Molecular formula	$C_{21}H_{24}FN_3O_4$	$C_{16}H_{18}FN_3O_3$	$C_{18}H_{18}N_8O_7S_3$
CAS number	354812-41-2	70458-96-7	73384-59-5
Abbreviation	MOX	NOR	CEF
Molecular structure	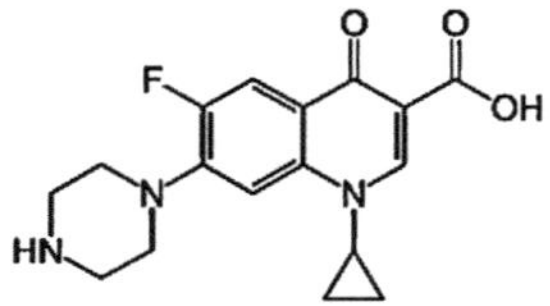	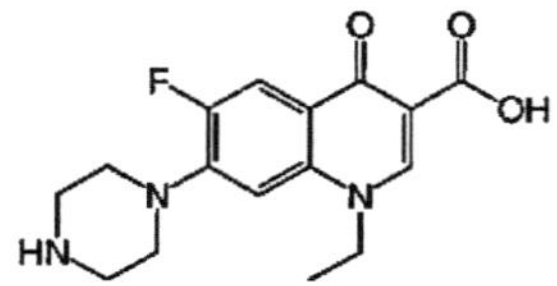	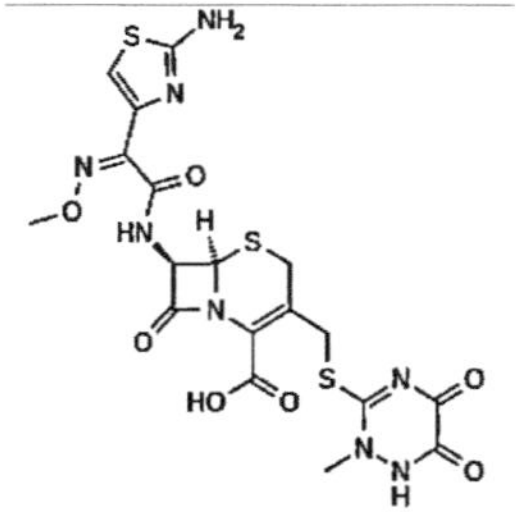
Molecular weight, g/mol	401.438	319.331	554.57
The acid dissociation constants, pKa	pKa1 = 6.43, pKa2 = 10.63	pKa1 = 0.16, pKa2 = 8.68	pKa1 = 1.72, pKa2 = 3.15, pKa3 = 4.34
Chemical name	(4aS-cis)-1-Cyclopropyl-6-fluoro-1,4-dihydro-8-methoxy-7-(octahydro-6H-pyrrolol[3,4-b]pyridin-6-yl)-4-oxo-3-quinolinecarboxylic acid	1,4-Dihydro-1-ethyl-6-fluoro-4-oxo-7-(1-piperazinyl)-3-quinolinecarboxylic acid	(6R,7R)-7-{[(2Z)-2-(2-amino-1,3-thiazol-4-yl)->2-(methoxyimino)acetyl]amino}-3-{[(2-methyl-5,6-dioxo-1,2,5,6-tetrahydro-1,2,4-triazin-3-yl)thio]methyl}-8-oxo-5-thia-1-azabicyclo[4.2.0]oct-2-ene-2-carboxylic acid
Water solubility, mg/mL at 25°C and pH 5–7.5	1.15–5	0.45–161	0.123
Concentration in surface water, μg/L	0.006–0.017	0.0023–0.12	59.5
Toxicity, oral LD50 (rat), g/kg	1.32	4	10

(NF) or nano-filtration, (RO) or reverse-osmosis ozonation, chlorination, filtration, and other materials.[21] Based on the published research work[22] the conventional biological processes (such as wastewater treatment plants) cannot remove antibiotic pollutants. Several methods have been used for treatment of antibiotic-rich effluents such as electrochemical oxidation,[23] biodegradation,[24] photodegradation,[12,25] catalytic degradation,[26,27] micro-extraction,[28] oxidation (catalytic degradation),[29] and adsorption.[4,30,31] It has been shown that adsorption is an effective, simple, and economical method to remove antibiotic pollutants at low concentration from waters.[14,16,25,32–34] Recently, many studies focused on antibiotics adsorption by porous materials, such as zeolites.[35,36]

For the removal of a broad range of antibiotic pollutants, adsorption is the extensively used technique because of its easy operation, simple design, and somewhat simple regeneration. Adsorption systems aimed at reducing environmental risk, reducing energy intensity of processes, refusal of special sorbents, etc., should be attributed to compliance with the principles of "green economy." This approach implies a high degree of minimization of expenses during water purification from toxic and harmful components. The economic aspects of this process play a very relevant role, allowing the purification and additional treatment of wastewater, thus reducing the anthropogenic load on water bodies. Adsorption characterizes with low cost and no by-product formation, and thus is being considered an effective way to remove antibiotics in water. This method is a well-controlled process that allows removing various contaminants up to almost any residual concentration, regardless of the chemical stability of pollutants, the absence of secondary contaminants, and the cost efficiency associated with the repeated use of the adsorbent.[37] Clearly, there is always a need for low-cost effective adsorbents for the removal of antibiotics from various contaminated media. While considering the protection of our living environment, the utmost and perspective direction of use of natural zeolites is their use for conditioning of potable water, treatment of domestic, industrial, and agricultural wastes from ammonia nitrogen, oil products, and toxic ions of non-ferrous and heavy metals.[38] The researches have shown that zeolite-containing rocks possess a unique spectrum of physicochemical, adsorption, and ion-exchange properties, due to which they find wide applications in practice of wastewaters treatment. Adsorption of antibiotics on activated carbons and other adsorbents including zeolites has been explained in many studies,[32,39–43] but the possibility of adsorption of moxifloxacin HCl, norfloxacin, and ceftriaxone on natural zeolite for wastewater treatment has not been addressed by researchers.

At present, 15 varieties of zeolite minerals have been identified in Georgia, but among them industrial significance may have only: analcime, clinoptilolite-heulandite, phillipsite, laumontite, and mordenite. The presence of large reserves of natural zeolites in Georgia contributed to the study of adsorption properties of both initial and nano-modified samples to create adsorption materials effective for the extraction and removal of pharmaceutical ingredients. Their complex evaluation allows us to fully use the mineral wealth and to increase their investment attractiveness.[44] Natural zeolites as hydrated aluminosilicates characterized by high surface area and leading cation exchange capacities can be presumably used to remove some antibiotics with actual charged moieties through cation exchange.[13,45,46]

The objective of the present book chapter is to describe research study carried out by authors during the last 3 years[4,46,47] regarding the investigation of adsorption and the possibility of adsorptive removal of the most commonly used and the poorly metabolized antibiotics FQs namely moxifloxacin HCl, norfloxacinfrom, and β-lactam broad-spectrum cephalosporin namely ceftriaxone from aqueous solution as a model of wastewater by natural zeolites—clinoptilolite, mordenite from the local regions, Georgia and their acid-modified H-forms. To study adsorption on the above-mentioned test natural zeolites and control each pollutant in test aqueous solutions during the adsorption experiment, it has been developed and validated new, selective, and sensitive analytical methods for quantitative determination of each antibiotic compound. A literature review revealed that various analytical HPLC methods for the quantitative determination of moxifloxacin HCl, norfloxacin, and ceftriaxon have been reported in several papers and pharmacopeias,[47] but adequate and suitable HPLC methods with method validation data in support of investigation of adsorption process for the quantitative determination of the above-mentioned antibiotics at very low concentration level in aqueous solutions were not found.

The present study is the first case where adsorptive removal of moxifloxacin HCl, norfloxacin, and ceftriaxone has been investigated using natural zeolite as a cheap adsorbent with simple treatments and to demonstrate that natural zeolite is a potential adsorbent of antibiotic removal of wastewaters.

4.2 EXPERIMENTAL

4.2.1 PREPARATION OF ADSORBENTS

Local natural zeolite with a zeolite phase content of 55–65%—clinoptilolite ($(Ca, Na_2,K_2)3Al_6Si_{30}O_{72}\cdot 24H_2O$) was obtained from the Khandaki region of Kartli, Georgia, natural mordenite (Ca, Na_2, K_2)$Al_2Si_{10}O_{24}$· •$7H_2O$) with a zeolite phase content of 50–60% from the Bolnisi-Ratevani region (Georgia). To obtain a modified H-form, clinoptilolite and mordenite were treated with an acidic solution (2N, 5N, and 10N HCl solution). The high value of the silicate module (Si/Al) determines the high thermal- and acid-resistance of the mineral.[4,46] Both obtained adsorbents were separated into fractions with a particle size of 0.5–1 mm.

4.2.2 INSTRUMENTATION

For the study of adsorption dynamic process there was used the specially constructed laboratory dynamic type equipment with fixed bed adsorption glass column and high-pressure pump at constant temperature of 20°C. The equipment consists of three parts in dynamic condition through the prepared zeolite samples placed in a glass adsorption column; the antibiotic test solution at different concentrations was passed. The experiment was carried out with different liquid flow rates. The scheme of this equipment is given in Figure 4.2.[4,46] The concentration of each test adsorbate in influent and effluent stream was determined using HPLC methods.

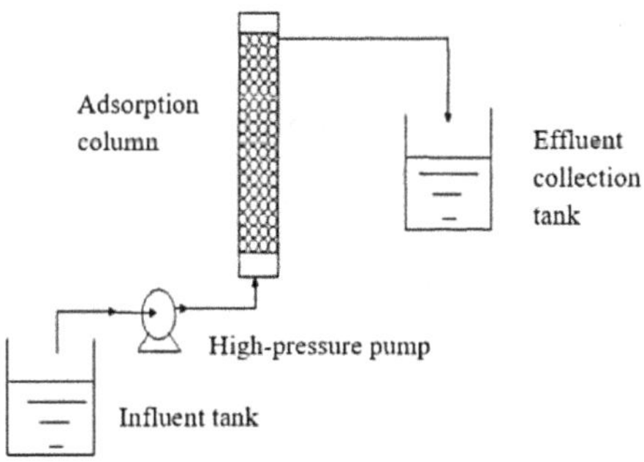

FIGURE 4.2 The laboratory dynamic type equipment for adsorption study.

The chromatographic analysis was carried out using LC-20AD Prominence Shimadzu HPLC System (Japan). Meanwhile, the HPLC grade water was prepared using Milli Q Advantage A10 purification system (France). Analytical balance ALX-210 (USA) and Hanna Instruments HI 2211 pH-meter (USA) were considered for standard preparation of sample. Infrared (IR) spectroscopic analysis was performed using Agilent Cary 630 Fourier Transform Infrared (FTIR) Spectroscopy (USA). The IR spectra of the zeolite were recorded in the region of vibrations of the aluminosilicate framework of the middle and far regions. "XRD" or X-ray diffraction analysis was carried out by X-ray diffractometer - Drone-4 (Russia). The X-ray diffractometer connected to a personal computer via the USB port multimeter AX-18B and the corresponding PC software-PC-Link, which allows the processing of experimental data in Excel format. All the measuring equipment were properly calibrated.

4.2.3 ANALYTICAL HPLC METHODS

HPLC analytical methods for the quantitative determination of MOX HCl, NOR, and CEF in test aqueous solutions were developed using HPLC columns—Agilent SB-C18 4.6 × 250 mm, 5 μm (USA). For the MOX determination a mixture of buffer pH 2.5 ± 0.1 and ACN 60:40 v/v, for NOR determination—a mixture of buffer–orthophosphoric acid solution (1:1000) and ACN 85:15 v/v; for CEF determination—a mixture of phosphate buffer pH 7 ± 0.1 (potassium hydrophosphate and potassium dihydrophosphate solution), citrate buffer pH 5 ± 0.1 (citric acid monohydrate and sodium citrate), and ACN 35:15:50 v/v were used as a mobile phase (MP) with isocratic elution. The flow rate of MP was 1.0 mL/min. The detector wavelengths were 293, 275, and 270 nm for MOX HCl, NOR, and CEF determinations, respectively. The injected volume was 20 μL. The column temperature was maintained at 40°C for MOX and NOR determinations, and at 25°C for CEF determination. The MP was used as a diluent for standard preparation by the following procedure: 20 mg of the analytical standard of each antibiotic compound was weighed and transferred to a 20-mL volumetric flask, dissolved in 15 mL of diluent and then diluted to volume with the same diluent, mixed well (standard stock solution – 1 mg/mL). The obtained solution was filtered through 0.45-μm polyvinylidene fluoride (PVDF) membrane syringe filter and 1 mL of this solution was transferred to 10 mL volumetric flask, diluted to volume with purified water, mixed well (0.1 mg/mL).

The developed HPLC methods were validated with respect to robustness (standard solution stability, PVDF membrane filter compatibility test, critical factors effect study using design of experiments—DoE) system suitability test (SST), specificity, linearity-range, accuracy, precision, limits of detection (LOD), and quantitation (LOQ) according to methodologies[17–19] and ICH Q2 guideline requirements.[20] Microsoft Excel was used for statistical assessment and graphical analysis.

4.2.4 ADSORPTION EXPERIMENT

For adsorption study the test stock solutions (adsorbate influent solution) from 0.2 mg/mL to 2.0 mg/mL concentration of each antibiotic were prepared by dissolving analytical standards in purified water. The effluent test solutions were diluted in the same way as the last step of dilution of standard preparation. The initial pH value of test solution was adjusted by adding 0.1 N NaOH and HCl solution.

To study adsorption by the static method, 0.2 g of zeolitic adsorbent was transferred to 50-mL Erlenmeyer flask and 20 mL of each antibiotic test stock solution was added at 0.2–2.0 mg/mL concentration range. Initially, zeolite sample with adsorbate antibiotic solution was left on an orbital shaker at 150 rpm for 15 min, then allowed to stand statically for the determined time interval, and then the adsorbent samples were centrifuged at 3000 rpm for 5 min. The taken aliquots from the obtained supernatants at different time intervals during the experiment were used to determine a concentration of each adsorbate using the corresponding HPLC method.

For the adsorption process study by the dynamic method the specially constructed laboratory dynamic type equipment was used. MOX HCl, NOR, and CEF influent solutions were added to a glass beaker separately and pumped with two different flow rates of the influent stream - 1.5, 2.5, and 5.0 mL/min into the adsorption column. The effluent samples were collected initially at different time intervals and the end of the adsorption experiment after the saturation state took place. We have then investigated the effect of working parameters such as the value of pH, the flow rate, and the initial concentration of test solution accordingly. Afterward, one additional glass flask was left under the same conditions. To be able to properly evaluate possible degradation of antibiotics with aliquots being removed analysis at every time interval and no significant losses were observed (<2 %). The obtained solutions were filtered through 0.45-μm PVDF membrane syringe filter, and 1 mL of these solutions were diluted to 10 mL with diluent, mixing well. The initial pH value of adsorbate effluent sample solution was adjusted by adding 0.1 M NaOH and HCl solution.

4.2.5 CALCULATIONS

The breakthrough curves for adsorption of each antibiotic—MOX HCl, NOR, and CEF on zeolite adsorbent in terms of the effluent to influent concentrations ratio, C/C_o, versus the contact time— τ, min were studied by carrying out a set of fixed bed experiments at constant temperature of 20°C.

The concentration of each antibiotic—C_u in an effluent test solution, expressed in mg/mL, was calculated by the following formula:

$$C_u = A_u \times W \times D_1 \times P/A_s \times D_2 \times 100 \quad (4.1)$$

where Au is the peak area of each antibiotic obtained with an influent/effluent test solution; As is the peak area of each antibiotic obtained with standard solution; W is the weight of analytical standard, mg; D_1 is the dilution factor of test solution; D_2 is the dilution factor of standard solution; and P is the purity of standard, %.

Using the following equation we have determined the removal efficiency—R, %:

$$R, \% = (C_0 - C_e) \times 100/C_0 \quad (4.2)$$

where C_e is the equilibrium concentration of the adsorbate test solution at the fixed time (contact time with adsorbent—τ, h), mg/mL; C_0 is the initial concentration in the adsorbate test solution, mg/mL;

The adsorption capacity—q, mg/g was calculated by the followed formula:

$$q = (C_0 - C_e) \times V/m \quad (4.3)$$

where V is the used volume of adsorbate solution, mL; m is the mass of adsorbent, g.

For the explanation of the mechanism of adsorptive removal of antibiotics and the interaction of zeolite adsorbent with antibiotics was studied using the adsorption isotherm. The adsorption isotherm data were interpreted using the most commonly used Langmuir isotherm model. The "Langmuir model" presumed mono-layer adsorption on the surface of the adsorbent. The experiment was carried out with different initial concentrations of each antibiotic over a range from 0.5 mg/mL to 2 mg/mL. Based on the adsorption experiment data the isotherm was plotted considering the relationship between the uptake of adsorbate and also the adsorbate equilibrium concentration (remaining in the solution after the system has attained the equilibrium state at a constant temperature of 20°C). The model is given in the following equilibrium equation:

$$1/q_e = 1/q_m K_L C_e + 1/q_m \quad (4.4)$$

where q_e is the uptake of the adsorbate – equilibrium adsorption capacity (mg/g), q_m is the monolayer adsorption capacity of adsorbent (The adsorption activity of the sorbent layer), mg/g; K_L is the Langmuir adsorption equilibrium constant, mL/mg.[4]

The length of the mass transfer zone (the height of the working layer), cm, was calculated by the formula

$$L_0 = L \times (t - t_b)/t_s - (1 - \phi) \times (t_s - t_b) \quad (4.5)$$

where L is the length of the sorbent layer, cm; $t_s = V_s/F$ is the time necessary for bed saturation, min; Vs is the volume of solvent flowing up to the instant of bed saturation, mL; F is the flow rate, mL/min; tb = Vb/F is the time for bed breakthrough, min; Vb is the volume of antibiotic solution flowing up to the instant of bed saturation, mL; $\phi = S_A/(S_A + S_B)$ is the symmetry coefficient of the breakthrough curve; S_A is the area above the breakthrough curve up to the moment of bed saturation; S_b is the area under the breakthrough curve up to the moment of bed saturation.[48]

4.3 RESULTS AND DISCUSSIONS

4.3.1 *IR SPECTROSCOPIC AND XRD ANALYSIS*

The crystal structure of clinoptilolite is characterized by quite open channels formed with 10- and 8-membered tetrahedral rings arranged in three directions: the channels parallel to the C axis have the following dimensions: 0.705 × 0.425 nm and 0.460 × 0.395 nm, respectively. 8-membered channels, arranged in parallel and at 500 angle to the A-axis, have diameters of 0.540 × 0.390 and 0.520 × 0.390 nm.[45] Mordenite structure is penetrated by a 2-dimensional system of 8- and 12-membered channels. Some cations are localized in 8-membered channels, and some—in large channels. Cations block narrow 8-membered channels, and only large channels with diameter of 0.67 nm are open for the diffusion of molecules; Na-, Ca-mordenites are unblocked after treatment with hot hydrochloric acid. The difference in the molecular-sieve properties is caused by the different arrangements of sodium cations, which in narrow-pore mordenite are located in large 12-membered channels (0.59–0.47 nm) and thus, narrow their effective diameter. In wide-pore ones they are localized in the depths of niches composed double eight-membered rings (0.39–0.47 nm) and do not prevent the penetration of large molecules into large channels.[46]

Based on the IR spectra of clinoptilolite, the zeolite structure is characterized by Si–O–Si(Al) inter-tetrahedral deformation and valence oscillations in the region of 774, 593.6, 520.9 cm^{-1}. After the treatment of the sample with 2N acid solution (HCl), various impurities are removed and the zeolite structure improvement takes place. This is indicated by the disappearance of bands in the spectrum of initial clinoptilolite in the areas of 1430.4 cm^{-1} and 875.0 cm^{-1}. These bends are typical for CO_3^{2-} and SO_4^{2-}. In addition, the dealumination of zeolite begins. This is indicated by the change of frequencies in the spectrum of aluminosilicate carcass of intra-tetrahedral valence vibrations Si–O–Si(Al) from 1040.9 cm^{-1} to 1052.0 cm^{-1}. It is worth mentioning that in the characteristic spectrum of clinoptilolite structure the intensity of Si–O–Si(Al) inter-tetrahedral bands of oscillations has changed insignificantly. Treatment of zeolite with 5N acid solution (HCl) has significantly reduced these vibration bends and accordingly, the intensity of intra-tetrahedral valence Si–O–Si(Al) vibrations has increased to 1065.1 cm^{-1}. This indicates an increase of module of SiO_2/Al_2O_3 in zeolite. Thus, acid treatment of clinoptilolite does not destroy the structure of zeolite.

Comparison of the IR spectra of natural mordenite with an acid-treated sample (5N HCl) showed the presence of all characteristic absorption bands of the aluminosilicon-oxygen framework of the zeolite. A somewhat different spectrum is observed for the sample modified with 10N HCl, which is probably caused by the deformation of the zeolite structure.[46] Along with the chemical and mineral compositions of natural zeolites, textural and structural characteristics are fundamental in the study and development of adsorbents involved in various adsorption processes. Modification of natural zeolites with solutions of hydrochloric acid shows a significant increase in the values of physical-mechanical and textural characteristics: the specific surface area of the zeolite increases on average up to five times, and the total pore volume doubles. To characterize the texture parameters of the studied samples (specific surface area, volume of mesopores, and total pore volume, pore-size distribution), the method of low-temperature adsorption-desorption of nitrogen at 77.2 K was used. The procedure was carried out on the equipment—Micromeritics Asap 2020 Plus Physisorption Analyzer (USA) by the Barrett-Joyner-Hallend (BJH) method. The natural mordenite sample was characterized by the lowest specific surface area, density, porosity, and total pore volume. Modification of samples with acid solution was accompanied by an increase in specific surface area (to 70.84 from 162.33 m^2/g), volumes (from 0.086 to 0.097 cm^3/g), pore diameters, porosity, and a decrease in true density. The most significant changes in the structure of the zeolite occurred during its treatment with 10N boiling HCl solution. By varying the type of modifying agent, it is possible to form a porous structure at the level of micro-, meso-, and macro-pores.[46]

Crystal structural changes of natural clinoptilolite after the acid treatment to obtain their acid-modified H-form were studied using X-ray diffraction analysis. The obtained diffractograms of the prepared adsorbent samples of clinoptilolite and clinoptilolite H-forms are given in Figure 4.3. From this figure we can observe that the intensities of the diffraction peaks are reduced for the sample of clinoptilolite H-forms in comparison with the sample of initial clinoptilolite. This decrease is characterized by the ratio of intensities (I) of characteristic peaks of clinoptilolite ($2\theta=9.9^\circ$, 22.4°, and 30.0°) to that for a quartz phase at $2\theta = 26.6^\circ$ (I_0). Reduction of intensity of the sample of clinoptilolite H-form (H-CL) is connected with the known removal of cations and aluminum from clinoptilolite caused by the acidic treatment with HCl. Meanwhile, it is quite clear that reduction in the relative intensity of diffraction peaks (I/I_0) in case of clinoptilolite H-form, especially, the diffraction peaks at 9.88° and 22.36° is associated with an increase in crystallinity compared to initial clinoptilolite.

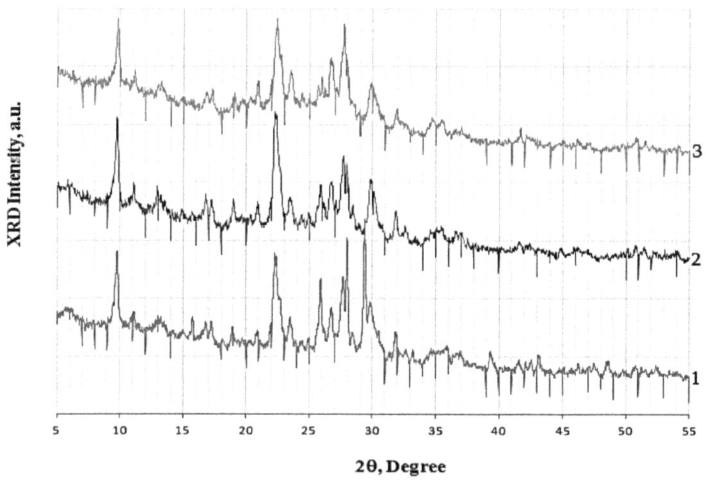

FIGURE 4.3 X-ray difractograms of initial CL (a), H-CL (modified with 2N HCl) (b), and H-CL (modified with 5N HCl) (c).

The diffractograms (Fig. 4.4) of the initial natural mordenite treated by HCl of different concentrations show that despite changes in the chemical composition, the crystal lattice of zeolite remains unchanged after acid treatment. The diffractograms of treated mordenite by 5M and 10M HCl are similar to the diffractogram of initial mordenite. This indicates an increase in the relative content of Si/Al ratio in zeolite framework and pore opening compared to natural zeolite.[48,49]

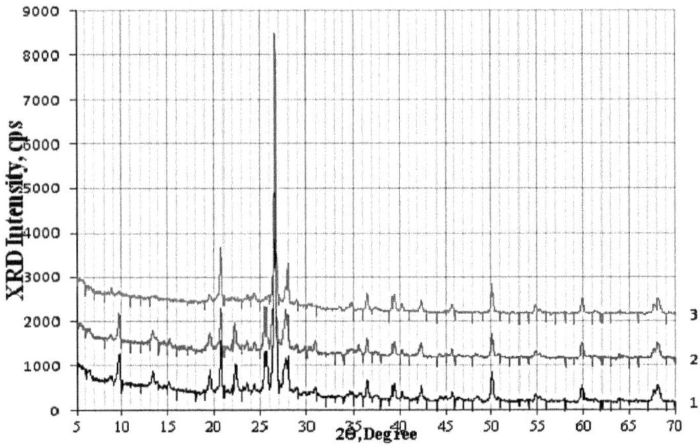

FIGURE 4.4 X-ray difractograms of initial MOR (a), H-MOR (modified with 5N HCl (b), H-MOR (modified with 10N HCl).

4.3.2 ADSORPTION OF FQ ANTIBIOTICS

In static conditions we study adsorption process of the selected adsorbates—MOX HCl and NOR with three different concentration solutions—0.5 mg/mL, 1.0 mg/mL, 2.0 mg/mL on clinoptilolite, mordenite, and their H-forms at the value of pH 7. The removal efficiency—R, % was calculated by formula (2) and plotted graphs describing the relationship between the amount of adsorbate and the contact time of adsorbate solution with the adsorbent. For all cases the same adsorption process occurred. Figures 5a and 5b depict the obtained graphs for NOR and MOX HCl, respectively. The fastest adsorption was observed at the low concentration—0.5 mg/mL in case of NOR. The removal efficiency increased with a decrease in the concentration of adsorbate solution. Accordingly, the amount of FQ antibiotic adsorbed:

- per unit mass of zeolite as a function of the contact time of each adsorbate with adsorbent,
- the structure,
- the molecular weight and the concentration of FQ antibiotic in solution.

The adsorption isotherm generally depends on the equilibrium equation providing remarkable information about the adsorbent surface properties by calculating the maximum adsorption capacity.[46] Based on the obtained data of the adsorption study by the static method the adsorption of FQ antibiotics on CL and H-CL was analyzed using the Langmuir model. The adsorption experiment was carried out using 0.2 g of zeolite adsorbent—CL and 20 mL of adsorbate solution with four different adsorbate concentrations over a range 0.5–2 mg/mL at constant pH value of 7 and temperature of 20°C. The adsorption isotherms are given in Figures 6a and 6b. The results show that isotherms have a good correlation (R^2>0.9) and define a monolayer adsorption. The adsorption mechanism is mainly composed of electrostatic interaction between the adsorbent surface and the adsorbate.

The adsorption process was investigated as well by the dynamic method using the laboratory dynamic type equipment with packed fixed bed adsorption column. The behaviors of breakthrough curves for MOX HCl and NOR adsorption on natural zeolites and modification forms at different flow rates of 1.5, 2.5, and 5.0 mL/min and inlet FQ antibiotic concentration of 0.2 mg/mL and 1.0 mg/mL at constant temperature of 20°C were observed. The experimental data obtained with the FQ antibiotics adsorption on the natural zeolites samples in dynamic conditions indicate an improvement in the process of adsorption of NOR and MOX HCl by the sample of zeolites modified with HCl solution. The efficiency of the equilibrium adsorption capacity of the sorbent layer (η) increases. The height of the working layer (L_0) decreases, which increases the repeatability of sorbent "elementary" volumes in the process of repeated acts of sorption-desorption. The obtained data demonstrate the influence of the antibiotic chemical nature and chemical structure on the process of its extraction from the aqueous solution by zeolite. The height of the working layer—L_0 was determined from the breakthrough curves which is an important characteristic of the dynamics of the adsorption process and calculated according to the formula (5). The efficiency equilibrium adsorption capacity of the adsorbent layer (the degree of utilization of the adsorbent)—η, % was determined by the ratio of the dynamic adsorption capacity of the layer to the equilibrium adsorption capacity of the sorbent layer.

The interaction mechanism between adsorbate and adsorbent which could be considered CL has a frame, open, and stable three-dimensional structure with negative charge, and FQ antibiotics have positively charged functional groups. Consequently, natural clinoptilolite would be able to maintain antibiotics via the cation exchange mechanism. Even so it is hypothesized

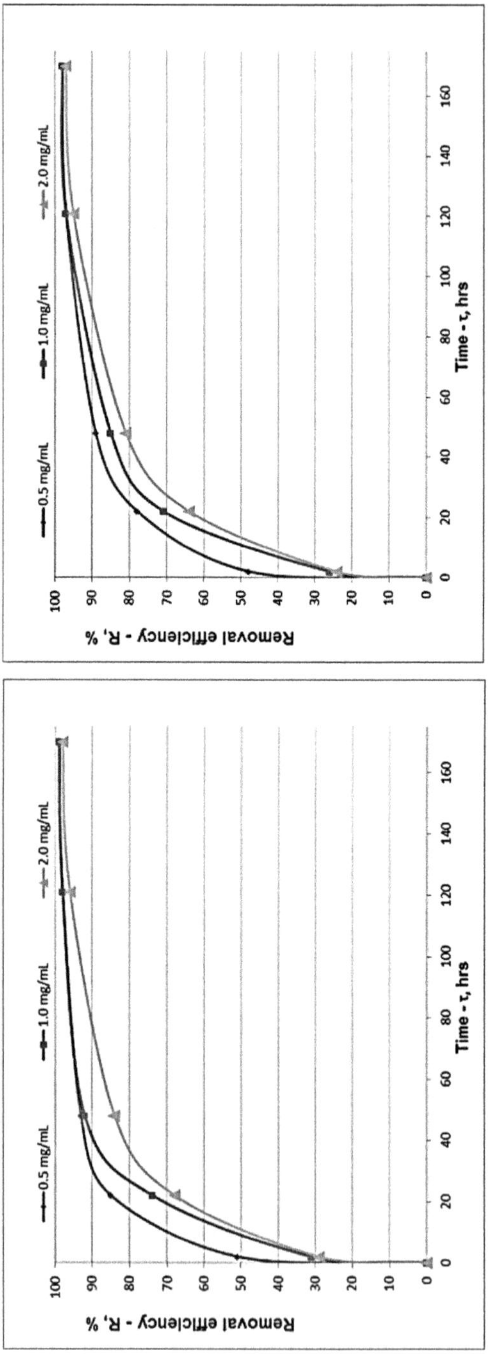

FIGURE 4.5 The graph of the removal efficiency, % versus the concentration of adsorbate solution for NOR (a) and MOX HCl (b) on CL at three different concentrations.

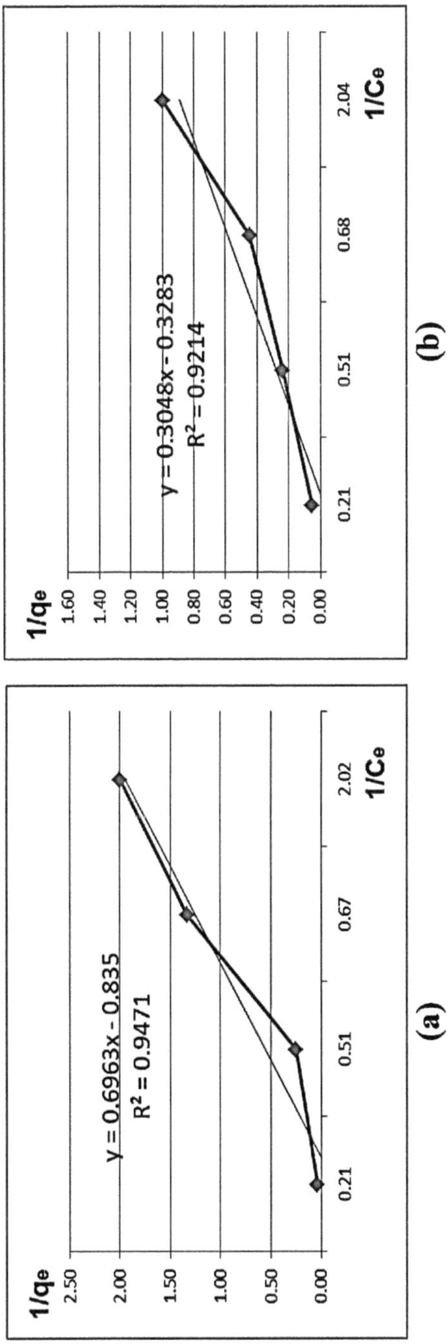

FIGURE 4.6 The Langmuir isotherm model for NOR (a) and MOX HCl (b) adsorption on CL.

that antibiotics are uptake by zeolite in associate with π–π electron–donor–acceptor interaction on moxifloxacin and norfloxacin as the principal mechanism of adsorption. π–π electron–donor–acceptor interaction between the benzene ring, carboxyl of FQs, and hydroxyl groups on zeolite adsorbent surface may be the significant factor. Augmentation of adsorption capacity in case of H-CL is linked with the effect of alteration including pore sizes. Its porous structure change furnishes a great surface area for adsorption.[4]

At different flow rates (1.5, 2.5, and 5.0 mL/min) the results show that the decrease in dynamic adsorption capacities of all the test antibiotics was originated by increasing the flow rate of inlet solution of each adsorbate. The high adsorption dynamic capacities were observed for NOR compared to second FQ antibiotic—MOX HCl which can be expressed with great water solubility and lower molecular weight of NOR relative to MOX HCl. The decrease in dynamic adsorption capacities was observed by increasing the inlet concentration of each adsorbate. This in reality could be connected with the enhancement of driving force for mass transfer across the liquid film along with acceleration of adsorption rate which generates an early saturation of the fixed-bed adsorbent column. The data show that the acid-modified H-forms of both natural zeolites are characterized with high dynamic adsorption capacity caused by increasing the Si/Al ratio in zeolite framework and pore opening in comparison with natural clinoptilolite and mordenite. The fact was confirmed by XRD analysis results. In the same conditions of experiment the obtained results indicate that the established dynamic adsorption capacities of mordenite and mordenite acid-modified forms are higher than the values for clinoptilolite and clinoptilolite H-forms.

4.3.3 ADSORPTION OF CEFTRIAXON BY DYNAMIC METHOD

The adsorption process was investigated for ceftriaxone by the dynamic method using the laboratory dynamic type equipment with packed fixed bed adsorption column. The behavior of breakthrough curves on two modified H-forms of clinoptilolite at different flow rates of 1.5, 2.5, and 5.0 mL/min and inlet ceftriaxone concentrations of 0.2 mg/mL and 1.0 mg/mL at constant temperature of 20°C was observed. Under the same dynamic conditions adsorption of ceftriaxone on initial clinoptilolite is ineffective; therefore, the dynamic adsorption capacity is very low (<0.009 mg/g); saturation was reached very rapidly. Analysis of the data obtained from the breakthrough curves (Figs. 7a and 7b) showed that the degree of use of the clinoptilolite layer activity in the case of modification with 5N HCl solution is a little bit higher than in the case of the sorbent modified with 2N HCl solution. The capacities of dynamic activity (q_d) of the adsorbent layer are improved. The breakthrough adsorption curve in the second case is elongated and has a different shape relative to the curve obtained on the H-CL (modified with 2N HCl) which supposes the monomolecular adsorption of ceftriaxone on the surface of the zeolite. At the time of 100 min the value of effluent to influent concentration ratio C/C_0 for CEF on H-CL (modified with 2N HCl) at the flow rate of 1.5 mL/min is at least 1.5 times higher than the value observed for CEF on H-CL (modified with 5N HCl). This can be related to low interaction between adsorbate and H-CL which accelerates breakthrough and saturation. Also, the early breakthrough of adsorbate is due to relatively low adsorption. Hence, the clinoptilolite acid-modified form is characterized by higher dynamic adsorption capacity caused by increasing the Si/Al ratio in the zeolite framework and pore opening in comparison with natural clinoptilolite.[13]

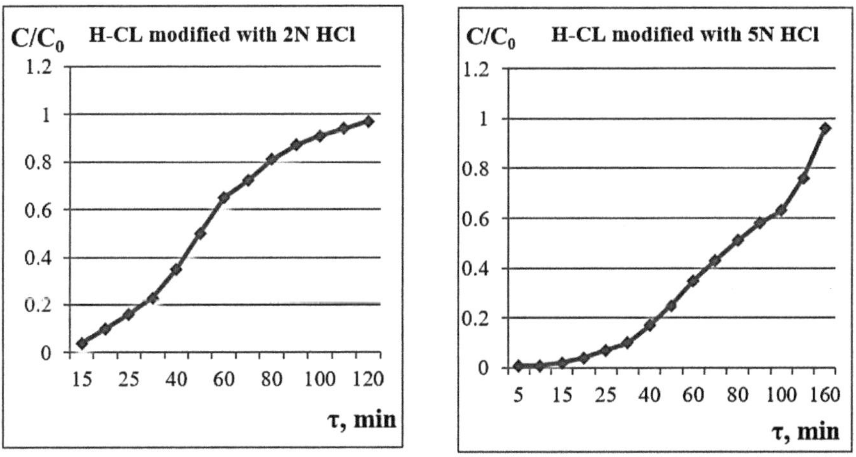

(a) *(b)*

FIGURE 4.7 The breakthrough curves of CEF on H-CL with 2N HCl (a) and H-CL with 5N HCl (b) at the flow rate of 1.5 mL/min and adsorbate concentration of 1.0 mg/mL at the neutral pH value.

4.3.4 VALIDATION OF ANALYTICAL HPLC METHODS FOR MOX HCL AND NOR DETERMINATION

Robustness study: The last chromatographic conditions were observed by improving the system operational parameters. These parameters are:

a) the wavelength for detection,

b) the composition of the mobile phase,
c) the flow rate,
d) the nature of stationary phase, and
e) checking the system suitability test parameters: theoretical plates, tailing factor, peak purity, etc.

The calibration curve showed very good linearity for a wide concentration range at 293 and 275 nm for MOX and NOR, respectively. For the robustness study, the qualitative critical factors were considered and selected, which are given with their levels summarized in Table 4.2. The SST parameters—column efficiency (theoretical plates—N), tailing factor (USP symmetry—As), relative standard deviation (RSD) of peak areas (RSD_A) and RSD of retention times (RSD_{RT}) (n = 6), and peak purity (PP) obtained from standard solution chromatograms were used as the response variable for the analytical HPLC method. The experiment was conducted in $2^{5-2} = 8$ runs for five two-level factors. Five critical factors were selected, and small variations (low and high levels) were induced in the nominal values of both methods. The 8-run design experiment was performed to assess the effect of each factor on the SST results. Table 4.3 shows the DoE results of the robustness test for the developed HPLC methods. The variability of the checked STT parameters is within the acceptance criteria, and the results show that these analytical procedures are robust.

TABLE 4.2 Robustness Factors and Design of Experiment.

No	Factor (Xi)	Unit	Low level (−)	Nominal level (0)	High level (+)
1	Flow rate of MP (X1)	mL/min	0.9	1.0	1.1
2	MOX/NOR standard stock solution (X2)	pH	4.0	7.0	–
3	ACN percentage in MP for MOX/NOR (X3)	%	35 10	40 15	45 20
4	Column temperature (X4)	°C	35	40	45
5	DAD wavelength for MOX/NOR (X5)	nm	291 273	293 275	295 277

The stability of the standard solution was studied initially, after 24 h, 3, 5, 7, 9 days stored at room temperature against a freshly prepared standard solution. The stability was checked using two standard solutions by calculating the percentage difference between peak areas of standard solutions stored and freshly prepared which should not be more than 5.0% and the bias in terms of peak area between two standard solutions should be within 0.98–1.02 (acceptance criteria). The percentage differences between peak areas obtained with standard solution stored at room temperature for 9 days and freshly prepared one are 4.4% and 3.9% for MOX HCl and NOR, respectively. This gives the confidence that FQ antibiotics residues are stable within 9 days and the concentration does not change in test solutions during the experiment of adsorption study.

The PVDF membrane syringe filter compatibility was checked using standard solution and by calculating the percentage difference between peak areas of standard solutions filtered and nonfiltered, which should not be more than 0.5% (acceptance criteria). The percentage difference between peak areas of standard solutions filtered and nonfiltered is 0.14% and 0.32% for MOX HCl and NOR, respectively, which gives the confidence that adsorption of each analyte does not occur on the used membrane syringe filter.

Specificity: The specificity parameter was checked by injecting standard solution and the background control (without FQ antibiotic) test solution—blank. The results show that there is no interference from blank and diluent at the retention time of analyte (adsorbate) peak. The MOX and NOR peaks were pure. Purity factor (975 for MOX and 980 for NOR) was more than purity threshold (950). Figures 8a and 8b show the 3D spectra and the chromatogram obtained from the standard solution of MOX HCl and NOR, respectively.

Linearity and range, limits of quantitation (LOQ), and detection (LOD): The standard working solutions were prepared at six different concentration levels ranging from 0.00005 mg/mL to 2.0 mg/mL from the standard stock solution of both FQ antibiotics—MOX HCl and NOR. Six replicate injections ($n = 6$) were carried out for each concentration level. The linearity over the given range was checked by the square of correlation coefficient—R^2 (acceptance criteria: >0.998), RSD of peak areas—RSD_A (acceptance criteria: <2.0 % for >1.0 mg/mL concentration levels and RSD_A<5.0 % for >0.00005 mg/mL concentration levels and for the last concentration <10 %), RSD of retention times—RSD_{RT} (acceptance criteria: <1.0 %).

The calibration curves (linearity graph) were constructed by plotting the peak area versus the corresponding concentration of the injected working standard solution. The value of the square of the correlation coefficient indicates very good linearity (R2>0.9996 for MOX HCl and R2>0.9999 for NOR). Figures 4.9 and 4.10 show the linearity graph for MOX HCl and NOR, respectively.

TABLE 4.3 The Results of Robustness Parameter (Critical Factors Effect Study).

No	Factors					Parameters							
						N		As		RSDA, %		RSDRT, %	
	X1	X2	X3	X4	X5	MOX	NOR	MOX	NOR	MOX	NOR	MOX	NOR
1	+	+	+	+	+	2865	3895	1.22	0.96	1.362	0.736	0.698	0.365
2	+	+	–	+	+	2563	3645	1.35	0.93	1.568	0.987	0.745	0.498
3	+		+	–	+	3256	3712	1.27	0.91	1.032	0.898	0.125	0.245
4	+		–	–	–	2463	3100	1.18	0.89	1.500	1.456	0.632	0364
5	–	+	+	–	–	3125	3564	1.23	0.93	1.952	1.365	0.455	0.522
6	–	+	–	–	+	3145	3666	1.09	0.86	1.300	1.112	0.245	0.601
7	–	–	+	+	–	2699	3056	1.11	0.87	0.856	1.354	0.112	0.471
8	–	–	–	+	+	2491	3444	1.13	0.89	1.505	1.866	0.733	0.326
Acceptance criteria						>2000		<2.0		<2.0		<1.0	

The limits of quantitation (LOQ) and detection (LOD) were established by injecting a series of stepwise diluted working solutions. The LOQ was estimated by calculating RSD of peak areas—RSD_A for six replicate injections which should be <10% and the signal-to-noise ratio—S/N - >10 (acceptance criteria). The LOD was estimated to be three times and more of s/N ratio (acceptance criteria). The determined LOQ and LOD for MOX HCl and NOR are presented in Table 4.4.

System suitability test parameters: The SST parameters were measured to check the chromatographic system performance. This test was carried out by six replicate injections (n = 6) of the standard solution of each FQ antibiotic. The main SST parameters including RSD of peak areas—RSD_A, RSD of the retention times—RSD_{RT}, peak tailing factor—A_s (USP coefficient of the peak symmetry $S = W_{0.05}/2f$) and column efficiency—the number of theoretical plates were measured. The results are summarized in Table 4.5.

TABLE 4.5 The System Suitability Parameters Results.

Parameter	MOX HCl	NOR	Acceptance criteria
N	3166	4532	>2000
RSD_A (n = 6), %	1.796	1.629	<2.0
RSD_{RT} (n = 6), %	0.762	0.866	<1.0
A_s	1.12	0.93	<2.0

Accuracy: The accuracy was expressed as the percentage of each FQ antibiotic recovered from a spiked test solution (effluent test solution + added standard solution) with the corresponding RSD. The average recovery should be within 95.0–105.0% and RSD of the percentage recoveries should be <3.0% for each concentration level of spiked solution (acceptance criteria). The recovery—Rec, % was calculated by the following formula:

$$Rec,\% = (C_1 - C_2) \times 100/C_s \tag{4.6}$$

where C_1—the concentration of each FQ antibiotic obtained with the spiked test solution, mg/mL, C_2—the concentration of each FQ antibiotic obtained with the effluent test solution, mg/mL and C_s—the concentration of each FQ antibiotic obtained with standard solution, mg/mL.[17–19] The results of the recovery are given in Table 4.6 which is well within the usually accepted limits indicating the accuracy of both methods.

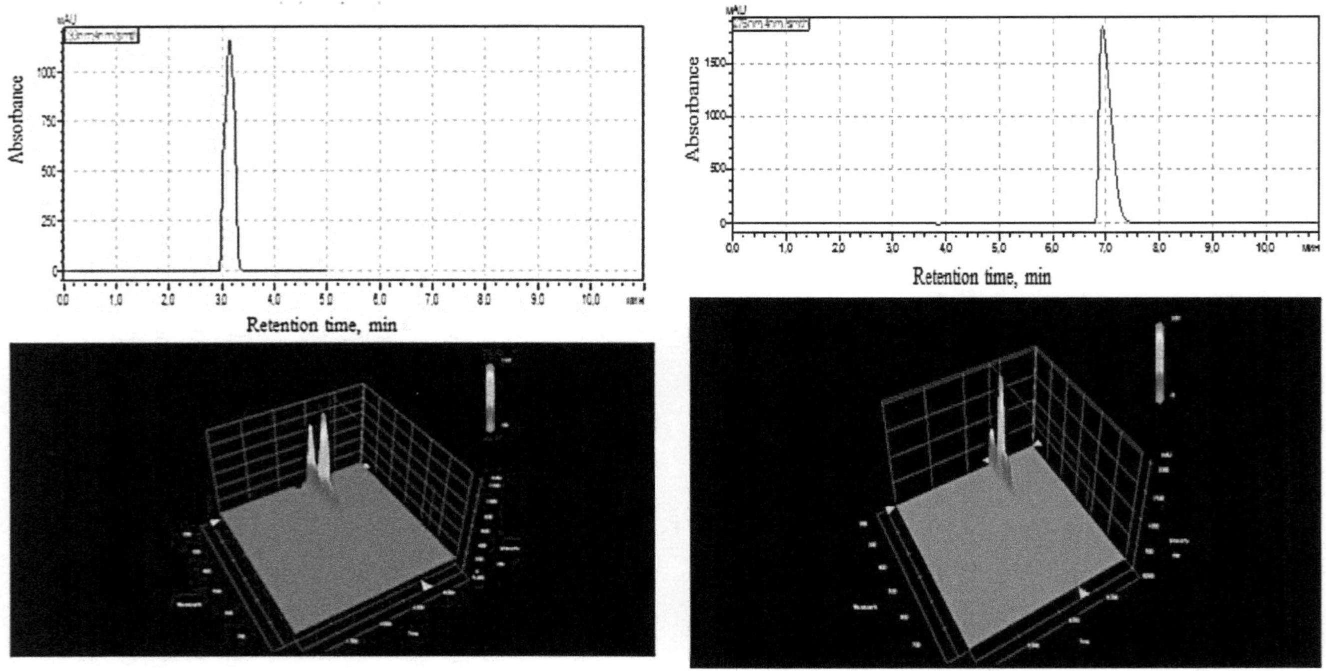

a) b)

FIGURE 4.8 The chromatograms and the 3D spectra of MOX HCl (a) and NOR (b) peaks obtained from standards solution with 0.1 mg/mL concentration.

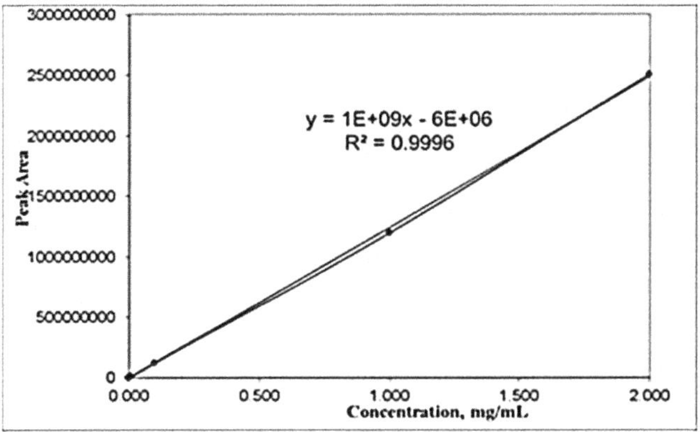

FIGURE 4.9 The calibration curve for MOX HCl over the concentration range 0.00005–2.0 mg/mL.

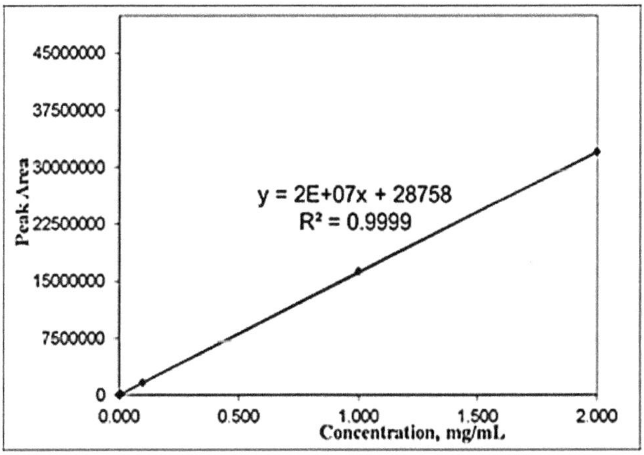

FIGURE 4.10 The calibration curve for NOR over the concentration range 0.00005–2.0 mg/mL.

TABLE 4.4 The LOQ and LOD of MOX HCl and NOR Methods.

Parameter	Value	
	MOX HCl	NOR
LOQ, mg/mL	0.00005	0.00005
LOD, mg/mL	0.00001	0.000008
RSD of peak areas for LOQ (n = 6)	3.561	2.956
RSD of retention times for LOQ (n = 6)	0.671	0.583
s/N for LOQ	25	21
s/N for LOD	4	7

TABLE 4.6 The Results of the Accuracy Study.

Added standard solution, mg/mL	Average concentration of spiked test solution, mg/mL (n = 3)	Concentration of effluent test solution, mg/mL	Acceptance criteria
MOX HCl			
0.0788	0.1319		97.5
0.0985	0.1527	0.05650	98.5
0.1182	0.1750		100.2
Average recovery - Rec, %			98.7
RSD, % of the percentage recoveries			1.383
NOR			
0.0797	0.1654		99.7
0.0996	0.0961	0.0862	96.5
0.1195	0.1222		102.3
Average recovery - Rec, %			99.5
RSD, % of the percentage recoveries			2.920

Precision: The precision of the developed analytical methods was estimated by measuring repeatability (intra-day) on six replicate injections of standard solution and on six individual determinations of MOX HCl, and NOR in test solution obtained using the adsorption static method. For sample preparation, 0.2–0.2 g of zeolite adsorbent samples were transferred separately to six 50 mL Erlenmeyer flasks and added 20 mL of each FQ antibiotic standard stock solutions at 1.0 mg/mL concentration. Initially, zeolite samples with adsorbate FQ antibiotic solutions were left on an orbital shaker at 150 rpm for 15 min, then statically for 3 h. After that zeolite samples were centrifuged at 3000 rpm for 5 min and 1–1 mL of the obtained supernatants were used and diluted to 10 mL with diluent, and mixed well. The precision was checked by calculating RSD of the concentration of each FQ antibiotic, mg mL^{-1} and retention times of MOX and NOR for six individual determinations which should not be more than 5.0% and 1.0%, respectively. The results are shown in Table 4.7. The results were within the acceptance criteria which indicate that this method has good precision.

TABLE 4.7 THE RESULTS OF THE PRECISION STUDY.

The number of test solution	MOX HCL		NOR	
	Concentration, mg/mL	Retention time, min	Concentration, mg/mL	Retention time, min
1	0.0351	3.326	0.0401	7.674
2	0.0385	3.388	0.0421	7.537
3	0.0355	3.390	0.0435	7.578
4	0.0341	3.347	0.0445	7.495
5	0.0377	3.345	0.0412	7.529
6	0.0366	3.352	0.043	7.480
Average	0.0363	3.358	0.0424	7.549
SD	0.002	0.026	0.002	0.070
RSD	4.582	0.762	3.774	0.931

4.3.5 VALIDATION OF ANALYTICAL HPLC METHOD FOR CEF DETERMINATION

System suitability test: In order to check the chromatographic system performance, the system suitability test was performed by using six replicate injections (n = 6) of the CEF standard solution at the concentration—1.0 mg/mL. The following parameters: the relative standard deviation (RSD) of peak areas—RSD_A, the RSD of the retention times—RSD_{RT}, the peak tailing factor (the USP coefficient of the peak symmetry), the column efficiency - the number of theoretical plates were measured.

Specificity: This parameter was checked by injecting the standard solution, the negative control solution as a blank (solution without CEF), and the model test solution. The CEF peak was pure. The purity factor (980) was greater than the purity threshold (950). Figure 4.11 shows the chromatogram obtained from the CEF standard solution, respectively.

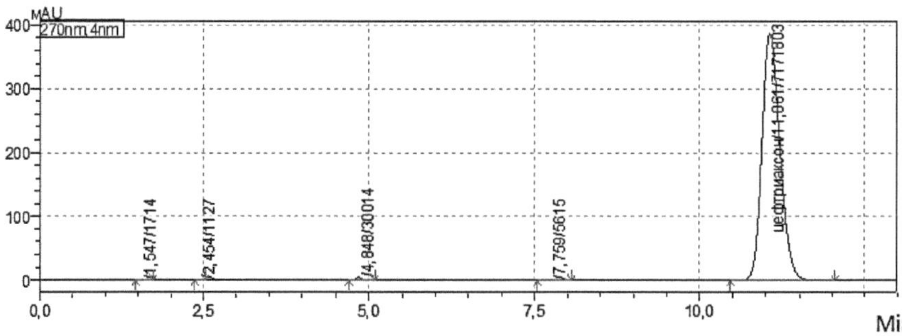

FIGURE 4.11 The chromatogram of the CEF standard solution (peaks with RT = 1.547, 2.454, 4.848, 7.759 belong to solvent; principal peak with RT = 11.061 belongs to CEF).

Linearity and range, limits of quantitation (LOQ), and detection (LOD): The calibration curve (plot of linearity) was constructed by plotting the average peak area of CEF versus the corresponding concentration of the injected working standard solution. The calibration curve (linearity graph) is linear and the square value of correlation coefficient ($R^2 = 0.99998$) indicates very good linearity in a wide range of concentrations of CEF at 270 nm. Figure 4.12 shows the linearity graph for CEF.

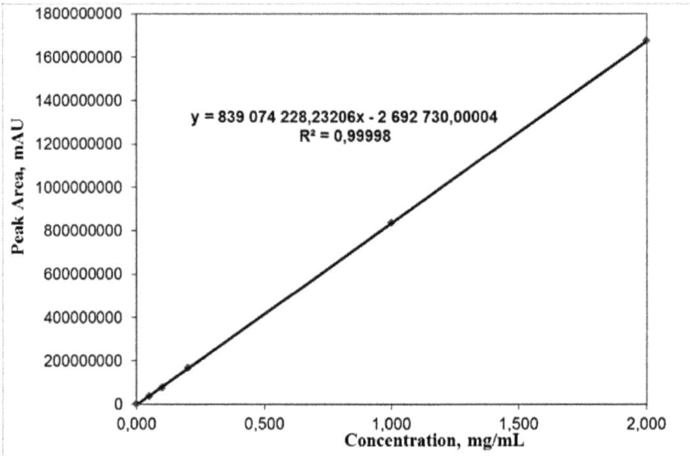

FIGURE 4.12 The calibration curve at the concentration range of 0.0001–2.0 mg/mL of CEF.

The LOQ was established for checking the sensitivity parameter of the method by injecting a series of stepwise-diluted working standard solutions. The LOQ was estimated by calculating the RSD_A for six replicate injections ($n = 6$), which should be < 10% and the ratio of signal to noise (s/N) - > 10 (acceptance criteria).

Accuracy: The accuracy was assessed by comparing the analyte amount determined versus the known amount spiked at three different concentration levels (80%, 100%, and 120% of the model test solution—1.0 mg/mL CEF) with three replicate injections ($n = 3$). The average percentage recovery should be within 95.0–105.0% and the RSD of percentage recovery rates for three individual determinations should be not more than 5.0% for each concentration level of spiked test solution (acceptance criteria).

Precision: The intraday precision of the analytical method was estimated by measuring repeatability on six individual determinations of CEF concentration in model test solution at the same concentrations (100% of the model test solution). The parameter was checked during the accuracy study. The precision was checked by the RSD_C for six individual determinations of CEF concentration (mg/mL) which should not be more than 3.0% (acceptance criteria).

The standard solution stability study indicated that the CEF solution was stable within 24 h at ambient temperature.

The results were obtained from SST, precision, and accuracy studies; also the values of LOQ are summarized in Table 4.8.

TABLE 4.8 The Method Validation Data.

Validation parameter	Characteristic	Result	Acceptance criteria
SST	Column efficiency	>3600	> 2000
	Symmetry	0.95–1.2	< 2.0
	RSD_A (n = 6), %	0.130	≤ 2.0
	RSD_{RT} (n = 6), %	0.261	≤ 1.0
Precision	RSD_C (n = 6), %	2.114	≤ 3.0
Accuracy	Recovery, % at 80%	98.5	
	Recovery, % at 100%	99.7	
	Recovery, % at 120%	101.1	95.0–105.0
	Main recovery, %	99.8	
LOQ	Concentration, mg/mL	0.0001	–
	s/N	> 23	> 10

Hence, the developed analytical HPLC methods were precise, accurate, specific, sensitive, and linear to determine quantitatively MOX HCl, NOR, and CEF in influents and effluent aqueous solutions during adsorption experiment.

4.4 CONCLUSIONS

The selected natural zeolites—clinoptilolite, mordenite, and their acid-modified H-form with hydrophilic pores have a greater affinity for adsorbing antibiotic pollutants in aqueous solution. The developed adsorption laboratory technology has been effectively applied for the removal of the frequently used fluoroquinolone antibiotics—moxifloxacin HCL, norfloxacin, and cephalosporin antibiotic namely ceftriaxone from aqueous solutions. The experimental data have shown that the high volumetric flow rate, the inlet concentration of antibiotic solution, and the pH value accelerate breakthrough of adsorbates and correspondingly reduce adsorption capacities.

Moreover, this book chapter provides functional information for the design of fixed bed adsorption column for removal studies of other antibiotics of different classes or contaminants. Hence, this research confirms that natural zeolites are competitive, eco-friendly, and efficient adsorbents in terms of cheapness, shape selectivity, and adsorption efficacy. Low cost, high adsorption capacity, readily available materials—clinoptilolite and mordenite with their acid-modified forms can be widely used for the removal of antibiotics from wastewaters and implemented in an industrial setting for separation and purification processes. Within the research new, specific, selective, and sensitive analytical HPLC procedures were developed and validated for quantitative determination of moxifloxacin HCl, norfloxacin, and ceftriaxone in aqueous solutions and matrices. Hence, the proposed and validated HPLC methods can be successfully used to measure adsorption of moxifloxacin HCl, norfloxacin, and ceftriaxone on the natural zeolites and their modified forms; also, they can be applied by other laboratories for routine analysis of wastewaters and pharmaceuticals.

ACKNOWLEDGMENT

The research was financially supported by the Shota Rustaveli National Science Foundation within the framework of the project #217138.

KEYWORDS

- **adsorption**
- **natural zeolite**
- **antibiotics**
- **wastewater**
- **HPLC**
- **validation**

REFERENCES

1. The Third National Program of Georgia on Environmental Protection, Working Version for 2017–2021, Tbilisi, 2017; p 161.
2. Water Resources Management and Protection, Approaches of Georgia and the European Union, Tbilisi, 2007; p 27.

3. Glassmeyer, S. T.; Hinchey, E. K.; Boehme, S. E.; Daughton, C. G.; Ruhoy, I. S.; Conerly, O.; Daniels, R. L.; Lauer, L.; McCarthy, M.; Nettesheim, T. G.; Sykes, K.; Thompson, V. G. Disposal Practices for Unwanted Residential Medications in the United States. *Environ. Int.* **2009,** *35* (3), 566–572.
4. Rubashvili, I.; Eprikashvili, L.; Kordzakhia, T.; Zautashvili, M.; Pirtskhalava, N.; Dzagania. M. Adsorptive Removal Study of the Frequently Used Fluoroquinolone Antibiotics – Moxifloxacin HCl and Norfloxacin from Wastewaters using Natural Zeolites. *Mediterranean J. Chem.* **2019,** *9* (2), 142–154. DOI: http://dx.doi.org/10.13171/mjc92190921700ar
5. Moussavi, G.; Alahabadi, A.; Yaghmaeian, K.; Eskandari, M. Preparation, Characterization and Adsorption Potential of the NH_4Cl-Induced Activated Carbon for the Removal of Amoxicillin Antibiotic from Water. *Chem. Eng. J.* **2013,** *217*, 119–128.
6. Chang, X.; Meyer, M. T.; Liu, X.; Zhao, Q.; Chen, H.; Chen, J. a.; Qiu, Z.; Yang, L.; Cao, J.; Shu, W. Determination of Antibiotics in Sewage from Hospitals, Nursery and Slaughter House, Wastewater Treatment Plant and Source Water in Chongqing Region of Three Gorge Reservoir in China. *Environ. Pollut.* **2010,** *158* (5), 1444–1450.
7. Brown, K. D.; Kulis, J.; Thomson, B.; Chapman, T. H.; Mawhinney, D. B. Occurrence of Antibiotics in Hospital, Residential, and Dairy Effluent, Municipal Wastewater, and the Rio Grande in New Mexico. *Sci. Total Environ.* **2006,** *366* (2-3), 772–783.
8. Watkinson, A. J.; Murby, E. J.; Kolpin, D. W.; Costanzo, S. D. The Occurrence of Antibiotics in an Urban Watershed: From Wastewater to Drinking Water. *Sci. Total Environ.* **2009,** *407* (8), 2711–2723.
9. Hernando, M.; Mezcua, M.; Fernandezalba, A.; Barcelo, D. Environmental Risk Assessment of Pharmaceutical Residues in Wastewater Effluents, Surface Waters and Sediments, *Talanta.* **2006,** *69* (2), 334–342. DOI: 10.1016/j.talanta.2005.09.037
10. Holstein, J. J.; Hübschle, C. B.; Dittrich, B. Electrostatic Properties of Ninefluoroquinolone Antibiotics Derived Directly from their Crystal Structure Refinements. *Cryst. Eng. Comm.* **2012,** *14*, 2520–2531.
11. Jia, A.; Wan, Y.; Xiao, Y.; Hu, J. Occurrence and Fate of Quinolone and Fluoroquinolone Antibiotics in a Municipal Sewage Treatment Plant. *Water Res.* **2012,** *46* (2), 387–394.
12. Khorsandi, H.; Teymori, M.; Aghapour, A. A.; Jafari, S. J.; Taghipour, S.; Bargeshadi, R. Photodegradation of Ceftriaxone in Aqueous Solution by using UVC and UVC/H_2O_2 Oxidation Processes. *Appl. Water Sci.* 2019, *9*, 81. https://doi.org/10.1007/s13201-019-0964-2;
13. Eprikashvili, L.; Kordzakhia, T.; Zautashvili, M.; Rubashvilia, I.; Pirtskhalava, N.; Dzagania, M.; Tsintskaladze, G.; Antia, G. Effect of Clino Ptilolite Acid Activation on Ceftriaxone Sorption from Wastewaters. *Res. J. Chem. Environ.* **2021,** *25* (5), 80–85.
14. Yu, F.; Li, Y.; Han, S.; Ma, J. Adsorptive Removal of Antibiotics from Aqueous Solution using Carbon Materials. *Chemosphere* **2016,** *153*, 365–385.
15. Choi, T. A. In *Side Effects of Drugs Annual*; Chapter 26 - Miscellaneous Antibacterial Drugs; Aronson, J. K., Ed.; Elsevier: Amsterdam, 2014; vol 35, pp 463–481.
16. Darweesh, M. T.; Ahmed, M. J. Adsorption of Ciprofloxacin and Norfloxacin from Aqueous Solution onto Granular Activated Carbon Infixed Bed Column. *Ecotoxicol. Environ. Safety.* **2017,** *138*, 139–145.
17. Suzuki, S.; Phuong Hoa, P. T. Distribution of Quinolones, Sulfonamides, Tetracyclines in Aquatic Environment and Antibiotic Resistance in Indochina. *Front. Microbio.* **2012,** *3* (67).

18. Carvalho, I. T.; Santos, L. Antibiotics in the Aquatic Environments: A Review of the European Scenario. *Environ. Int.* **2016,** *94*, 736–757.
19. US Pharmacopeia National Formulary USP 41-NF-36, Moxifloxacin, The United States Pharmacopeial Convention; United Book Press, Inc: Baltimore, 2018; pp 8361–8363.
20. US Pharmacopeia National Formulary USP 41-NF-36. Norfloxacin, The United States Pharmacopeial Convention; United Book Press, Inc: Baltimore, 2018; pp 2976–2977.
21. Le-Minh, N.; Khan, S. J.; Drewes, J. E.; Stuetz, R. M. Fate of Antibiotics during Municipal Water Recycling Treatment Processes. *Water Res.* **2010,** *44* (15), 4295–4323.
22. Torres-Pérez, J.; Gérente, C.; Andrès, Y. Sustainable Activated Carbons from Agricultural Residues Dedicated to Antibiotic Removal by Adsorption. *Chin. J. Chem. Eng.* **2012,** *20*, 524–529.
23. Zhu, L.; Santiago-Schübel, B.; Xiao, H.; Hollert, H.; Kueppers, S. Electrochemical Oxidation Offluoroquinolone Antibiotics: Mechanism, Residual Antibacterial Activity and Toxicity Change. *Water Res.* **2016,** *102*, 52–62.
24. Cˇvancˇarová, M.; Moeder, M.; Filipová, A.; Cajthaml, T. Biotransformation of Fluoroquinolone Antibiotics by Ligninolytic Fungi–metabolites, Enzymes and Residual Antibacterial Activity. *Chemosphere* **2015,** *136*, 311–320.
25. Maraschi, F.; Sturini, M.; Speltini, A.; Pretali, L.; Profumo, A.; Pastorello, A. TiO_2-Modified Zeolites for Fluoroquinolones Removal from Wastewaters and Reuse after Solar Light Regeneration. *J. Environ. Chem. Eng.* **2014,** *2*, 2170–2176.
26. Feng, M.; Wang, X.; Chen, J.; Qu, R.; Sui, Y.; Cizmas, L.; Wang, Z.; Sharma, V. K. Degradation of Fluoroquinolone Antibiotics by Ferrate (VI): Effects of Water Constituents and Oxidized Products. *Water Res.* **2016,** *103*, 48–57.
27. Zhao, Y.; Liang, X.; Wang, Y.; Shi, H.; Liu, E.; Fan, J.; Hu, X. Degradation and Removal of Ceftriaxone Sodium in Aquatic Environment with Bi_2WO_6/g-C_3N_4 Photocatalyst. *J. Coll. Interface Sci.* **2018,** *523*, 7–17 https://doi.org/10.1016/j.jcis.2018.03.078
28. Ebrahimpour, B.; Yamini, Y.; Moradi, M. Application of Ionic Surfactant as a Carrier and Emulsifier Agent for the Microextraction of Fluoroquinolones. *J. Pharm. Biomed. Anal.* **2012,** *66*, 264–270.
29. Guo, H.; Gao, N.; Yang, Y.; Zhang, Y. Kinetics and Transformation Pathways on Oxidation of Fluoroquinolones with Thermally Activated Persulfate. *Chem. Eng. J.* **2016,** *292*, 82–91.
30. Ferreira, V. R. A.; Amorim, C. L.; Cravo, S. M.; Tiritan, M. E. Fluoroquinolones Biosorption onto Microbial Biomass: Activated Sludge and Aerobic Granular Sludge. *Int. Biodeterior. Biodegrad.* **2016,** *110*, 53–60.
31. Shi, X.; Karachi, A.; Hosseini, M.; Yazd, M. S.; Kamyab, H.; Ebrahimi, M.; Parsaee, Z. Ultrasound Wave Assisted Removal of Ceftriaxone Sodium in Aqueous Media with Novel Nano Composite g-C_3N_4/MWCNT/Bi_2WO_6 based on CCD-RSM Model. *Ultrason. Sonochem.* **2020,** *68*, 104460. https://doi.org/10.1016/j.ultsonch.2019.01.018;
32. Tan, F.; Sun, D.; Gao, J.; Zhao, Q.; Wang, X.; Teng, F.; Quan, X.; Chen, J. Preparation of Molecularly Imprinted Polymer Nanoparticles for Selective Removal of Fluoroquinolone Antibiotics in Aqueous Solution. *J. Hazard. Mater.* **2013,** *244–245*, 750–757.
33. Braschi, I.; Gatti, G.; Paul, G.; Gessa, C. E.; Cossi, M.; Marchese, L. Sulfonamide Antibiotics Embedded in High Silica Zeolite Y: A Combined Experimental and Theoretical Study of Host-Guest and Guest-Guest Interactions. *Langmuir* **2010,** *26* (12), 9524–9532.

34. Cao, J.; Lei, C.; Yang, B.; Li, Z.; Lei, L.; Hou, Y.; Feng, X. Zeolitic Imidazolate Frameworkderived Core-Shell-Structured CoS_2/CoS_2-N-C Supported on Electrochemically Exfoliated Graphene Foil for Efficient Oxygen Evolution. *Batt. Supercaps* **2018,** *1*, 1–8.
35. Zukal, A.; Shamzhy, M.; Kubů, M.; Čejka, J. The Effect of Pore Size Dimensions in Isoreticular Zeolites on Carbon Dioxide Adsorption Heats. *J. CO_2 Util.* **2018,** *24*, 157–163.
36. Chen, Z.; Maa, W.; Lub, G.; Menga, F.; Duana, S.; Zhanga, Z.; Weic, L.; Pana, Y. Adsorption of Levofloxacin onto Mechanochemistry Treated Zeolite: Modeling and Site Energy Distribution Analysis. *Sep. Purif. Technol.* **2019,** *222*, 30–34.
37. Jiang, N.; Shang, R.; Heijman, S. G. J.; Rietveld, L. S. High-Silica Zeolites for Adsorption of Organic Micro-Pollutants in Water Treatment: A Review. *Water Res.* **2018,** *144*, 145–161.
38. Abu-Lail, L.; Bergendahl, J. A.; Thompson, R. W. Adsorption of Methyl Tertiary Butyl Ether on Granular Zeolites: Batch and Column Studies. *J. Hazard Mater.* **2010,** *178*, 363–369.
39. Liang, Z.; Zhao, Z.; Sun, T.; Shi, W.; Cui, F. Adsorption of Quinolone Antibiotics in Spherical Mesoporous Silica: Effects of the Retained Template and its Alkyl Chain Length. *J. Hazard. Mater.* **2016,** *305*, 8–14.
40. Liu, H.; Ning, W.; Cheng, P.; Zhang, J.; Wang, Y.; Zhang, C. Evaluation of Animal Hairs-based Activated Carbon for Sorption of Norfloxacin and Acetaminophen by Comparing with Cattail Fiber-based Activated Carbon. *J. Anal. Appl. Pyrolysis.* **2013,** *101*, 156–165.
41. Patiño, Y.; Díaz, E.; Ordóñez, S. Pre-concentration of Nalidixic Acid through Adsorption–Desorption Cycles: Adsorbent Selection and Modeling. *Chem. Eng. J.* **2016,** *283*, 486–494.
42. Martucci, A.; Pasti, L.; Marchetti, N.; Cavazzini, A.; Dondi, F.; Albertia, A. Adsorption of Pharmaceuticals from Aqueous Solutions on Synthetic Zeolites. *Micropor. Mesopor. Mater.* **2012,** *148* (1), 174–183.
43. Ašperger, D.; Varga, I.; Babić, S.; Ćurković, L. Adsorption of Norfloxacine on Natural Zeolite – Clinoptilolite. *Holistic App. Environ.* **2014,** *4* (1), 3–15.
44. Eprikashvili, L. G.; Kordzakhia, K. T.; Kordzakhia, T. N.; Zautashvili, M. G.; Pirtskhalava, N. V.; Dzagania, M. A. Condition of the Atmospheric Air of Georgia and Some Methods of its Protection against Pollution. *APARATURA BADAWCZA I DYDAKTYCZNA* **2013,** *v. XVIII* (3), 195–204.
45. Eprikashvili, L.; Zautashvili, M.; Kordzakhia, T.; Pirtshkalava, N.; Dzagania, M.; Rubashvili, I.; Tsitsishvili, V. Intensification of Bioproductivity of Agricultural Cultures by Adding Natural Zeolites and Brown Coals into the Soil. *Ann. Agrarian Sci.* **2016,** *14*, 67–71. https://doi.org/10.1016/j.aasci.2016.05.013
46. Eprikashvili, L.; Kordzakhia, T.; Zautashvili, M.; Rubashvili, I.; Dzagania, M.; Pirtskhalava, N.; Tsintskaladze, G. Feasibility Study for the use of Natural Mordenite in Purification Processes of Wastewaters from Pharmaceutical Pollutants. *Int. J. Adv. Res.* **2021,** *9* (03), 683–691. https://www.journalijar.com/article/36739/feasibility-study-for-the-use-of-natural-mordenite-in-purification-processes-of-wastewaters-from-pharmaceutical-pollutants/
47. Rubashvili, I.; Zautashvili, M.; Kordzakhia, T.; Eprikashvili, L. Development and Validation of Quantitative Determination HPLC Metods of the Fluoroquinolone

Antibiotics – Moxifloxacin Hydrochloride and Norfloxacine in Support of Adsorption Study on Natural Zeolites. *PERIÓDICO TCHÊ QUÍMICA* **2019,** *16* (3), 10–20. http://deboni.he.com.br/Periodico33.pdf

48. Eprikashvili, L.; Kordzakhia, T.; Andronikashvili, T. In *Zeolites – The Unique Desiccation Agent for Organic Liquids*. LAP LAMBERT Academic Publishing, 2015; p 91.
49. Tsitsishvili, G. V.; Andronikashvili, T. G.; Kirov, G. R.; Filizova, L. D. In *Natural Zeolites*; Ellis Horwood Limited: Chichester, 1991; p 297.

CHAPTER 5

Investigation of the Impact of Epoxy Compounds Based on Environmentally Friendly and Renewable Raw Materials in Bitumen Modification Processes

VOLODYMYR GUNKA[1], YURII HRYNCHUK[1], YURIY PRYSIAZHNYI[1], YURIY DEMCHUK[1], IURII SIDUN[2], VOLODYMYR REUTSKYY[1], and MICHAEL BRATYCHAK[1]

[1]*Institute of Chemistry and Chemical Technology, Lviv Polytechnic National University, Ukraine*

[2]*Institute of Building and Environmental Engineering, Lviv Polytechnic National University, Ukraine*

ABSTRACT

The paper is devoted to the research of the possibility of creating a new complex modifier for bitumen based on epoxy compounds of plant origin. Crude rapeseed oil was used as a raw material for the production of epoxy bitumen modifier. The effect of adding modifiers (epoxidized compounds) on the basis of renewable raw materials with and without initiators-hardeners (adipic and formic acid, maleic anhydride, polyethylene polyamine) on the performance properties of the obtained bitumens, such as penetration, temperature with the surface of gravel and glass, the temperature of fragility according to Fraas, ductility, elasticity, and others, was researched. The design of the optimal composition of asphalt concrete mixtures obtained from bitumens modified with epoxy modifiers on the basis of renewable raw materials is provided in the paper. Samples of asphalt concrete obtained with the use of bitumen modified with epoxides based on crude rapeseed oil were obtained and tested.

Environmental Technology and Sustainability: Physical, Chemical, and Biological Technologies for Environmental Protection. Tamara Tatrishvili, Ann Rose Abraham, & A. K. Haghi (Eds.)

5.1 INTRODUCTION

As for today, the most common road surface remains hot asphalt concrete (HMA). Despite the high demand, such coverage is not perfect. The problem lies in the binder—bitumen, the quality of which can be largely improved. Various methods and modifiers are used to improve the properties of bitumen.[2,3,10,11, 17,18,21,24,25,31] Usually, modifiers are divided into: polymers, adhesion promoters, waxes, fibers, natural bitumen, rubber crumb, and various chemical reagents (polyphosphoric acid, ferric chloride, and organometallic complexes). The widely used modifiers of these are polymers, which are also divided into thermoplastics, thermoplastic elastomers, terpolymers (reactoplasts or thermosetting polymers), and latexes.

Among these modifiers, one of the less studied are thermosetting polymers which include epoxy resins with their hardeners. However, current research indicates that epoxy-containing additives can radically change the properties of bitumen and asphalt concrete and the effect of their use may be much more interesting and higher in comparison with other modifiers. Hot-mix asphalt concrete of different types, which include epoxy resins, is called epoxy asphalt concrete.

Studies[16,23,26–28,32] have shown that epoxy asphalt concrete has high strength and track resistance at elevated temperatures, high resistance to cracking from prolonged transport and low temperatures, and resistance to fuels and lubricants, increasing (2–4) times the service life of pavements.

Epoxy asphalt concrete can be used for the installation of asphalt pavements on congested sections of roads, concrete, and reinforced concrete foundations of bridges.[30]

Epoxy compounds are monomers or short-chain copolymers having an epoxy group at either end that is polymerized by hardeners. In epoxy bituminous binder systems, polymer chains are formed by the reaction of an epoxy modifier with initiators with reactive functional groups in a solid matrix. This phenomenon is called polymerization, which involves the formation, branching, and crosslinking of epoxy polymer chains. During polymerization, when all the chains are interconnected in one network, a higher molecular weight of the composite is achieved.[19]

In Refs. [14,15] it is claimed that the possible formation of epoxy asphalt concrete is accompanied by polymerization reactions of epoxidized vegetable oil with a hardener in the structure of bitumen, which allows changing the properties of the resulting product. Commodity amines and amides for petroleum epoxy resins, dibasic carboxylic acids, etc., can be used as hardeners.

In recent years, researchers at Lviv Polytechnic National University are conducting research on the possibility of using epoxidized vegetable oils to improve bitumen and asphalt concrete based on them.[13–15,22]

Particular attention should be paid to the plant origin of raw materials for modifier road bitumen, which is environmentally friendly, renewable, and promising given the current state of affairs and with environmental aspects of hydrocarbon use and their availability in general.

Studies[20,29] also confirm that epoxy resins can significantly improve the properties of asphalt binder. However, the reviewed studies also indicate that modern epoxy resins made from crude oil are quite expensive and nonenvironmentally friendly. Therefore, the use of epoxy compounds based on vegetable origin, in particular vegetable oils, for the modification of road bitumen and asphalt concrete based on them is promising. These are cheap to produce, affordable, environmentally friendly, and made from relatively cheap renewable raw materials, which are also produced in sufficient quantities by Ukraine.

Taking into account previous experience,[13,15] that shows the feasibility of using epoxidized rapeseed oil to modify road bitumen and HMA, the authors propose to modify the bitumen with so-called bio-based epoxy rape oil (ERO-In). What is ERO-In? Under this term we understand the use as a modifier of epoxy rapeseed oil (EPO) obtained from renewable and environmentally friendly raw materials, rapeseed oil with known hardeners (AA, MA, or PEPA) to initiate the polymerization reaction in bitumen.

Therefore, our research was aimed at developing a new model that can predict the behavior of modified bitumen bio-based epoxy rape oil in different conditions and ways to intensify this process using hardener initiators. It is important for us to understand the mechanism of modification and curing and how it affects the physical and mechanical properties of epoxy bitumen systems. This paper presents these results.

5.2 EXPERIMENTAL

5.2.1 MATERIALS

OB1—oxidized bitumen brand BND 70/100, obtained at the plant for the production of oxidized bitumen PJSC "Ukrtatnafta" (Kremenchuk, Poltava region, Ukraine).

TABLE 5.1 Characteristics of Bitumen OB1.

Index	Value	Standard or Ref.
Penetration at 25°C (dmm)	71	[8]
Softening point (°C)	46	[9]
Ductility at 25°C (cm)	>100	[4]
Adhesion to gravel (mark)	2.5	[11]
Adhesion to glass (%)	33	[11]
Fraas breaking point (°C)	−10	[7]
Plasticity interval (°C)	56	PI = SP – FBP
Resistance to hardening at 163°C (RTFOT method)		
Mass change (wt%)	0.03	[5]
Softening point after RTFOT (°C)	52.2	
Penetration at 25°C after RTFOT (dmm)	55	
Softening point change (°C)	6.2	
Residual penetration (%)	77.5	

Rapeseed oil epoxide (ERO) was obtained according to reference,[12] the characteristics of the raw material of the epoxidation process (RO) and the product of this process (ERO) are given in Table 5.2.

TABLE 5.2 Characteristics of RO and ERO.

Index	RO	ERO
Epoxy number (wt%)	0.28	6.72
Acid number (g KOH/100g)	0.50	0.45
Bromine number (g Br/100g)	80.30	31.9

Granite crushed stone (S1—fraction 8/11 mm; S2—fraction 20/40 mm) selected at LLC Novograd-Volyn Stone Crushing Plant. Meets the requirements of the Ukrainian standard.[6]

5.2.2 TECHNIQUES OF EXPERIMENTS

5.2.2.1 MIXING BITUMEN WITH ERO-IN

After epoxidation, ERO was mixed with the initiator, and this composition (ERO-In) was added to the heated bitumen. A thermostated container with a stirrer (speed 1000 rpm) was used for mixing. The ERO-In mixture was charged to the heated bitumen reactor and modified with constant stirring for some time. The block diagram of the modification is shown in Figure 5.1.

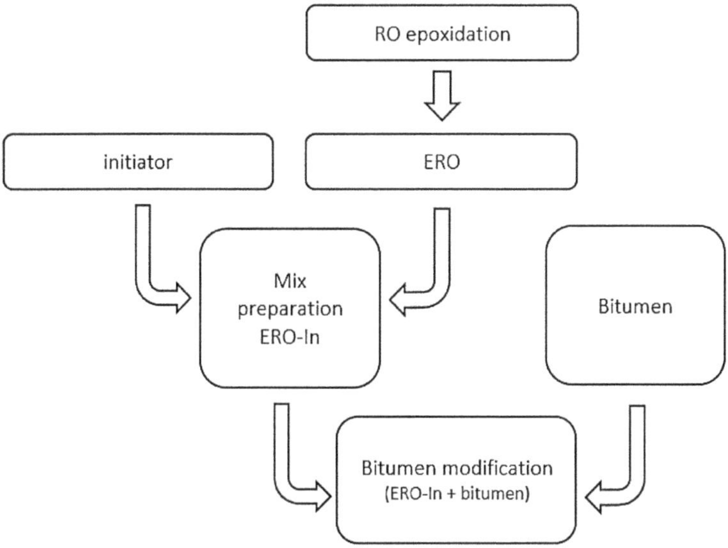

FIGURE 5.1 ERO-In bitumen modification scheme.

5.2.2.2 *BITUMEN ANALYSES*

The main performance characteristics of bitumen were determined according to the methods given in Table 5.1.

Rolling bottle tests were conducted according to Ref. [33]. We used crushed stone of 8/11 fraction obtained from natural stone (S1) and selected at LLC Novograd-Volyn Stone Crushing Plant.

5.2.2.3 *FTIR SPECTROSCOPY*

The FTIR spectra samples were recorded on a Spectrum Two spectrometer (PerkinElmer, USA) by using diamond U-ATR single reflection accessory. PerkinElmer Spectrum software was used to draw the spectra. The spectra (16 scans per spectrum) of the samples were collected in the mid-infrared wavenumber range from 4000 to 400 cm^{-1}, with a spectral resolution of 4 cm^{-1}.

5.2.2.4 *PREPARATION AND TESTING OF HMA*

Physico-mechanical properties of HMA were determined on samples obtained by compaction of the mixture in steel molds.

The HMA study in the form of cylindrical specimens (diameter and height 71.4 mm and weight 655.0 g) was performed according to Ukrainian research methods. The average density of HMA was determined by hydrostatic weighing. Residual porosity was determined by pore volume in HMA based on pre-set average density of cylindrical samples and actual density of HMA mix. Water-saturation was determined by the amount of water, which is absorbed by a sample at pre-set mode of saturation in vacuum unit. The compression tensile strength at 20 and 50°C of HMA was determined on mechanical presses with press-plate movement speed of (3.0 ± 0.1) mm/min. For testing, HMA samples are placed under the press plates with flat upper and lower cylinder faces. Before testing the samples are thermostat-conditioned in a vessel with water during (60 ± 5) min at the temperature: (50 ± 1)°C, (20 ± 1)°C. The samples for testing compression tensile strength at 50°C are placed (before the thermostat-conditioning) into the tight polyethylene bags—to prevent their contact with water. The compressive strength after water saturation at a temperature of 50°C was determined on samples after their water saturation.

FIGURE 5.2 HMA samples for testing.

5.3 RESULTS AND DISCUSSION

Initially, the feasibility of adding different types of hardener initiators to the ERO to obtain a bitumen modifier was investigated.

5.3.1 INFLUENCE OF EXPEDIENCY OF USING THE INITIATOR IN THE PROCESS OF MODIFICATION

According to the proposed modification scheme (Fig. 5.1) was prepared a mixture of ERO with initiators, in which the content of the initiator was 10 wt%. Polyethylene polyamine (PEPA), maleic anhydride (MA), formic acid (FA), and adipic acid (AA) were used as initiators.

The parameters of bitumen modification are given in Table 5.3.

TABLE 5.3 Bitumen Modification Parameters.

Parameter	Value
Temperature (°C)	180
Duration (min)	300
Expense of ERO-In (kg/100 kg bitumen)	3.0
The content of the initiator in ERO-In (wt%)	10.0

The test results of the obtained bitumen–polymer compositions are presented in Table 5.4.

TABLE 5.4 Characteristics of the Obtained Bitumen–Polymer Compositions.

Bituminous composition	Softening point (°C)	Penetration at 25°C (dmm)	Ductility at 25°C (cm)	Adhesion to glass (%)
OB1	46	70	>100	33
OB1 + ERO	48	68	>100	38
OB1 + ERO-AA	47	66	>100	65
OB1 + ERO-FA	46	95	>100	35
OB1 + ERO-MA	49	85	>100	55
OB1 + ERO-PEPA	47	95	>100	95

The results in Table 5.4 show that the use of AA, MA, and PEPA can improve the adhesion to glass. When modifying bitumen OB1 ERO without the use of the initiator, its conduct properties do not change. In the presence of PEPA as an initiator, the adhesion of the obtained modified bitumen increases from 33 to 95%, AA—from 33 to 65%, MA—from 33 to 55%. Negative results with the use of formic acid as an initiator (ERO-FA) make it inappropriate to use it in subsequent studies.

5.3.2 DETERMINATION OF OPTIMAL CONDITIONS FOR MODIFICATION OF ERO-IN BITUMENS

Next, studies were conducted to determine the conditions (bitumen/epoxy/initiator ratio, temperature, duration) for modification of bitumen with rapeseed epoxy and production of modified road bitumen that meets all the requirements for modified road bitumen and current regulations.

5.3.2.1 INFLUENCE OF INITIATOR CONTENT IN ERO-IN COMPOSITION ON MODIFICATION PROCESS

To determine the optimal content of the initiator in the ERO-In composition in road bitumens, the content of the initiators in the ERO-In was changed, and the remaining modification parameters were fixed. The content of the initiator in ERO-In was varied from 1 to 20 wt% for AA and PEPA and from 1 to 50 wt% for MA.[1] The parameters of bitumen modification are given in Table 5.5, and the obtained results in Figure 5.3–5.5.

TABLE 5.5 Bitumen Modification Parameters.

Parameter	Value
Temperature (°C)	160
Duration (min)	120
Consumption of ERO-In (kg/100 kg bitumen)	3.0

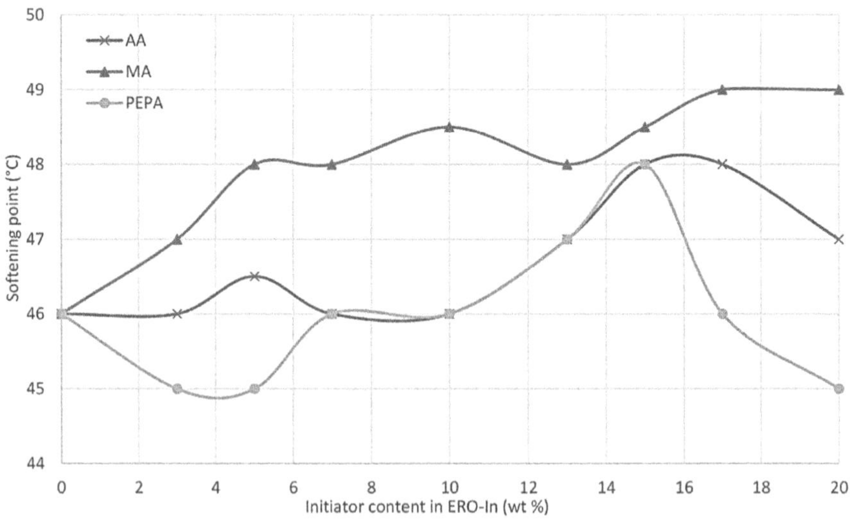

FIGURE 5.3 Influence of initiator content in ERO-In on softening temperature of modified bitumen.

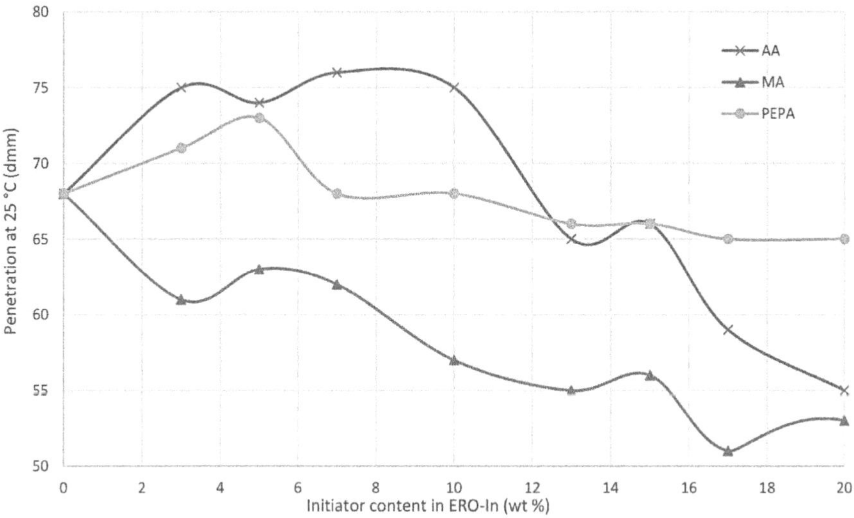

FIGURE 5.4 Influence of initiator content in ERO-In on penetration of modified bitumen.

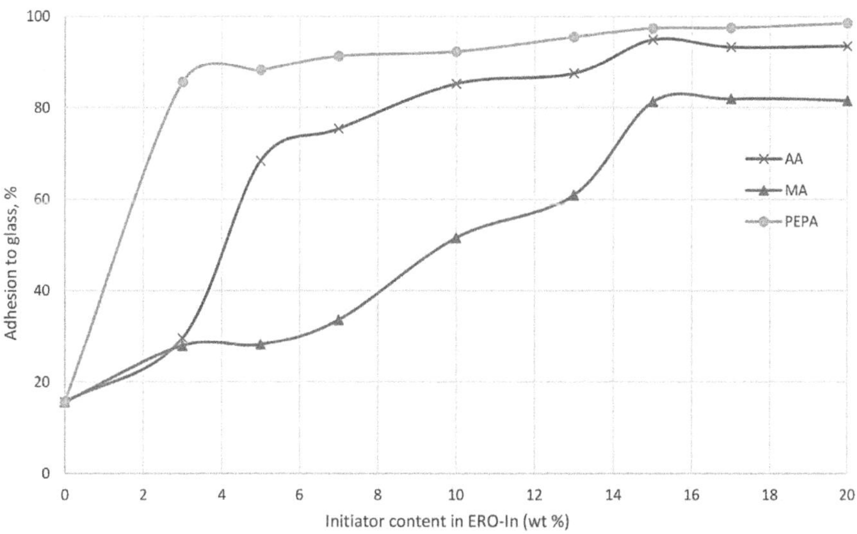

FIGURE 5.5 Influence of initiator content in ERO-In on adhesion to modified bitumen to glass.

Analysis of the results (Fig. 5.3–5.5) showed that increasing the content of the initiator in the composition of ERO-In from 1 to 20 wt% improves the properties of modified bitumen in almost all respects. Thus, at 15 wt% of the AA initiator in

ERO-In, the softening temperature increases by 2°, the penetration rate decreases by 5–10 dmm, and the adhesion rate to the glass relative to the original bitumen increases 3–4 times. Also, the content of 15–20 wt% MA in the composition of the modifier allows us to obtain modified bitumens with a softening temperature of 50°C and penetration of 55 dmm, but such an important indicator as adhesion has increased only 2 times. PEPA in the composition of ERO-In allows us to obtain modified bitumens with the highest adhesion to glass > 98%, other indicators are virtually unchanged. It is worth mentioning that in the case of maleic anhydride, a study was conducted with the content of MA also at 45% and 50%, but the desired result at these ratios could not be obtained. Other mass ratios of initiators with ERO-In should not be used due to unsatisfactory properties of the obtained modified bitumens or due to too high content of initiator, which significantly increases the cost of bitumen–polymer composition.

Based on the obtained results, it was assumed that the optimal content of all initiators in the ERO-In composition is 15 wt%.

5.3.2.2 INFLUENCE OF TEMPERATURE ON THE MODIFICATION PROCESS

To study the effect of temperature on the properties of modified ERO-In bitumen, all process parameters were fixed (Table 5.6) except temperature. The modification process was carried out in the temperature range of 140–180°C. The results of the experiments are shown in Figure 5.6–5.8.

TABLE 5.6 Bitumen Modification Parameters.

Parameter	Value
Duration (min)	120
Consumption of ERO-In (kg/100 kg bitumen)	3.0
The content of the initiator in ERO-In (wt%)	15.0

The results (Fig. 5.6–5.8) show that with increasing the temperature of the process of modification of oxidized bitumen from 130 to 180°C, its performance properties improve. This trend is maintained for all ERO-In compositions in the temperature range of 130–160°C. Further increase of temperature to 180°C practically does not change the reached indicators at temperature of 160°C that does inexpediency of carrying out modification at higher temperatures. Taking into account the obtained results and the operating temperature of asphalt concrete formation, it was assumed that the optimum temperature of the modification process was 160°C.

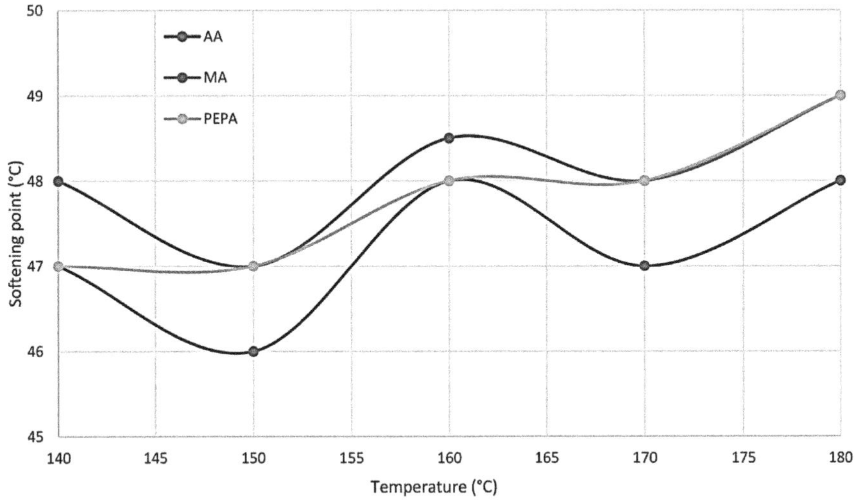

FIGURE 5.6 Influence of temperature on softening temperature of modified ERO-In bitumen.

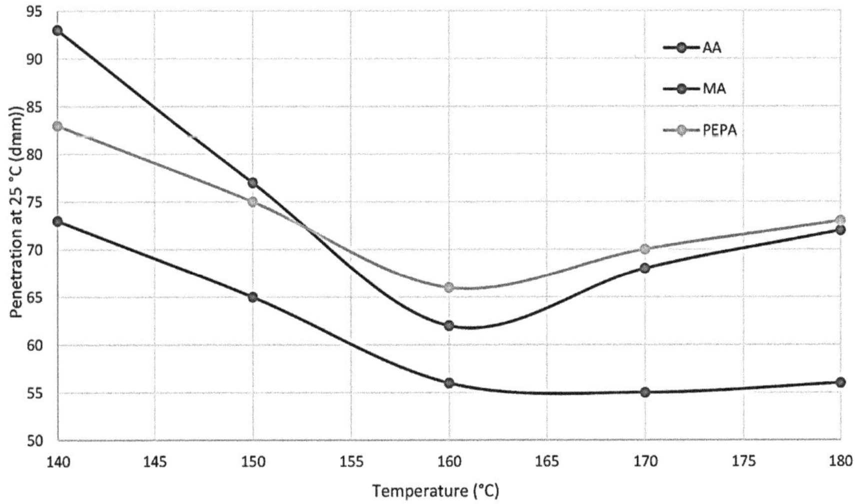

FIGURE 5.7 The effect of temperature on the penetration of modified bitumen ERO-In.

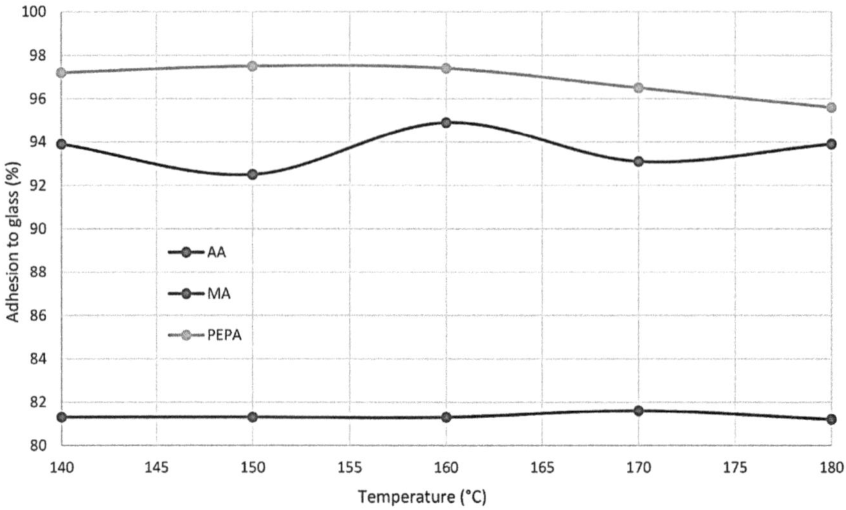

FIGURE 5.8 Influence of temperature on adhesion to bitumen glass modified ERO-In.

5.3.2.3 INFLUENCE OF DURATION ON THE MODIFICATION PROCESS

To study the effect of duration on the properties of modified bitumen ERO-In conducted a study with a change in the duration of the modification from 0 to 300 min. The results and conditions of the experiments are shown in Figure 5.9–5.11.

TABLE 5.7 Bitumen Modification Parameters.

Parameter	**Value**
Temperature (°C)	160
Consumption of ERO-In (kg/100 kg bitumen)	3.0
Initiator content in ERO-In (wt%)	15.0

Comparing the obtained results (Fig. 5.9–5.11), it is seen that with increasing of the duration of the process of modification of road bitumen to 120 min, its indicators reach maximum values and change little during further modification. Namely, the softening temperature increases by 4–5°C, and the penetration, which characterizes their hardness, decreases slightly or does not change. The rate of adhesion to the glass is also virtually unchanged, which makes it impractical to modify road bitumen for more than 120 min.

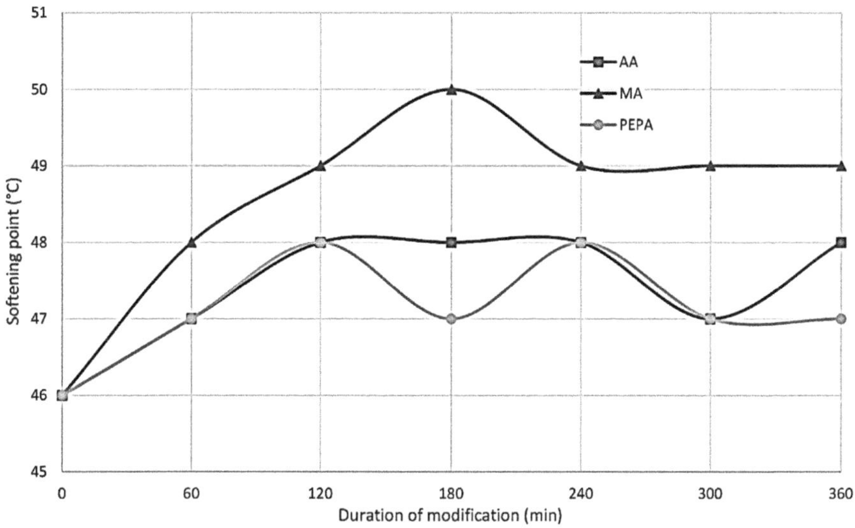

FIGURE 5.9 Influence of process duration on softening temperature of modified bitumen.

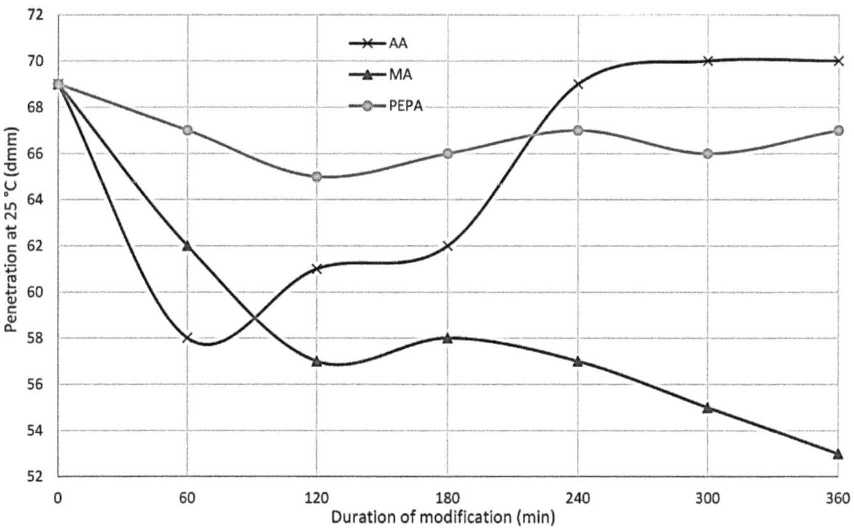

FIGURE 5.10 Influence of process duration on penetration of modified bitumen.

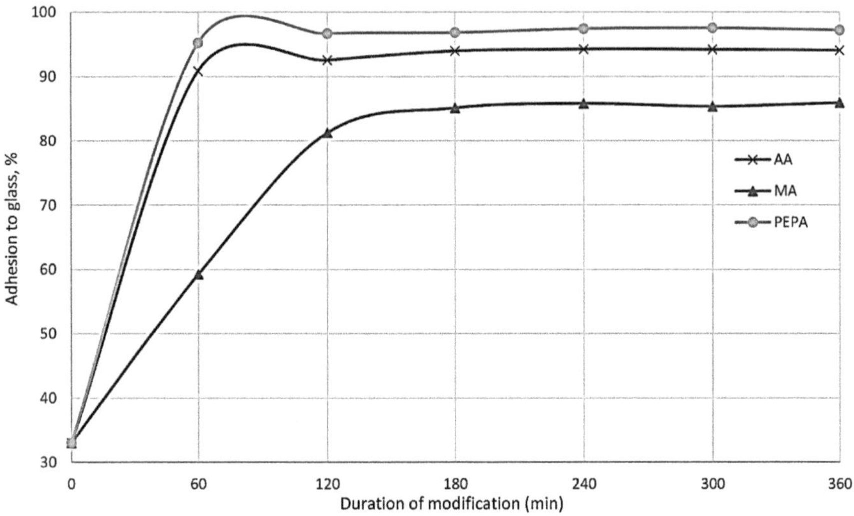

FIGURE 5.11 Influence of process duration on adhesion of modified bitumen.

5.3.2.4 THE IMPACT OF ERO-IN CONSUMPTION ON THE MODIFICATION PROCESS

The study of the impact of ERO-In consumption was performed in the range from 0.3 to 7.0 kg/100 kg bitumen. The process conditions are given in Table 5.8, and the results obtained in Figures 5.12–5.14.

TABLE 5.8 Bitumen Modification Parameters.

Parameter	Value
Temperature (°C)	160
Duration (min)	120
Initiator content in ERO-In (wt%)	15.0

From the results we can observe that the increase in the consumption of ERO-In in bitumen from 1 to 3 kg/100 kg bitumen allows us to obtain optimal softening temperature, increased by 2°–3° and adhesion by 2–3 times (Fig. 5.12–5.14). With a further increase in the consumption of ERO-In more than 3 kg/100 kg bitumen, the change in the performance of modified bitumen is unsatisfactory. The penetration of bitumen increases with increasing ERO-In content in the bituminous composition; similarly, the softening temperature decreases slightly. This change in the parameters allows us to obtain good indicators of

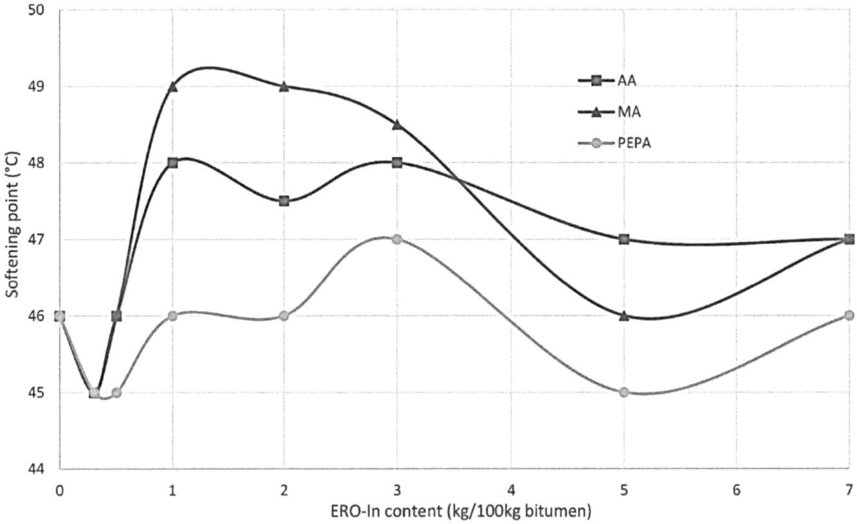

FIGURE 5.12 The effect of ERO-In consumption on the softening temperature of modified bitumen.

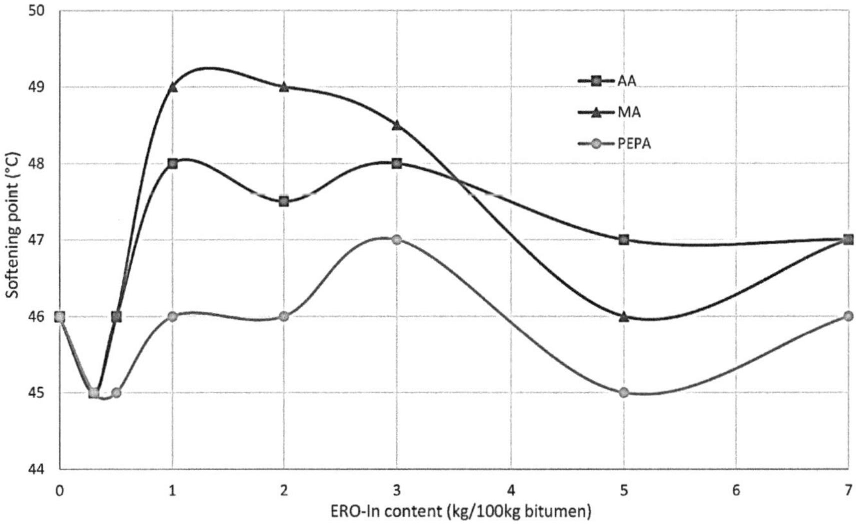

FIGURE 5.13 The effect of ERO-In consumption on the penetration of modified bitumen.

the qualitative characteristics of road bitumens and can be considered for further studies. The best results were achieved at the cost of ERO-In in the composition with bitumen 1–3 kg/100 kg bitumen, in terms of adhesion it is enough less than 1 kg/100 kg bitumen ERO-In to achieve positive results.

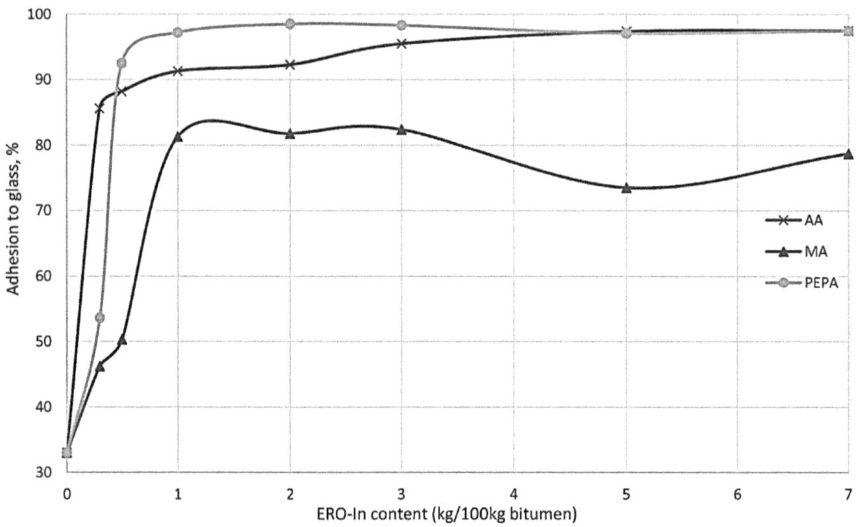

FIGURE 5.14 The effect of ERO-In consumption on the adhesion of modified bitumen.

Changing the adhesive properties of modified bitumen ERO-In allows us to study the proposed additive as well as the adhesive one. After all, bitumen modified with ERO-In meets the requirements for bitumen modified with adhesives.

5.4 STUDY OF THE ADHESIVE PROPERTIES OF BITUMEN MODIFIED WITH ERO-IN

OB1 bitumen and bitumen modified with rapeseed oil epoxy were used as starting materials for research, which will show the difference between the properties of modified bitumen with and without the use of additives.

The parameters for obtaining a bituminous composition were selected on the basis of previous studies and are shown in Table 5.9.

TABLE 5.9 Bitumen Modification Parameters.

Parameter	Value
Temperature (°C)	160
Duration (min)	120
Consumption of ERO-In (kg/100 kg bitumen)	1.0
Initiator content in ERO-In (wt%)	15.0

The main characteristics of the original and modified oxidized bitumen brand BND 70/100 bitumen are given in Table 5.10.

TABLE 5.10 Characteristics of Bitumen Modified by ERO-In.

Bitum	Penetration at 25°C (dmm)	Softening point (°C)	Adhesion to glass (%)	Adhesion to gravel (mark)
OB1	71	46	26.3	2.5
OB1 + ERO-AA	73	47	92.4	4.5
OB1 + ERO-MA	62	49	81.3	4.5
OB1 + ERO-PEPA	67	47	97.2	5.0

Analyzing the results of Table 5.10 it is seen that when adding 1 kg/100 kg bitumen ERO-In, the softening temperature increases by 1° (from 46 to 47°C), penetration increases by 2 points (from 71 to 73 dmm), ERO- MA softening temperature increases by 3° (from 46 to 49°C), and penetration decreases by 9 dmm (from 71 to 62 dmm). When adding 1% of the mass on the original bitumen ERO-In (PEPA), the softening temperature increases by 1° (from 46 to 47°C), the penetration decreases by 4 dmm (from 71 to 67 dmm). However, it is interesting to note the significant increase in the adhesion of petroleum bitumen to the glass surface and gravel after adding ERO-In to the composition. Figures 5.15 and 5.16 show the photos of the studied samples of adhesive properties to glass and stone.

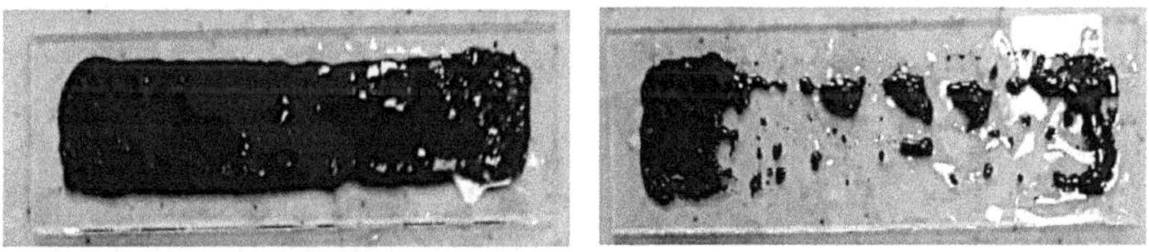

OB1 + ERO-PEPA OB1

FIGURE 5.15 Adhesion of bitumen to the glass surface.

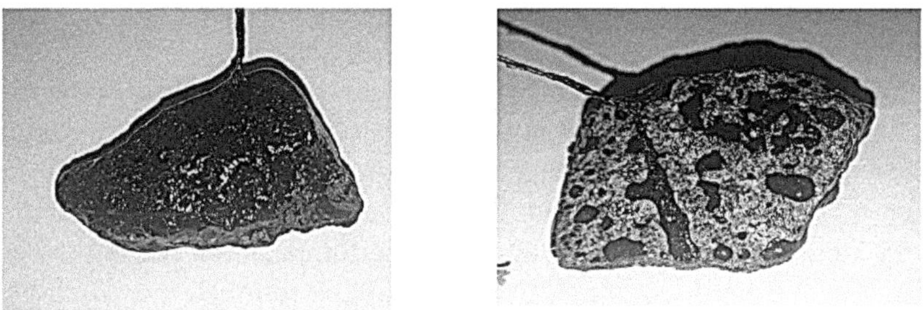

OB1 + ERO-PEPA OB1

FIGURE 5.16 Adhesion of bitumen to the surface of rubble S2.

To confirm the adhesion properties of the modified bitumen ERO-In, the tendency of bitumen to peel off from the aggregate for a long time, the so-called rolling bottle test, was studied (Fig. 5.17).

As can be seen from the results in Figure 5.17, the degree of coating by bitumen ERO-In for 6 h is 80–82% depending on the initiator, while for the original bitumen (OB1) binder coating—76%, bitumen coating modified ERO-In for 72 h—40–53% depending on the initiator and OB1—33%. This indicates that the bitumen composition with ERO-In is less prone to peeling and has a higher bond strength of bitumen with crushed stone compared to the original unmodified bitumen.

If we analyze the effect of initiators on the adhesion properties of bitumen to mineral fillers modified by ERO-In, the best adhesion properties with mineral materials showed bitumen modified with ERO-PEPA composition, while ERO-AA and ERO-MA compositions showed satisfactory and almost identical results.

The obtained results show that bitumen modified with ERO in a mixture with initiators-hardeners has better adhesion properties with mineral materials in comparison with the original unmodified bitumen.

5.5 FTIR SPECTROSCOPY

FTIR spectroscopic studies were performed to confirm the interaction of epoxy bonds with bitumen molecules. For comparison, the spectra of ERO, source oxidized bitumen (OB1), and modified ERO-In bitumen were obtained. The obtained FTIR spectra are shown in Figures 5.18–5.22.

The presence of double bonds in bitumen is confirmed by the absorption band of 1590 cm^{-1}. Free $-CH_3$ groups are proven by the absorption band at 2918 cm^{-1}. Groups -CH2- present in bitumen are confirmed by the absorption band at 2850 cm^{-1}. The presence of benzene rings—absorption bands at 1457 and 1375 cm^{-1}. One-, two-, and three-substituted benzene rings - 782, 720, and 864 cm^{-1}, respectively (Fig. 5.18).

Bands 1250 and 910–920 cm^{-1} (Fig. 5.19) are characteristic of epoxy cycle fluctuations, which are absent in the spectra of modified bitumen samples, which indicates the interaction of epoxy groups with bitumen molecules (Figs. 5.20–5.22). That is, in our opinion, to accelerate the opening of the epoxy ring and give bitumen improved performance properties (such as elasticity) in the process of modifying ERO-In oil residues, it is advisable to use the proposed initiator-hardener.

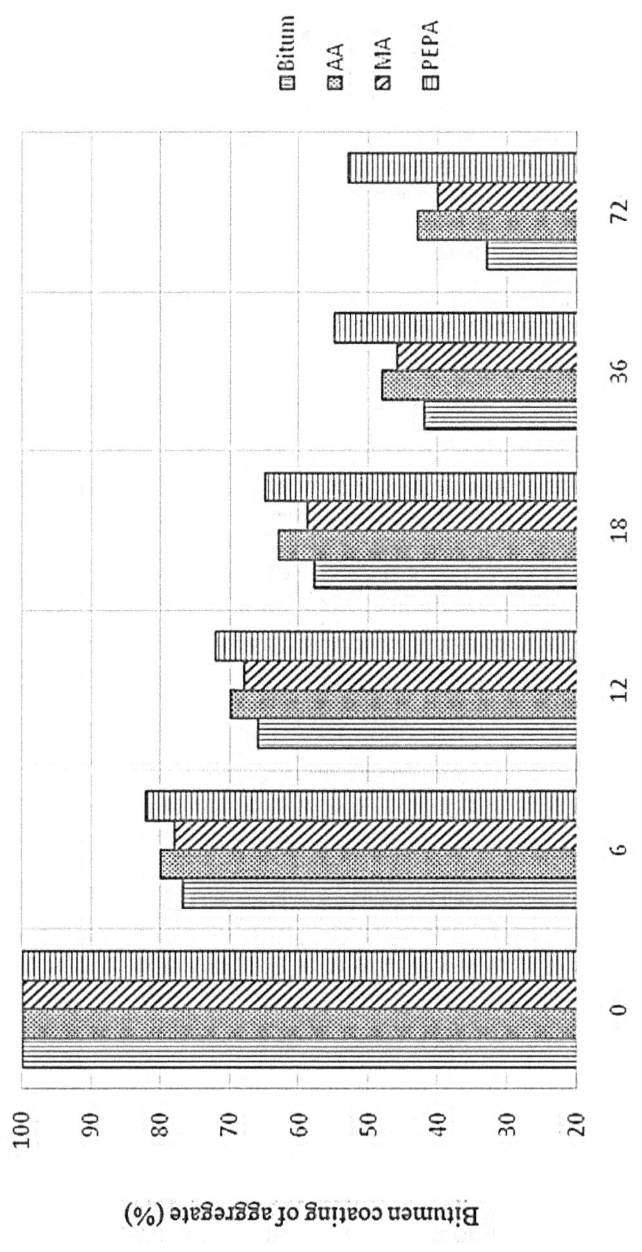

FIGURE 5.17 Rolling-bottle-tests.

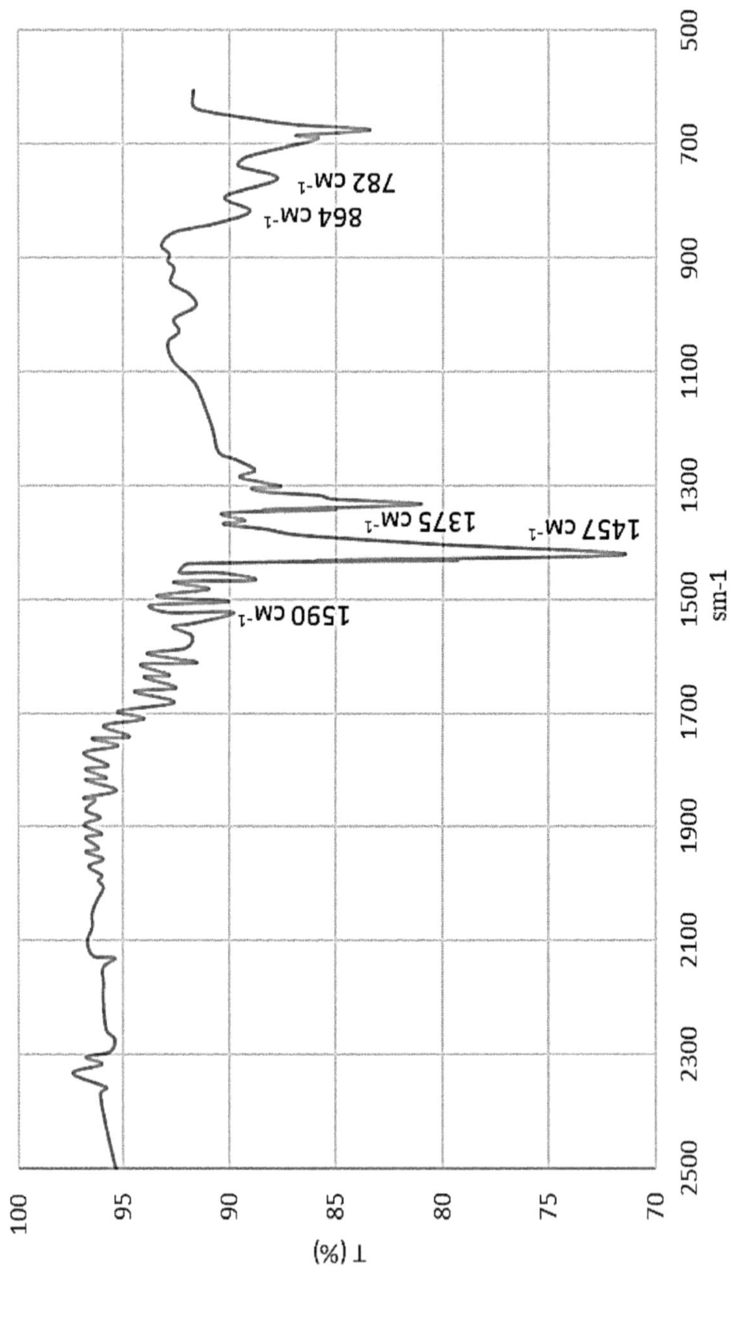

FIGURE 5.18 FTIR spectra OB1.

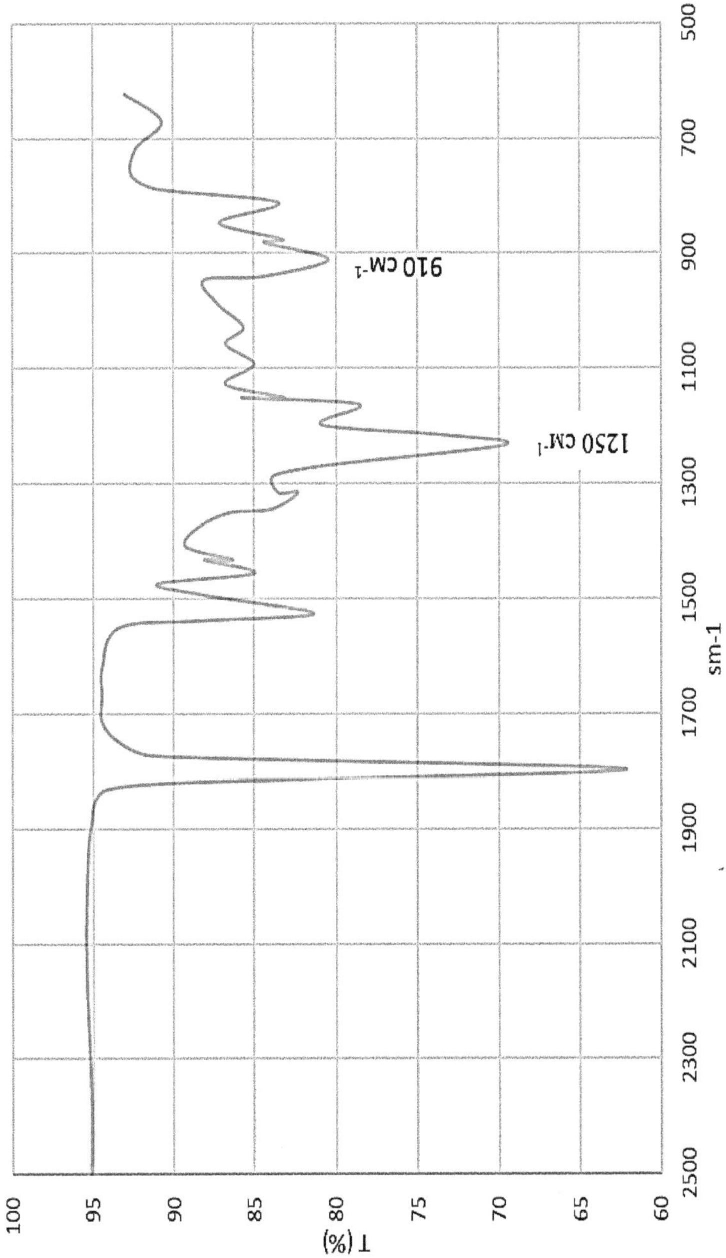

FIGURE 5.19 FTIR spectra ERO.

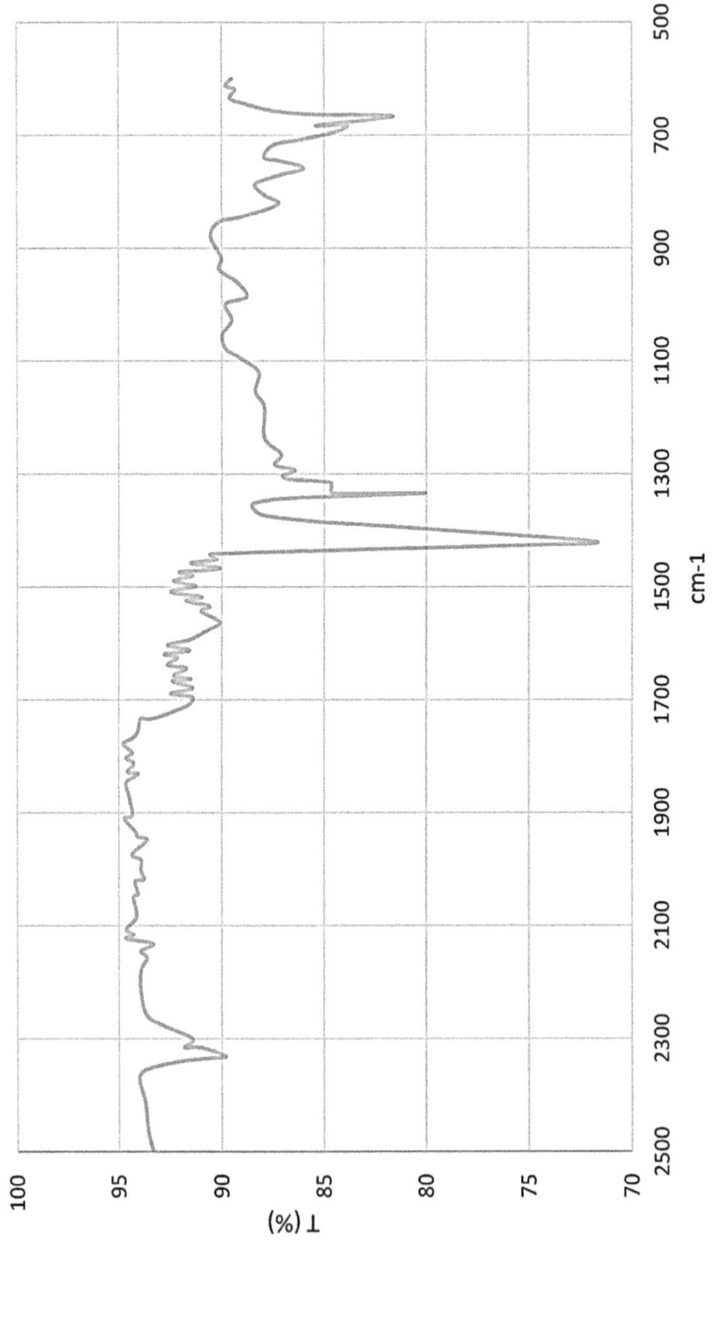

FIGURE 5.20 FTIR spectra OB1 + ERO-MA (3.0 kg/100 kg bitumen).

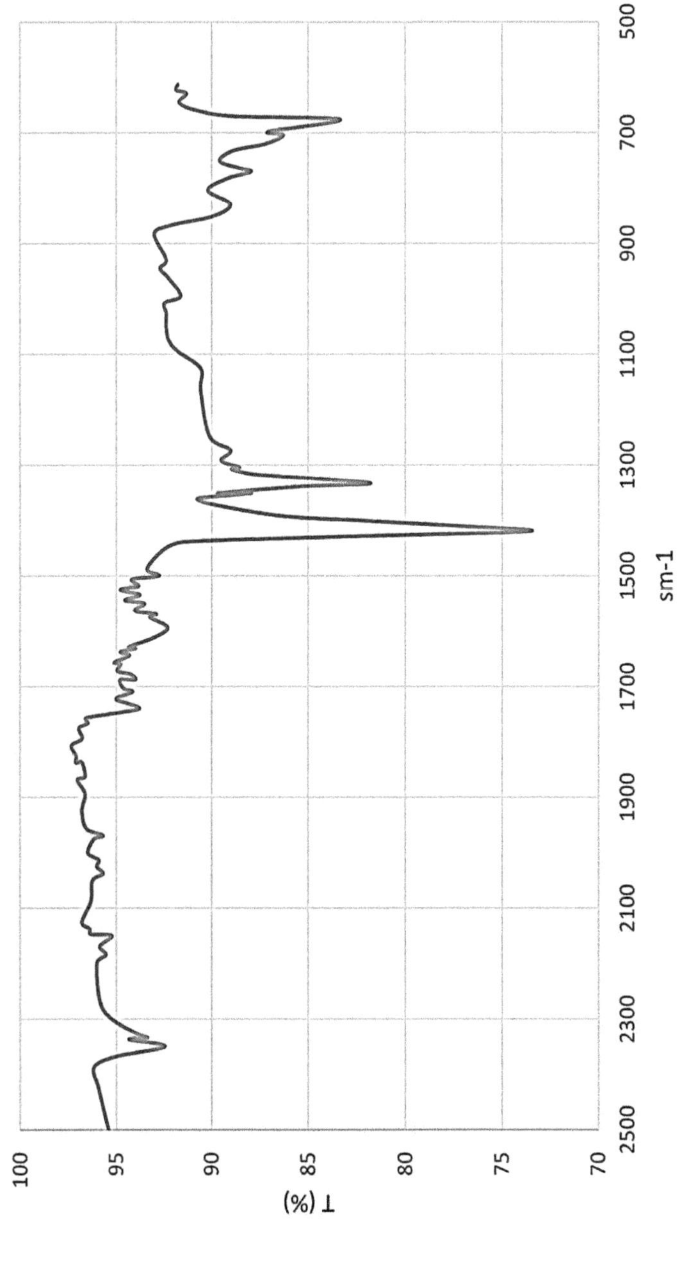

FIGURE 5.21 FTIR spectra OB1 + ERO-AA (3.0 kg/100 kg bitumen).

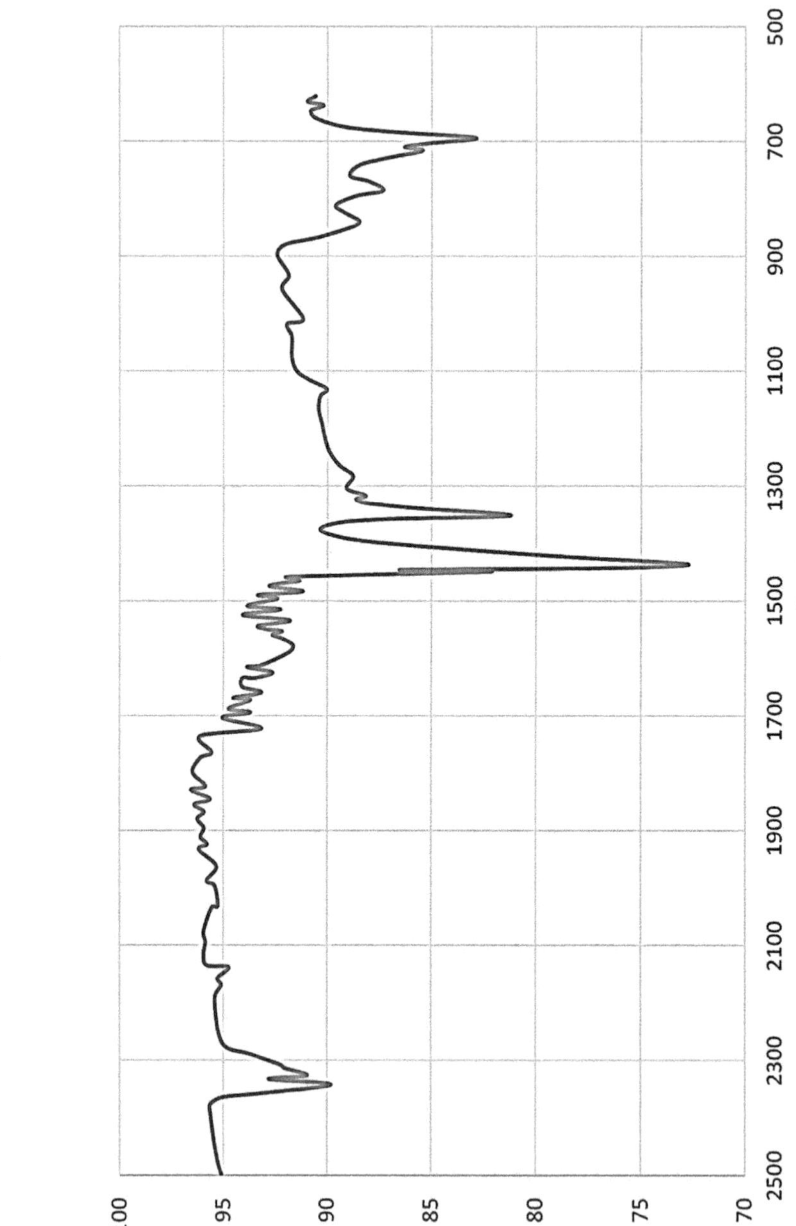

FIGURE 5.22 FTIR spectra OB1 + ERO-PEPA (3.0 kg/100 kg bitumen).

For example, the anhydride groups contained in maleic anhydride react easily with the free epoxy groups contained in the epoxidized oil and with the petroleum residue, forming a crosslinked structure in the mixture that is not related to the bitumen structure.

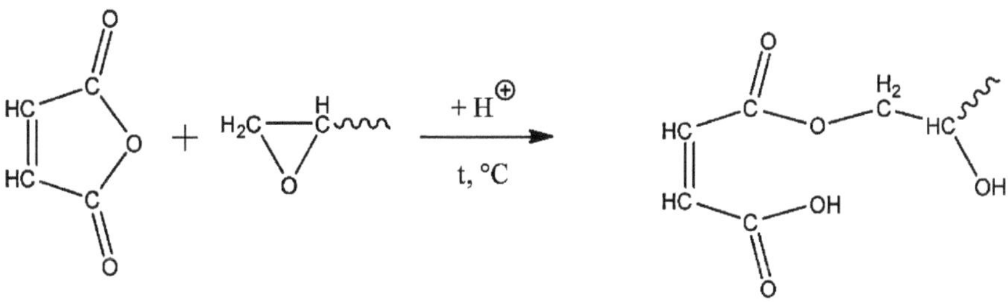

5.6 THE IMPACT OF ERO-IN ON THE PROPERTIES OF HMA

Given the positive effect of ERO-In on the properties of bitumen, the effect of this modifier on HMA prepared on the basis of source bitumen (OB1) and modified bitumen (OB1 + ERO-In) was also studied.

To study the effect of modified ERO-In bitumen on the properties of HMA, the physical and mechanical properties of modified and unmodified HMA were compared. To be able to assess the impact of ERO-In, it was necessary to create similar in particle size distribution of asphalt concrete. Therefore, to ensure equivalent conditions for comparison and objective assessment of the impact of ERO-In on the characteristics of asphalt concrete, a mixture with the same particle size distribution was prepared for each series of experiments.

Both pure OB1 bitumen and modified ERO-In in the amount of 3 kg/100 kg bitumen at a temperature of 160°C for 2 h were used as starting materials. (AA) or Adipic acid, (MA) or maleic anhydride, and (PEPA) or polyethylene polyamine can be considered initiators. This selection could show the distinction between the properties of modified HMA with and without additives.

The composition of the HMA is given in Table 5.11.

To determine the physical and mechanical properties of asphalt concrete, we used cylindrical samples, which are made by compaction of mixtures and prepared in the laboratory.

Thereafter by forming asphalt concrete samples on the basis of standard and modified bitumen, their physical and mechanical properties were

TABLE 5.11 Composition HMA.

Name of material	Content of material in asphalt concrete (% wt)
Aggregates	90
including factions	
15/20 mm	6.0
10/15 mm	15.0
5/10 mm	24
0.071/5 mm	45
Mineral powder	10
Bitumen (OB1, OB1 + ERO-AA, OB1 + ERO-MA, and OB1 + ERO-PEPA)	5.5

obtained and development in the properties of asphalt concrete prepared using modified ERO-In bitumen was studied. All tests were conducted in the laboratory based on the Ukrainian standard.[34] The test results are presented in Table 5.12.

TABLE 5.12 PHYSICAL AND MECHANICAL PROPERTIES OF HMA.

Sample	Average density (g/cm3)	Water-saturation (% vol)	Compression strength (MPa):	
			20°C	50°C
OB1	2.37	2.8	5.6	2.2
OB1 + ERO-AA	2.37	2.3	6,5	2.2
OB1 + ERO-MA	2.37	3.2	6,6	2.6
OB1 + ERO-PEPA	2.37	3.0	6.7	2.4
Requirements to HMA according to the Ukrainian standard	–	≤ 3.5	≥ 2.7	≥ 1.2

The effect on the strength of HMA under compression at a temperature of 20°C is nearly the same for all modified bitumens. However, the compressive strength of HMA at a temperature of 50°C so far varies depending on the initiator used. Analyzing the observed results, we conclude that ERO-In is appropriate for use as a modifier of bitumen used for HMA, in particular at 20°C the highest strength index shows HMA modified ERO-PEPA, although at 50°C preferable strength results show composition ERO-MA. In general, ERO-In modified HMA has greater compressive strength at 20°C and 50°C (compared to HMA based on unmodified bitumen). It can be concluded that the cohesive strength of the mixture also augments. This could possibly show that they are specified by a substantial heat resistance, and asphalt pavements equipped with their use will have greater track resistance in operating conditions.

It is worth mentioning that the modification of ERO-In bitumen is practically simple and does not need specific equipment. Meanwhile, it is performed by standard technology. Thus this additive is not expensive and made on the basis of domestic, environmentally friendly, and renewable raw materials.

5.7 CONCLUSIONS

1. The possibility of creating a new complex modifier for bitumen based on epoxy compounds of plant origin was studied. Adipic acid, maleic anhydride, and polyethylene polyamine were studied as initiators and hardeners of rapeseed oil epoxy in bitumen. The efficiency of using all the studied compositions on the qualitative properties of bitumen, in particular, the adhesion with mineral fillers of asphalt concrete while maintaining all other indicators within the norms was studied.
2. The influence of the main parameters of bitumen epoxy modification with rapeseed oil using initiators was studied. The optimal content of the initiator in a mixture with epoxy is wt 15%. The dependencies of the operational properties of the modified bitumen on the process temperature are established, and the optimum temperature of the modification process is set at – 160°C. It is proved that the optimal amount of modifier (epoxy/initiator) in bitumen is 1–3 kg/100 kg bitumen, in which the adhesion increases by 2–3 times, the softening temperature is almost the same compared to unmodified bitumen, penetration is reduced by ~ 10–15%.
3. The design of the optimal composition of asphalt concrete mixtures obtained from bitumens modified with epoxy modifiers based on renewable raw materials (rapeseed oil) using initiators (adipic acid, maleic anhydride, and polyethylene polyamine) was made. It is established that this mixture has a positive effect on the characteristics of asphalt concrete made of modified bitumen, namely the limit of its compressive strength increases, which increases the durability of the road surface.

ACKNOWLEDGMENT

The research was performed on the equipment of the Scientific Equipment Collective Use Center: "Laboratory of Advanced Technologies, Creation and Physico-Chemical Analysis of a New Substances and Functional Materials" Lviv Polytechnic National University (https://lpnu.ua/ckkno).

KEYWORDS

- **bitumen**
- **asphalt**
- **epoxy**
- **rapeseed oil**
- **maleic anhydride**
- **polyethylene polyamine**

REFERENCES

1. Al Fuhaid, A.; Lu, Q.; Luo, S. Laboratory Evaluation of Biobased Epoxy Asphalt Binder for Asphalt Pavement. *J. Mater. Civ. Eng.* **2018,** *30* (7), 0002383. https://doi.org/10.1061/(ASCE)MT.1943-5533.0002383
2. Bratychak, M.; Gunka, V.; Prysiazhnyi, Y.; Hrynchuk, Y.; Sidun, I.; Demchuk, Y.; Shyshchak, O. Production of Bitumen Modified with Low-Molecular Organic Compounds from Petroleum Residues. 1. Effect of Solvent Nature on the Properties of Petroleum Residues Modified with Folmaldehyde. *Chem. Chem. Technol.* **2021,** *15* (2), 274–283. https://doi.org/10.23939/chcht15.02.274
3. Demchuk, Y.; Sidun, I.; Gunka, V.; Pyshyev, S.; Solodkyy, S. Effect of Phenol-cresol-formaldehyde Resin on Adhesive and Physico-mechanical Properties of Road Bitumen. *Chem. Chem. Technol.* **2018,** *12* (4), 456–461. https://doi.org/10.23939/chcht12.04.456
4. DSTU 8825:2019. Bitumen and Bituminous Binders. Determination of the Ductility; Kyiv, Ukraine, 2019.
5. DSTU B EN 12607-1:2015. Bitumen and Bituminous Binders. Deterination of the Resistance to Hardening under Influence of Heat and Air – Part 1: RTFOT Method; Kyiv, Ukraine, 2015.
6. DSTU B V. 2.7-75-98. Solid Natural Crushed Stone and Gravel for Building Materials, Products, Structures and Construction Works. Specifications; Kyiv, Ukraine, 1998.
7. DSTU EN 12593:2018. Bitumen and Bituminous Binders. Determination of the Fraas Breaking Point; Kyiv, Ukraine, 2018.
8. DSTU EN 1426:2018. Bitumen and Bituminous Binders. Determination of Needle Penetration; Kyiv, Ukraine, 2018.
9. DSTU EN 1427:2018. Bitumen and Bituminous Binders. Determination of the Softening Point. Ring and Ball Method; Kyiv, Ukraine, 2018.

10. Gunka, V.; Bilushchak, H.; Prysiazhnyi, Y.; Demchuk, Y.; Hrynchuk, Y.; Sidun, I.; Shyshchak, O.; Bratychak, M. Production of Bitumen Modified with Low-Molecular Organic Compounds from Petroleum Residues.4. Determining the Optimal Conditions for Tar Modification with Formaldehyde and Properties of the Modified Products. *Chem. Chem. Technol.* **2022,** *16* (1), 142–149. https://doi.org/10.23939/chcht16.01.142
11. Gunka, V.; Demchuk, Y.; Sidun, I; Miroshnichenko, D.; Nyakuma, B. B.; Pyshyev, S. Application of Phenol-cresol-formaldehyde Resin as an Adhesion Promoter for Bitumen and Asphalt Concrete. *Road Mater. Pavement Design* **2021,** *22* (12), 2906–2918. https://doi.org/10.1080/14680629.2020.1808518
12. Hrynchuk, Y. M.; Starchevskyy, V. L.; Nykypanchuk, M. V. Patent UA №83020, Byul. №16, 27.08.2013 Ukraine, 2013.
13. Nykypanchuk, M.; Hrynchuk, Y.; Olchovyk, M. Effect of Modified Bitumen on Physico-mechanical Properties of Asphalt Concrete. *Chem. Chem. Technol.* **2013,** *7* (4), 467–470. https://doi.org/10.23939/chcht07.04.467
14. Hrynchuk, Y.; Gunka, V.; Reutskyy, V.; Koval, I.; Matcipura, P.; Sidun, I.; Mosiuk, M. Possibility Improvement Technology of Modification Road Bitumen by the Green Epoxy Rapeseed Oil on the Basis of Renewable Raw Material. *Pet. Coal* **2020,** *62* (4), 1566–1571.
15. Hrynchuk, Y.; Sidun, I.; Gunka, V.; Prysiazhnyi, Y.; Reutskyy, V.; Mosiuk, M. Epoxide of Rapeseed Oil-modifier for Bitumen and Asphalt Concrete. *Pet. Coal* **2019,** *61* (4), 836–842.
16. Kang, Y.; Song, M.; Pu, L.; Liu, T. Rheological Behaviors of Epoxy Asphalt Binder in Comparison of Base Asphalt Binder and SBS Modified Asphalt Binder. *Constr. Build. Mater.* **2015,** *76*, 343–350. https://doi.org/10.1016/j.conbuildmat.2014.12.020.
17. Koval, I. Correlation between Diameter of Microorganisms and Efficiency of Microorganisms Destruction under Gas/Cavitation Conditions. *Chem. Chem. Technol.* **2021,** *15* (1), 98–104. https://doi.org/10.23939/chcht15.01.098
18. Koval, I.; Starchevskyy, V. Gas Nature Effect on the Destruction of Various Microorganisms under Cavitation Action. *Chem. Chem. Technol.* **2020,** *14* (2), 264–270. https://doi.org/10.23939/chcht14.02.264
19. Martynyuk, M. I.; Sirenko, H. O.; Boyko, L. YA. Epoksydni smoly i kompozytsiyni materialy na yikh osnovi (ohlyad). Visnyk Prykarpat·s'koho natsional'noho universytetu imeni Vasylya Stefanyka; Seriya: Khimiya, 2014; vol 18, pp 115–132. http://nbuv.gov.ua/UJRN/vpnu_chem_2014_18_16.
20. Pan, L.; Wang, Y. T.; Wang, C. S.; Wang, Z. L.; Xie, H. F. Characterization of Thermosetting Epoxy Modified Asphalt Adhesives. *Thermosetting Resin.* **2011,** *26* (4), 33–37.
21. Shevchuk, L.; Strogan, O.; Koval, I. Equipment for Magnetic-cavity Water Disinfection. *Chem. Chem. Technol.* **2012,** *6* (2), 219–223. https://doi.org/10.23939/chcht06.02.219
22. Starchevskyy, V.; Hrynchuk, Y.; Matcipura, P.; Reutskyy, V. Influence of Initiators on the Adhesion Properties of Bitumen Modified by Natural Origin Epoxide. *Chem. Chem. Technol.* **2021,** *15* (1), 142–147. https://doi.org/10.23939/chcht15.01.142
23. Vyrozhemskyi, V.; Kopynets, I.; Kischynskyi, S.; Bidnenko, N. In *Epoxy Asphalt Concrete is a Perspective Material for the Construction of Roads*, IOP Conference Series: Materials Science and Engineering; IOP Publishing, 2017; vol 236 (1), p 012022. DOI: 10.1088/1757-899X/236/1/012022

24. Wei, J.; Zhang, Y. Study on the Curing Process of Epoxy Asphalt. *J. Test. Evaluat.* **2012,** *40* (7), 1169–1176. https://doi.org/10.1520/JTE20120136
25. Wręczycki, J.; Demchuk, Y.; Bieliński, D. M.; Bratychak, M.; Gunka, V.; Anyszka, R.; Gozdek, T. Bitumen Binders Modified with Sulfur/Organic Copolymers. *Materials* **2022,** *15* (5), 1774. https://doi.org/10.3390/ma15051774
26. Xu, P. J.; Cong, P. L.; Ye, H.; Chen, S. F. Modification of Epoxy Asphalt by Hyperbranched Polyester. *Adv. Mater. Res.* **2013,** *716*, 379–382. https://doi.org/10.4028/www.scientific.net/AMR.716.379
27. Yin, H.; Zhang, Y.; Sun, Y.; Xu, W.; Yu, D.; Xie, H.; Cheng, R. Performance of Hot Mix Epoxy Asphalt Binder and its Concrete. *Mater. Struct.* **2015,** *48* (11), 3825–3835. https://doi.org/10.1617/s11527-014-0442-0
28. Yu, J.; Cong, P.; Wu, S. Laboratory Investigation of the Properties of Asphalt Modified with Epoxy Resin. *J. Appl. Polym. Sci.* **2009,** *113* (6), 3557–3563. https://doi.org/10.1002/app.30324
29. Yu, L. Application of Epoxy Asphalt to Adhesive Layer for Deck of Long-Span Steel Bridge, Highway. 2015**;** vol *3*, pp 56–59. https://doi.org/10.1155/2021/3454029
30. Zelenovsky, V.; Kopinets, I.; Onishchenko, A. Experience of Application of epoxн Asphalt Concrete Pavement on Road Bridges. *Dorogi i mosti* **2019,** *19–20*, 78–92. https://doi.org/10.36100/dorogimosti2019.19.078
31. Zhang, W.; Ding, L.; Jia, Z. Design of SBS-Modified Bitumen Stabilizer Powder based on the Vulcanization Mechanism. *Appl. Sci.* **2018,** *8* (3), 457. https://doi.org/10.3390/app8030457
32. Zhou, X.; Wu, S.; Liu, G.; Pan, P. Molecular Simulations and Experimental Evaluation on the Curing of Epoxy Bitumen. *Mater. Struct.* **2016,** *49* (1), 241–247. https://doi.org/10.1617/s11527-014-0491-4
33. DSTU EN 12697-11:2018 Bituminous Mixtures - Test Methods for Hot Mix Asphalt - Part 11: Determination of the Adhesion Between Bitumen and Mineral Aggregate
34. DSTU B V.2.7-319:2016 Mixtures for Asphalt Concrete and Road Asphalt Concrete. Test Methods.

CHAPTER 6

Assessment of the Durability of Geopolymer Materials Based on Thermally Modified Polymineral Clay Rocks

ELENA SHAPAKIDZE[1], MARINA AVALIANI[2], MARINE NADIRASHVILI[1], VERA MAISURADZE[1], IOSEB GEJADZE[1], TAMAR PETRIASHVILI[1], and EVGENI KHUCHUA[1]

[1]*Iv. Javakhishvili Tbilisi State University, Al. Tvalchrelidze Caucasian Institute of Mineral Resources, Tbilisi, Georgia*

[2]*Iv. Javakhishvili Tbilisi State University, R. Agladze Institute of Inorganic Chemistry and Electrochemistry, Tbilisi, Georgia*

ABSTRACT

The cement production sector is one of the very high energy-intensive and polluting industries in the world economy, although demand for Portland cement is constantly growing worldwide.

When clinker is fired as a result of limestone decarbonization, a huge amount of greenhouse gases, mainly CO_2, is released into the atmosphere, which accounts for about 8% of all global emissions. Therefore, any reduction in the use of clinker will have a major impact on global warming. To achieve this goal, new technologies for obtaining clinker-free binders are being developed, one of which is geopolymer materials.

Geopolymer materials (GPM) can become an alternative to Portland cement (OPC), as carbon emissions are reduced by about 80%. In addition, given that geopolymer concretes are generally stronger and more durable,

Environmental Technology and Sustainability: Physical, Chemical, and Biological Technologies for Environmental Protection. Tamara Tatrishvili, Ann Rose Abraham, & A. K. Haghi (Eds.)

they offer a longer service life, which reduces the need for raw materials in the time ahead.

For the widespread use of geopolymers in the construction industry, it is necessary, along with other properties, to study their durability.

This chapter describes the technology for obtaining geopolymer materials using thermally modified polymineral clay rocks and assessing their durability. Studies have shown that geopolymer materials have good durability and higher resistance to aggressive solutions compared to OPC.

6.1 INTRODUCTION

The trend of environmental degradation is a growing concern worldwide. While the cement industry is one of the most energy-intensive and polluting industries in the global economy, new materials are emerging in civil engineering as an alternative to reduce the environmental impact of this industry.

During clinker burning, as a result of limestone decarbonization, a huge amount of greenhouse gases, mainly CO_2, is released into the atmosphere, which accounts for about 8% of all global emissions. Therefore, any reduction in the use of clinker will have a major impact on global warming. To achieve this goal, new technologies for obtaining clinker-free binders are being developed, one of which is geopolymer materials.

Geopolymer materials (GPM) can become an alternative to Portland cement (OPC), as carbon emissions are reduced by about 80%. In addition, given that geopolymer concretes are generally stronger and more durable, they offer a longer service life, which reduces the need for raw materials in the future.

A geopolymer is a material obtained by alkaline activation of aluminosilicates at an ambient temperature or slightly elevated temperature, having an amorphous or semicrystalline polymer structure with Si^{4+} and Al^{3+} cations, tetrahedrally coordinated and connected by oxygen bridges.[1,2] When hydrated, the geopolymers give a product that is predominantly calcium silicate hydrate.

Using modern research methods, it has been proven that geopolymer concrete was an ancient form of concrete that was rediscovered in the 20th century by Joseph Davidovich. He developed the concept of geopolymerization and gave new insights into this class of inorganic polymers. Davidovich evolved the concept of geopolymer (Si/Al inorganic polymer) to explain these chemical processes and the resulting material properties much better.

But even earlier, at the end of the 1950s, V. D. Glukhovsky[3] was the first to discover the possibility of making binders from low-basic calcium or calcium-free aluminosilicates (clays) and solutions of alkali metals. He named these binders "soil cements" and "soil silicates" to reflect their resemblance to natural minerals.

Fundamental studies in this direction were carried out by V. D. Glukhovsky and his colleagues,[4–6] as a result of which a new class of alkaline or alkali-activated types of cement (AAC) appeared. In Ukraine, many large infrastructure projects have been built using "alkali cement."

Geopolymers have many engineering properties in contrast to the Portland cement.

These engineering properties are:

- high compressive strength,
- high flexural strength,
- low shrinkage,
- high resistance to fire,
- low permeability, and
- high resistance to aggressive environments.

Due to these properties, geopolymers can be used in many areas: ceramics,[7] concrete,[8] insulation,[9] stabilization and sorption of hazardous wastes,[10] etc. They rely on many factors such as the nature and chemical mineralogical composition of raw materials.

Studies have proven that significant advantages of geopolymers are their high strength, density, water resistance, heat and heat resistance, and corrosion resistance.[11–13] However, today, the advantage of using these materials is only the possibility of using a huge amount of accumulated industrial waste around the world.

For the widespread use of geopolymers in construction, it is necessary, along with other properties, to study their durability and behavior in aggressive environments.

Sulfate and chloride corrosion of building materials is one of the most significant types of corrosion. It manifests itself when building structures are exposed to mineralized sea and ground waters, atmospheric air with a high content of sulfur compounds, and acid rain.[14]

According to Ref. [15], geopolymer concrete based on low-calcium fly ash has high sulfate resistance. Samples exposed to sodium sulfate solution for 1 year showed no visible signs of surface deterioration, cracking, or flaking. The compressive strength values remained the same as before the samples were immersed in an aggressive environment.

According to Ref. [16,17], the leading cause of the destruction of reinforced concrete structures is the destruction under the action of carbonization and chloride ions, which cause corrosion of the reinforcement. They do not have a direct effect on the concrete but contribute to the corrosion of reinforcement in concrete. The atmospheric carbon dioxide reacts with calcium hydroxide, causing a decrease in pH in the pore space. As a result, the protective properties of concrete in relation to reinforcing steel are reduced. Chloride ions can penetrate concrete through aggregate, mixing water, or accelerators. However, this rarely occurs in practice due to severe restrictions on the chloride content of concrete. As a rule, chlorides penetrate concrete from the outside, either from seawater or also due to the use of deicing salts. Carbon dioxide and chlorine compounds, as well as other aggressive substances, can only cause concrete degradation in the presence of water.

The mechanism of chloride penetration and diffusion of CO_2 from the environment, together with the movement of water in concrete, plays an important role in the demolishing of concrete. These phenomena are the key factors that determine the durability of concrete based on Portland cement and geopolymer concrete.

This chapter describes the production of geopolymer materials synthesized based on thermally modified clay rocks and their assessment of durability.

6.2 EXPERIMENTAL PART

6.2.1 MATERIALS

For the study, a geopolymer material was used, made based on clay rocks of Georgia (argillite and low-melting clay) modified at 700°C and granulated blast furnace slag of the Rustavi Metallurgical Plant.

An alkaline activator NaOH, Na_2CO_3, and Na_2SiO_3 was used as a mixing liquid.

6.2.2 METHODS

Temperature modification of argillite and clay was carried out by heating the material in a muffle furnace to a temperature of 700°C with exposure at a maximum temperature for 1 h.

Geopolymer materials were prepared as follows: Granulated blast furnace slag and modified argillite or clay in a ratio of 80:20 were ground together in a laboratory ball mill to a specific surface area of 8000–10000 cm^2/g. Dry substances were added to the resulting powder in a certain amount: NaOH or Na_2CO_3, or Na_2SiO_3, or their mixture, which were mixed well for 5 min. Water was added to the dry mixture to obtain a dough of normal consistency. Samples were molded with a size of 2 × 2 × 2 cm. The molds, together with the samples, were wrapped in a plastic film to prevent the binder from drying out and were immediately placed in a heat treatment chamber. Heat treatment of geopolymer materials was carried out at 80°C for 24 h. The samples were randomly cooled in the chamber until they reached room temperature.

In order to test for corrosion resistance, geopolymer materials were immersed in aggressive solutions: H_2SO_4 (concentration 2% and 5%), HCl (concentration 2% and 5%) and Na_2SO_4 (concentration 5%).

6.3 RESULTS AND DISCUSSION

Table 6.1 shows the chemical compositions of clay rocks: Argillite from the Teleti deposit (No. 1) and clay from the Gardabani deposit (No. 2).

TABLE 6.1 Chemical Compositions of Clay Rocks, Mass. %.

No.	LOI	SiO_2	Al_2O_3	Fe_2O_3	Mn_2O_3	CaO	MgO	SO_3	Na_2O	K_2O
1	7.01	47.19	15.90	13.36	0.10	6.30	4.10	1.39	2.86	1.30
2	10.60	52.84	15.07	6.47	–	7.06	2.49	1.36	1.19	2.17

The assessment of the corrosion resistance of geopolymer materials was carried out according to the change in the mass and strength of the samples after 180 days of their immersion in aggressive solutions, which were 2% and 5% solutions of H_2SO_4 and HCl, as well as a 5% solution of Na_2SO_4.

For comparison, OPC was tested, sealed with ordinary water, which was immersed in the same aggressive solutions.

As shown by the test results (Table 6.2), geopolymer materials have a higher acid resistance and sulfate resistance compared to OPC, which agrees well with the data of other studies.[18,19] According to the authors of Ref. [20], the high corrosion resistance of geopolymer materials is explained by the absence of $Ca(OH)_2$ in their composition, a compound that is the main cause of the destruction of Portland cement concrete.

TABLE 6.2 Compositions of Geopolymer Materials and the Results of Their Corrosion Resistance.

No.	Composition of the geopolymer (%)		Alkaline activator composition (%)	Weight loss of samples (%) after 180 days immersion in solution					Strength before testing, MPa	Loss of strength of samples (%) after 180 days immersion in solution				
				H_2SO_4		HCl		Na_2SO_4		H_2SO_4		HCl		Na_2SO_4
				2%	5%	2%	5%	5%		2%	5%	2%	5%	5%
11*	Slag 80	Clay 20	NaOH (4) + Na_2SiO_3 (10)	1.49	2.17	2.67	4.77	0.62	97.5	6	8	9	11	1.8
27*	Slag (80)	Argillite (20)	NaOH (4) + Na_2SiO_3 (10)	0.25	1.32	0.15	3.91	0.63	92.5	7	11	10	15	1.9
27**	Slag (80)	Argillite (20)	NaOH (4) + Na_2SiO_3 (10)	7.34	9.50	4.53	12.9	0.64	65.0	12	25	17	29	2.6
15*	Slag (80)	Argillite (20)	Na_2CO_3 (15)	1.15	2.49	1.72	5.5	0.22	88.0	7	12	11	17	1.5
25*	Slag (80)	Argillite (20)	Na_2CO_3 (7)	1.38	2.92	0.75	5.88	0.87	82.0	8	11	10	17	2.1
6*	Slag (100)	-	Na_2SiO_3 (10)	4.53	6.97	5.34	9.94	2.29	77.0	15	26	24	37	3.3
7*	Slag (100)	-	NaOH (4) + Na_2SiO_3 (10)	5.81	9.18	3.22	9.06	2.53	75.0	19	23	22	31	4.2
Ordinary Portland Cement (OPC)				18.11	***	27.7	***	32.8	73.0	62	***	70	***	80

*Curing mode: Heat treatment at 8°C for 24 h.

**Curing mode: In air at 20°C for 28 days.

***Samples destroyed.

Composition No. 27** showed lower results in corrosion resistance, which is clearly due to the fact that this composition was not subjected to heat treatment. It is known that the heat treatment of geopolymers accelerates the processes of polymerization and pozzolanization,[21,22] which increase their resistance to aggressive solutions.

Compositions No. 6 and 7 are also characterized by lower mechanical strength and corrosion resistance, since they do not include a clay component, a source of metakaolin, which provides the geopolymer structure with a high degree of polymerization.

According to the authors of Ref. [20,23], the corrosion resistance of geopolymer materials is also because there are no high-base calcium hydroaluminates in their hardening products that cause sulfate corrosion, and there is also no free lime, leaching of which occurs in soft waters. Therefore, geopolymer concretes are superior in corrosion resistance even to sulfate-resistant concretes based on Portland cement.

According to Ref. [19], if metakaolin is mixed with a certain amount of NaOH solution or NaOH + Na_2SiO_3 solution and then cured at temperatures below 100°C, it is possible to obtain a solid substance having a lattice structure of aluminosilicate resembling a zeolite material having high mechanical strength. This is what happens when geopolymers are hardened according to the curing mode: heat treatment at 80°C for 24 h (Table 6.2—compositions No. 11*, 27*, 15*, and 25*).

The durability of the geopolymer was determined using the composition: slag (80%) + argillite (20%). The composition of the alkaline activator is NaOH + Na_2SiO_3. The results of tests for strength during storage in air are given in Table 6.3. As the results show, the dynamics of changes in strength over time show stable indicators.

TABLE 6.3 Dynamics of Changes in the Strength of GPM Over Time.

Compressive strength, MPa				
The original	**After 30 days**	**After 90 days**	**After 180 days**	**After 360 days**
92.5	95.1	89.7	93.6	91.9

For X-ray phase analysis, composition No. 27* was chosen, which was synthesized from 80% slag and 20% argillite (Table 6.2) and activated with a solution of NaOH + Na_2SiO_3.

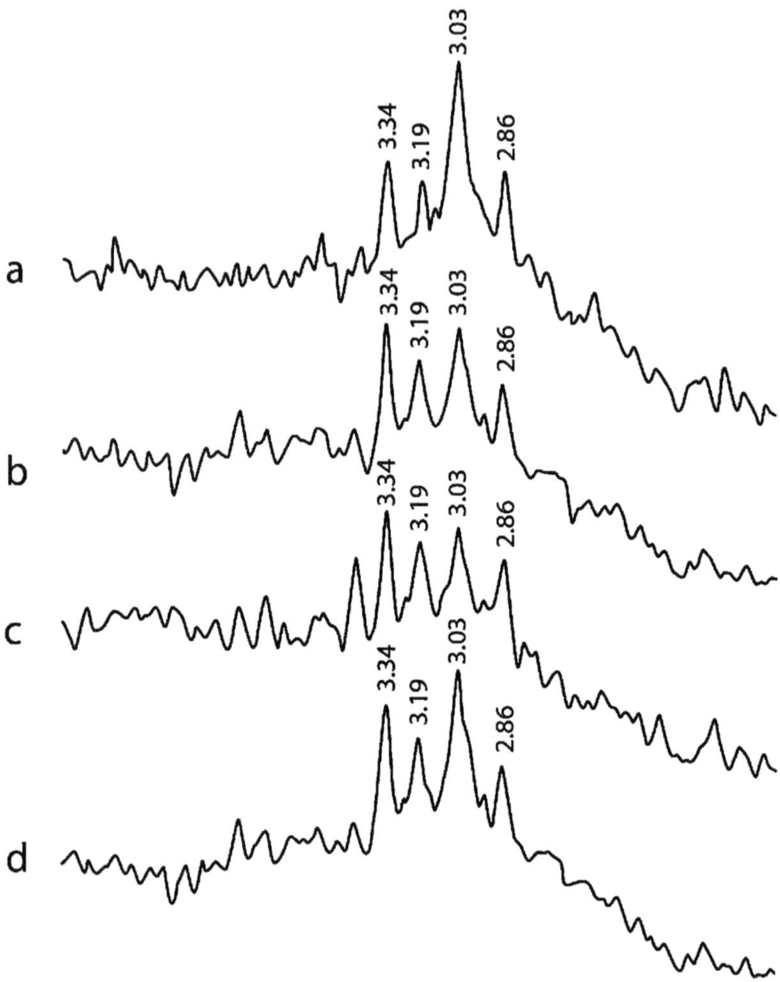

FIGURE 6.1 X-ray patterns of sample No. 27*: (a) before immersion in an aggressive solution; (b) after 180 days of immersion in 5% H_2SO_4 solution; (c) after 180 days of immersion in 5% HCl solution; (d) after 180 days of immersion in 5% Na_2SO_4 solution.

The X-ray diffraction patterns of the geopolymer material show a certain process. The main phases of the material before its immersion in an aggressive solution (Fig. 6.1a) are: X-ray amorphous phase and diffraction lines of quartz (3.33 Å), feldspar (3.19 Å), calcite (3.03 Å), and slag (2.86 Å).

After 180 days immersing of the material in a 5% H_2SO_4 solution (Fig. 6.1b), the phase composition remains the same, but the amount of the X-ray amorphous phase decreases, as does the intensity of the diffraction lines of calcite (3.03 Å). The same picture is observed after immersion of the material in a 5% HCl solution: a decrease in the amount of the X-ray amorphous phase and calcite (Fig. 6.1c).

As the authors of Ref. [24] believe, the main reason for the loss of concrete strength in an acid solution is the degradation of the geopolymer matrix, which is reflected in a decrease in the intensity of the X-ray amorphous phase in X-ray patterns 1b and 1c.

Another process is observed when the material is immersed in a 5% Na_2SO_4 solution (Fig. 6.1d). Here, the phase composition and phase ratios remain almost on par with the initial material.

So, as our studies have shown, geopolymers obtained from local raw materials have a high resistance to aggressive solutions, which is in good agreement with the data of other researchers whom we have cited.

Geopolymers are environment-friendly materials, and the technology for their production belongs to green technologies.

If geopolymer materials displace or at least reduce the production of OPC, this will allow to obtain resource-saving and environmental effects by replacing carbonate raw materials, which during firing lose almost half of their mass when emitting carbon dioxide, polluting the environment.

6.4 CONCLUSIONS

1. GPM synthesized on the basis of thermally modified clay rocks have good durability: The dynamics of changes in strength over time show stable indicators.
2. Corrosion resistance of GPM depends on the composition of the aggressive solution: (a) the higher the concentration of the aggressive solution, the greater the loss of mass and strength; (b) they are less stable in HCl solutions than in H_2SO_4, and rather stable in Na_2SO_4 solution.
3. The indicators of corrosion resistance of GPM (change in mass and strength) after immersion in various aggressive solutions correspond to changes in their phase compositions.
4. Corrosion resistance of GPM is always higher than OPC under the same conditions of aggression.

ACKNOWLEDGMENT

The authors are grateful to the Shota Rustaveli National Science Foundation of Georgia, with the financial support of which this work was carried out [grant number FR-18-783].

KEYWORDS

- **geopolymer material**
- **aggressive solutions**
- **corrosion resistance**
- **durability**
- **thermally modified clay rocks**

REFERENCES

1. Davidovits, J. In *Carbon-Dioxide Greenhouse-Warming: What Future for Portland Cement*. Proceedings, Emerging Technologies Symposium on Cement and Concretes in the Global Environment/Portland Cement Association, Chicago, Illinois, March 1993; p 21.
2. Duxon, P. Fernández-Jiménez, A.; Provis, J.; et al. Geopolymer Technology: The Current State of the Art. *J. Mater. Sci.* **2007,** 42, 2917–2933.
3. Glukhovsky, V. D. Soil Silicates (in Russian). Publish. Gosstroyizdat, Kiev, 1959.
4. Glukhovsky, V. D.; Krivenko, P. V.; Ryumina, P. V.; Gerasimchuk, V. L. Production of Concrete and Structures Based on Slag-alkali Binders. Publish. *Budivelnik, Kiev* **1988,** 144.
5. Krivenko, P. In *Alkaline Cements*, Proceedings of the First International Conference on Alkaline Cements and Concrete, VIPOL Stock Company, Kiev, Ukraine, 1994; Vol. 1, pp 11–129.
6. Palomo, A.; Krivenko, P.; Garcia-Lodeiro, I.; Kavalerova, E.; Maltseva, O.; Fernández-Jiménez, A. A Review on Alkaline Activation: New Analytical Perspectives. *Materiales de Construcción* **2014,** *64*, 315.
7. Lemougna, P. N.; Wang, K.; Tang, Q.; Cui, X. Synthesis and Characterization of Low Temperature (<800°C) Ceramics from Red Mud Geopolymer Precursor. *Constr. Build. Mater.* **2017,** *131*, 564–573.
8. Zhang, H. Y.; Kodur, V.; Wu, B.; Yan, J.; Yuan, Z. S. Effect of Temperature on Bond Characteristics of Geopolymer Concrete. *Constr. Build. Mater.* **2018,** *163*, 277–285.
9. Colangelo, F.; *et al.* Mechanical and Thermal Properties of Lightweight Geopolymer Composites. *Cement Concrete Compos.* **2018,** *86*, 266–272.

10. El-Naggar, M. R.; Amin, M. Impact of Alkali cations on Properties of Metakaolin and Metakaolin/Slag Geopolymers: Microstructures in Relation to Sorption of 134Cs Radionuclide. *J. Hazard. Mater.* **2018,** *344*, 913–924.
11. Fernando, P. T.; Said, J. Resistance to Acid Attack, Abrasion and Leaching Behavior of Alkali-activated Mine Waste Binders. *Mat. Struct.* **2011,** *44*, 487–498.
12. Bakharev, T. Resistance of Geopolymer Materials to Acid Attack. *Cement and Concrete Research* **2005,** *35*, 658–670.
13. Eroshkina, N. A.; Korovkin, M. O. Geopolymer Building Materials Based on Industrial Waste. Publishing House of the Penza State University of Architecture and Construction (PGUAS). *Penza* **2014,** 128.
14. Eroshkina, N. A.; Korovkin, M. O. Geopolymer Building Materials Based on Industrial Waste. Publishing House of the Penza State University of Architecture and Construction (PGUAS). *Penza* **2014,** 128.
15. Davidovits, J. *Geopolymer Chemistry and Applications*, 3rd ed. France, Saint-Quentin: Institute Geopolymer, 2011; p 614.
16. Siddiqui, K. S. Strength and Durability of Low-Calcium Fly Ash-based Geopolymer Concrete. Final Year Honours Dissertation. The University of Western Australia, Perth, 2007.
17. Shaikh, F. U. A. Effects of Alkali Solutions on Corrosion Durability of Geopolymer Concrete. *Adv. Concr.* **2014,** *2* (2), 109–123.
18. Bakharev, T. Resistance of Geopolymer Materials to Acid Attack. *Cem. Concr. Res.* **2005,** *35*, 658–670.
19. Palomo, A.; Blanco-Varela, M. T.; Granizo, M. L.; Puertas, F.; Vazquez, T.; Grutzeck, M. W. Chemical Stability of Cementitious Materials Based on Metakaolin. *Cem. Concr. Res.* **1999,** *29*, 997–1004.
20. Kim, Y. Y.; Lee, B. J.; Saraswathy, V.; Kwon, S. J. Strength and Durability Performance of Alkali-Activated Rice Husk Ash Geopolymer Mortar. *Sci. World. J.* **2014,** *2014*, 1–11.
21. Hewlett, P.; Liska, M. *Lea's Chemistry of Cement Concrete*, 5th ed.; Imprint: Butterworth-Heinemann. Elsevier, 2019; p 896.
22. Bankharev, T. Durability of Geopolymer Materials in Sodium and Magnesium Sulfate Solutions. *Cem. Concr. Res.* **2005,** *35* (6), 1233–1246.
23. Eroshkina, N. A.; Korovkin, M. O. Methods for Assessing and Increasing the Durability of Geopolymer Building Materials Based on Industrial Waste. Tutorial. Publishing house of the Penza State University of Architecture and Construction (PSUAC). *Penza* **2014,** 115.
24. Lingyu, T.; Dongpo, H.; Jianing, Zh.; Hongguang, W. Durability of Geopolymers and Geopolymer Concretes: A Review. *Rev. Adv. Mater. Sci.* **2021,** *60*, 1–14. https://doi.org/10.1515/rams-2021-0002.

CHAPTER 7

Synthesis Process Research of Urea–Formaldehyde Linear Oligomers when Carrying Out Polycondensation in a Solution to Obtain Biodegradable Nitrogen Fertilizers

G. PAPAVA, M. GURGENISHVILI, I. CHITREKASHVILI, N. DOKHTURISHVILI, E. GAVASHELIDZE, N. GELASHVILI, and K. ARCHVADZE

Petre Melikishvili Institute of Physical and Organic Chemistry of Ivane Javakhishvili Tbilisi State University, Tbilisi, Georgia

ABSTRACT

To rise the yield of grain crops, ecologically pure and economically efficient biocomposites were developed. Therefore, we have considered polymerized nitrogen fertilizers of linear structure acting by the prolongation mechanism. And also selected and studied microorganisms capable of destroying such fertilizers. To achieve this goal, a novel process for creation of polymerized fertilizers is reported. It is observed that at the 110–125°C range, up to deep conversion of fertilizers, reaction rate constants keep constant values, while they are determined based on the Arrhenius's second-order equation.

The results observed from infrared (IR) spectroscopy revealed that at the carbamide–formaldehyde interaction (at the first stage of the reaction) when methylol derivatives are formed (specter wave 1030 cm^{-1}), conversion of methylol groups into dimethylene ether groups (specter wave 1085 and 1110 cm^{-1}) was concluded.

Environmental Technology and Sustainability: Physical, Chemical, and Biological Technologies for Environmental Protection. Tamara Tatrishvili, Ann Rose Abraham, & A. K. Haghi (Eds.)

The change of the rate constants from the inverse absolute temperature occurs in accordance with the Arrhenius equation.

With the application of polymerized nitrogenous fertilizers the fixed hectare norm of nitrogenous fertilizers reduces by 40%, then productivity rises by 15–20% and the environment is protected from pollution by nitrogenous fertilizers.

7.1 INTRODUCTION

The high growth of the population calls for a rise in the production of grain crops. Annually arable lands reduce due to intense urbanization and rapid industrialization. To be able to resolve the issue, intense technologies must be assimilated, especially the application of nitrogen-containing fertilizers in raised doses. Based on the current data, every year more than 200 million ton nitrogenous fertilizers enter into the soil, but because of its good water solubility the major part of it (i.e., more than 50%) is lost due to its dryness and washing off, which results in considerable economic loss. At the same time, it contributes to global pollution, causing many diseases.[1,2] Therefore, our research aimed to receive, by chemical synthesis, a polymer that is biodegradable in soil, which would contain an amine group and peptide bond and would be characterized by prolongation action, to isolate from soil its destructor, including urease activity microorganisms, and to create nitrogen-containing biocomposites easily biodegradable in soil.

7.2 EXPERIMENTAL METHODS AND MATERIALS

The subject of our study was the linear structure nitrogenous fertilizer of prolongation action synthesized by us, which was obtained in the solution as well as in the melt, via carbamide and formaldehyde polycondensation.

To study biodegradation of nitrogenous fertilizers with prolongation effect, we isolated from soil freely residing nitrogen-fixing microorganisms (*Azotobacter* sp. and *Clostridium* sp.) and various strains of urobacteria (*Urobacillus* sp.).

Isolation of microorganisms from the soil and their identification was done according to the methods of M. Burger and T. Dobrovolskaya.[6–8] Nitrogen was determined by the I. Kjeldahl method.

7.3 RESULTS AND DISCUSSION

The process of creation of a polymer in the melt was investigated. To simplify penetration into the polymer, we synthesized linear polymers. High efficiency was achieved at definite molar ratio of starting components. The study of the process of creating a carbamide-based linear structure polymer was implemented at the following molar ratios of the initial components—carbamide and formaldehyde: 1:0.9, 1:1, 1:1.1, and 1:1.2, correspondingly. Reaction temperature was 60, 70, 80, 90, 100, and 135°C. Macromolecule formula was expressed as follows (eq 7.1):

$$HOH_2C[-NH-CO-NH-CH2]_n - NHCONH_2 \quad (7.1)$$

The IR spectral study showed that at the first stage, at the creation of methylol derivatives of carbamide (specter wave 1030 cm^{-1}), we also observe conversion of methylol groups to dimethylene ether groups (specter wave 1085 and 1110 cm^{-1}).

In the process of synthesis of the carbamide–formaldehyde oligomer, we considered it expedient to study some kinetic regularities of the process within the 110–125°C range at the carbamide–formaldehyde molar ratio of 1:1.2. Control over reaction progress was carried out according to the changes in formaldehyde concentration. It turned out that in the 110–125°C temperature range, until deep conversion, reaction rate constants during the reaction keep constant values when they are computed by the second-order equation.

Rectilinear dependence of the reaction rate constant of logarithm alteration to the inverse absolute temperature refers to the fact that reaction rate constants undergo change according to the Arrhenius equation.

The second order is proved also by the ratio of 1/a–x and reaction duration. In this temperature range the rectilinear relation is preserved. Activation energy computed according to the Arrhenius equation equals to 1554 kcal/mol.

TABLE 7.1 Values of Reaction Rates and Activation Energy* at 110–125°C Range at the Interaction of Formaldehyde and Carbamide.

No	Reaction duration, s.	Conversion degree, %			Reaction rate constant K · 10−2, l.mol−1, s−1		
		110°C	120°C	125°C	110°C	120°C	125°C
1	30	71.66	74.50	89.16	2.81	3.24	3.38
2	60	83.60	85.48	94.25	2.83	3.27	3.34
3	90	88.39	89.83	96.00	2.80	3.27	3.38
4	120	91.14	92.15	97.01	2.85	3.26	3.34
5	150	92.83	93.53	97.60	2.87	3.21	3.38
6	180	93.97	94.60	98.01	2.88	3.24	3.37
7	240	95.33	95.86	98.50	2.80	3.22	3.36
8	300	96.25	96.66	98.27	2.85	3.22	3.37

*Carbamide and formaldehyde molar ratio 1:1.2.

The process of degradation of linear structure polymers at the impact of bacteria is simplified because of the linear structure of molecules. Urease ferments easily penetrate macromolecules and biodegrade them.

Nitrogen concentration was determined in carbamide and fertilizers with prolongation action. Besides, the dynamics of changes in nitrogen content were defined in soil. The transition of nitrogen into soluble form in the case of fertilizer characterized by prolongation action, takes place purposefully slowly.

We have studied the degradation of linear structure polymer (fertilizer) by the emission of ammonia. With this in view, microorganisms apt to destruct the tested polymer were isolated from definite type soils (chestnut, alluvium acidic, and brown carbonaceous) of east Georgia.

As a source of nitrogen, various quantities of the polymer (to adapt microorganism) were added to the medium of cultivation of microorganisms (Christensen medium, where initially starting quantity of peptone was used as nitrogen source); when the adapted microorganisms were obtained, experiments were carried out according to the following scheme: only polymer without inoculation (control) and only with urea and polymer with inoculation (in the nutrient medium starting quantity of peptone and adapted microorganisms).

Analysis of the obtained results showed that at the fourth hour of the cultivation of microorganisms, ammonia content in the medium was approximately 0.6 mg/mL and by the 10th hour of the cultivation, it fell to 0. At this period, source of ammonia in cultivation medium is peptone (starting dose); by the 10th hour of the cultivation, peptone reserve is exhausted and microorganisms start to use carbamide as a nitrogen source (it destructs peptide bond and releases ammonia). As a result, from the 10th hour of the cultivation amount of ammonia in cultivation medium increases, that is, polymer is destructed with the emission of ammonia. These experiments determined active strains of *Urobacillus* sp. capable of destructing polymers.

We created the biocomposite by the selection of adequate ratio of lyophilized urobacterium mycelium and nitrogenous polymer fertilizers. Besides, activity of definite nitrogen fixers was defined in field conditions. Field experiments were carried out according to nine variants scheme with the four repetitions. Nitrogen fixers were applied on seed material. Experiments were carried out on autumn wheat breed (Bezostaya-1). Plant development phases such as germination, tillering, booting, heading, flowering, grain filling, and maturity were studied. The development of all phases proceeded normally. Indices of productivity in the studied variants at the application of nitrogen fixers on the seed material (*Azospirillum brasilense*) together with fertilizer characterized by the prolongation action showed that hectare norms of nitrogen compared with the control (obtained by agro rules) were reduced by more than 40–45%, while productivity was increased by 15–20%, correspondingly. This effect was achieved by joint application of fertilizers of prolongation action and nitrogen fixers. Such technology practically ensures reduction of nitrogenous fertilizers wash off and evaporation to the minimum, while the plant is provided with nitrogen along the whole vegetation period, which ensures growth of productivity and environment protection from pollution. Finally, it will give substantial economy and ecological effect.

7.4 CONCLUSION

New nitrogenous fertilizer of linear polymer structure was synthesized and kinetic regularities of its creation were studied.

Active strains of urobacteria able to destruct nitrogen-containing polymer of linear structure were isolated from the soil.

Biocomposites containing polymerized carbamide and its destructor microorganisms were obtained.

Biodegradable nitrogenous fertilizer of linear polymer structure increases production of grain crops by 15–20% and decreases hectare norms of nitrogenous fertilizers by 40–45%.

As a conclusion, in this chapter, the technology of producing polymerized nitrogenous fertilizers and polymer biodegradable biocomposites has been reported for the first time.

KEYWORDS

- **prolongation mechanism**
- **microorganisms**
- **biocomposites**
- **technology**
- **fertilizers**

REFERENCES

1. US Congress, Office of Technology Assesment Impacts of Technology on US Cropland and Rangeland Productivity, US Government Printing Office, Washington, D.C., 1982.
2. Steinhart, J. S.; Steinhart, C. E. *Science* **1974,** *184*, 307.
3. Lai, T. M.; Eberl, D. D. *Zeolites* **1986,** *6*, 129.
4. Russel, E. W. *Soil Conditions and Plant Growth*, 10th ed.; Longman: London, 1973.
5. Barbarick, K. A.; Lai, T. M.; Eberl, D. D. *Soil Sci. Soc. Am. J.* **1990,** *54*, 911.
6. Mishustin, E. N.; Fintsev, V. T. *Mikrobiologiya (Microbiology)*, Moscow: Agropromizdat, 1987.
7. *Reference Manual for Microbiologic and Virology Methods of Investigation.* Burger, M. O. Ed; Moscow: "Medicine", (Rus.), 1982.
8. Dobrovolskaya, T.; Skvortsova, I.; Lysak, V. *Methods for Isolation and Identification of Soil Bacteria*, MSU, Ed.; Moscow, 1989; pp 21–35 (Rus.).

CHAPTER 8

Calculation of the Stability Constant of Copper Fulvate, Using an Average Molecular Weight of Active Associate of Fulvic Acids at pH = 8 with Particular Application for Evaluation of the Ecological Condition of Water Reservoirs

TAMAR MAKHARADZE[1], NAZI GOLIADZE[1], TEONA MAKHARADZE[2], and SESILI ROGAVA[1]

[1]*Ivane Javakhishvili Tbilisi State University, Tbilisi, Georgia*

[2]*The State Laboratory of Agriculture, Tbilisi, Georgia*

ABSTRACT

The complexes of heavy metals with fulvic acids (FA) are identified with less bioavailability and consequently with less toxicity. Disregarding the most recent research works, empirical evidence on stability constants of complex compounds of FA with heavy metals (among them copper) is diversified, and they vary in many lines from each other. One of the reasons for such condition is neglecting an average molecular weight that is connected to fulvic acids, which finally creates the wrong results.

In this chapter, we have studied the complex formation process among copper (II) and FA by the solubility approach at pH = 8.0. In this regard, as a solid phase, $Cu(OH)_2$ suspension is applied. FA was then separated from lake Paravani by the adsorption chromatographic technique. The moderate molecular weight of the "active associate" (Mw = 2191) is considered for

Environmental Technology and Sustainability: Physical, Chemical, and Biological Technologies for Environmental Protection. Tamara Tatrishvili, Ann Rose Abraham, & A. K. Haghi (Eds.)

obtaining the composition of copper fulvate complex, the concentration of free ligand, and stability constant. β $[Cu(OH)_2FA] = 1.18 \times 10^7$; lg $\beta = 7.07$. Through the obtained results, it will be possible to calculate the migration forms of copper in natural waters, determine toxicity and bioavailability, and evaluate the ecological condition of water reservoirs.

8.1 INTRODUCTION

Copper in natural waters may be present as free ions (Cu^{2+}) and also as inorganic ($CuOH^+$, $Cu(OH)_2^0$, $CuHCO_3^+$, $CuSO_4^0$, $CuCl^+$, and others) and organic complexes (mainly humates and fulvates).[1–4] There is a consensus that Cu^{2+} is more toxic in comparison with other chemical forms.[5–7] The complexes of heavy metals with macromolecular organic substances are characterized with less bioavailability and accordingly with less toxicity.[8,9]

"FA" is one of the first macromolecular organic substances, which was discovered in natural waters. It makes up the fundamental part of dissolved organic matter of natural waters.[2,3,9–11] Fulvic acids take an active part in complexion and sorption processes, proceeding in plain waters, organize stable complexes with heavy metals, radionuclides, and stipulate migration forms involved in natural waters.[12–19]

Empirical data on stability constants (β) of complex compounds of fulvic acids with copper (II) are heterogeneous and they diverge in several lines from each other.[20–39] This condition is mainly specified by neglecting the moderate molecular weight (Mw) of the related FA, whose value in its turn relies upon the amount of pH and at last brings the wrong result.

Consequently, it's hard to study complex formation processes taking place in plain waters, identify migration forms of copper, and classify and assess the chemical–ecological condition of plain waters.

In this chapter, we worked on complex formation process between the pure samples of fulvic acids, separated from plain water and Cu(II), to calculate the composition and the average stability constant of copper fulvate complex. Complex formation process was studied at pH = 8.0 by the solubility method. $Cu(OH)_2$ suspension was used as a solid phase.

8.2 MATERIALS AND METHODS

To be able to obtain pure samples of fulvic acids for subsequent filtration through membrane filters (0.45 μm pore size), the water of lake Paravani was concentrated by a frozen procedure. It should be noted that the concentrated water samples were acidified with 6 M HCl to pH 2. Meanwhile, the sample was placed for 2 h on water bath at 60°C for coagulation of humic acids. Afterward, the solution was centrifuged for 10 min at 8000 rpm (Centrifuge T-23). Nevertheless, for isolation of fulvic acids from centrifuge, we have used the adsorption chromatographic method as well.

According to published research study presented in Ref. [40], charcoal was considered as a sorbent.[40] Desorption of amino acids and carbohydrates was also considered (i.e., 90% acetone water solution was used with 0.1 M HCl for desorption of polyphenols). The elution of the fraction of fulvic acids was performed with 0.1 N NaOH solutions.

We prepared alkaline solution of fulvic acids for the purification. This was passed through a cation exchanger (KU-2-8). To be able to attain the concentration of fulvic acids in obtained solution using gravimetric method, the part of the solution was dried under vacuum until the constant weight was achieved. Thus, we have prepared the model solutions of Fulvic acids.

It is worth mentioning that the solution of fulvate complexes was prepared by the solubility method. In such process, 0.1 ml suspension of copper hydroxide and increasing quantity of standard solution of fulvic acids were placed in 15 ml capacity fluoroplastic cylinders [pH = 8.0, ionic strength μ = 0.01 (KNO_3), and v = 10 ml]. In this research study, the concentrations of hydrogen ions were regulated by the addition of 0.01 M HNO_3 acid or 0.01 M NaOH (pH meter pH 2006). Afterward, it was stirred in order to achieve a reliable balance by using a mechanical mixer for 60 h. This suspension was then filtered through the membrane filters (considering 0.45 μm pore size). In the filtering procedure, the atomic absorption spectrophotometer (Perkin Elmer 200) was used to measure the concentration of copper.

8.3 RESULTS AND DISCUSSION

From Table 8.1 we can observe the following:

- In line with the augmenting of concentration of fulvic acids in solution, the concentration of Cu(II) raised many times because of formation of fulvate complex.
- During the determination of molar concentrations of "FA," the fact that they form correlations in water solutions was taken into consideration.

- During the experimental procedure for water solutions of fulvic acids, gel chromatographic method was carried out (in the interval pH 4–11).
- There is a line dependence between the average molecular weight of fulvic acids (Mw) and the amount of pH, which is shown in Ref. [2];
- Mw = 1350 at pH-4540,
- At pH = 8, Mw (FA) = 6260.
- In diluted water solutions, copper dihydroxy complex $Cu(OH)_2^0$ is the dominant form at pH = 8.0.[36]

Now, it should be noted that the reaction between copper dihydroxy complex and the anions of fulvic acids is written as:

$$Cu(OH)_2^0 + mFA \leftrightarrows Cu(OH)_2\,(FA)m \tag{8.4}$$

$$\beta = [Cu(OH)_2\,FAm]/\{[Cu(OH)_2^0]\,[FA]^m\} \tag{8.5}$$

In solution, the concentration of copper dihydroxy fulvic acid complex matches with the difference between final $[Cu(II)_{total}]$ and initial $[Cu(II)_{free}]$ concentrations of copper received after formation of complex:

$$[Cu(OH)_2FAm] = [Cu(II)_{total} - Cu(II)_{free}] \tag{8.6}$$

Putting these values in eq 8.5, we got eq 8.7,

$$\beta = [Cu(II)_{total} - Cu(II)_{free}]/\{[Cu(II)_{free}][FA]^m\} \tag{8.7}$$

From eq 8.7,

$$\beta[Cu(II)_{free}] = [Cu(II)_{total} - Cu(II)_{free}]/[FA]^m \tag{8.8}$$

At the fixed pH, the left part of the eq 8.8 is invariable and shown as K′.

$$K' = [Cu(II)_{total} - Cu(II)_{free}]/[FA]^m \tag{8.9}$$

The logarithm of this eq 8.9 is:

$$lgK' = lg[Cu(II)_{total} - Cu(II)_{free}] - mlg[FA] \tag{8.10}$$

The value (*m*) of the stoichiometric coefficient or the number of ligands in inner coordination sphere of complex equals to tangent of tilt angle of straight line built in coordinates.

$$lg[Cu(II)_{total} - Cu(II)_{free}] — lg[FA_{total}] \tag{8.11}$$

To be able to determine the exact value of tangent tilt angle of straight line, we used the least square method:

$$m = tg\alpha = (n\Sigma x_i y_i - \Sigma x_i \Sigma y_i)/(n\Sigma x_i^2 - (\Sigma x_i)^2) \quad (8.12)$$

where

$$x_i = \lg [FA_{total}] \text{ and } y_i = \lg [Cu(II)_{total} - Cu(II)_{free}] \quad (8.13)$$

TABLE 8.1 Solubility of Copper Hydroxide Dependence on the Concentration of Fulvic Acids and the Indispensable Data to Carry Out the Composition of Copper Dihydroxy Fulvate Complexes pH = 8.0; Mw(FA) = 6260; [Cu(II)free] = 3.30 × 10^{-6} mol/L.

Mol/L			$[Cu_{total}]$: $[FA_{total}]$	Log $[FA_{tota}]$	Log $[CuFA_m]$
$[FA_{total}]$	$[Cu_{total}]$	$[CuFA_m]$			
1.4 × 10^{-5}	4.22 × 10^{-5}	3.89 × 10^{-5}	1:0.33	–4.8539	–4.4100
2.1 × 10^{-5}	6.00 × 10^{-5}	5.67 × 10^{-5}	1:0.35	–4.6778	–4.2464
2.8 × 10^{-5}	8.02 × 10^{-5}	7.69 × 10^{-5}	1:0.35	–4.5528	–4.1141
3.5 × 10^{-5}	10.10 × 10^{-5}	9.77 × 10^{-5}	1:0.35	–4.4559	–4.0101
4.2 × 10^{-5}	12.03 × 10^{-5}	11.70 × 10^{-5}	1:0.36	–4.3767	–3.9318
4.9 × 10^{-5}	13.93 × 10^{-5}	13.60 × 10^{-5}	1:0.35	–4.3098	–3.8665
5.6 × 10^{-5}	15.43 × 10^{-5}	15.10 × 10^{-5}	1:0.36	–4.2518	–3.8210

TABLE 8.2 Determination of the Composition of Copper Fulvate Dihydroxy Complex, Using Least Square Method, Mw (FA) = 6260; pH = 8.0. Xi = lg $[FA_{total}]$, Yi = lg$[CuFA_m]$.

Xi	Yi	XiYi	Xi²
–4.8539	–4.4100	21.4057	23.5603
–4.6778	–4.2464	19.8638	21.8818
–4.5528	–4.1141	18.7307	20.7280
–4.4559	–4.0101	17.8686	19.8550
–4.3767	–3.9318	17.2083	19.1555
–4.3098	–3.8665	16.6638	18.5774
–4.2518	–3.8210	16.2461	18.0778

$\Sigma Xi = -31.4787$; $(\Sigma Xi)^2 = 990.9085$; $\Sigma Yi = -28.3999$; $\Sigma Xi^2 = 141.8358$; $\Sigma XiYi = 127.987$; m = 0.98.

The value of m (Mw(FA) = 6260), which equals 0.98 (Tables 8.1 and 8.2), obtained at pH = 8 in system $Cu(OH)_2$(solid)-Cu(II)(solution)-FA-H_2O is the copper dihydroxy fulvate, with the structure of 1:1. Therefore, the complex formation reaction (pH = 8.0) could be written as:

$$Cu(OH)_2^0 + FA = Cu(OH)_2FA. \quad (8.14)$$

Thus, for the stability constant:

$$\beta = [Cu(OH)_2FA]/\{[Cu(II)_{free}]\,[FA]\} = [Cu(II)_{total} - Cu(II)_{free}]/\{[Cu(II)_{free}]\,[FA]\} \quad (8.15)$$

The complexation of fulvic acids to metal ions cannot be explained in a very strict frame. It is mainly due to the incomplete defined nature of fulvic acids in contrast with the complexation of single ligands. Therefore, to be able to determine stability constants of fulvate complexes, it is needed to make some logical assumptions. In balanced solutions, correlation $[Cu(II)_{total}]:[FA_{total}]$ on average equals to 1:0.35 (please refer to Table 8.1). This means that during the complex formation process for fulvic acids, the value of Mw at pH = 8 equals to 6260 (divides and every 0.35 part carried out into copper's inner coordination sphere, as an integral ligand).

TABLE 8.3 Composition of Copper Dihydroxy Fulvate Complexes pH = 8.0; Mw (FA) = 2191; [Cu(II)free] = 3.30×10^{-6} mol/L.

Mol/L			Log $[FA_{tota}]$	Log $[CuFA_m]$
$[FA_{total}]$	$[Cu_{total}]$	$[CuFA_m]$		
4.0×10^{-5}	4.22×10^{-5}	3.89×10^{-5}	–4.3979	–4.4100
6.0×10^{-5}	6.00×10^{-5}	5.67×10^{-5}	–4.2218	–4.2464
8.0×10^{-5}	8.02×10^{-5}	7.69×10^{-5}	–4.0969	–4.1141
10.0×10^{-5}	10.10×10^{-5}	9.77×10^{-5}	–4.0000	–4.0101
12.0×10^{-5}	12.03×10^{-5}	11.70×10^{-5}	–3.9208	–3.9318
14.0×10^{-5}	13.93×10^{-5}	13.60×10^{-5}	–3.8539	–3.8665
16.0×10^{-5}	15.43×10^{-5}	15.10×10^{-5}	–3.7959	–3.8210

Thus, we can consider that Mw related to fulvic acids equals to 2191 when considered in the complex formation process. Now according to Ref. [40], these related parameters of fulvic acids in general are known as an "active associate." The Mw of the "active associate" of fulvic acids (Mw = 2191) is considered for calculation of the concentration of free ligand ($[FA_{free}]$) along with the average stability constant. It is worth mentioning that in case of using the moderate molecular weight of the associate (6260), it will not be possible to determine the concentration of free ligand. And of course, without access to this value, it is not possible to determine the stability constant of copper fulvate complex. As it is shown from data (Tables 8.3 and 8.4), using of active associate doesn't cause the changing of the composition of the complex, m = 1.01.

TABLE 8.4 The Calculation of the Composition of Copper Fulvate Dihydroxy Complex by the Least Square Method, Mw (FA) = 2191; pH = 8.0. Xi = lg $[FA_{total}]$, Yi = $lg[CuFA_m]$.

Xi	Yi	XiYi	Xi²
–4.3979	–4.4100	19.3995	19.3415
–4.2218	–4.2464	17.9274	17.8236
–4.0969	–4.1141	16.8550	16.7846
–4.0000	–4.0101	16.0404	16.0000
–3.9208	–3.9318	15.4158	15.3726
–3.8539	–3.8665	14.9011	14.8525
–3.7959	–3.8210	14.5041	14.4088

$\Sigma Xi = -28.2872$; $(\Sigma Xi)^2 = 800.1657$; $\Sigma Yi = -28.3999$; $\Sigma Xi^2 = 114.5836$; $\Sigma XiYi = 115.0433$; m = 1.01.

TABLE 8.5 The Required Data to Study Conditional Stability Constants of Copper Fulvate Complex by the Leden Method, pH = 8.0; Mw (FA) = 2191; Cu(II)free = 3.30×10^{-6} mol/L.

Mol/L				F (FA)
[FA] total	$[Cu(II)_{total}]$	[CuFA]	$[FA_{free}]$	
4.0×10^{-5}	4.22×10^{-5}	3.89×10^{-5}	0.11×10^{-5}	1.08×10^{7}
6.0×10^{-5}	6.00×10^{-5}	5.67×10^{-5}	0.33×10^{-5}	5.2×10^{6}
8.0×10^{-5}	8.02×10^{-5}	7.69×10^{-5}	0.31×10^{-5}	7.69×10^{6}
10.0×10^{-5}	10.10×10^{-5}	9.77×10^{-5}	0.23×10^{-5}	1.22×10^{7}
12.0×10^{-5}	12.03×10^{-5}	11.70×10^{-5}	0.30×10^{-5}	1.18×10^{7}
14.0×10^{-5}	13.93×10^{-5}	13.60×10^{-5}	0.4×10^{-5}	1.03×10^{7}
16.0×10^{-5}	15.43×10^{-5}	15.10×10^{-5}	0.90×10^{-5}	5.08×10^{6}

TABLE 8.6 Determination of Conditional Stability Constant of Copper Fulvate Complex by the Least Square Method, Mw (FA) = 2191; pH = 8.0; Xi = $[FA_{free}]$; Yi = F (FA).

Xi	Yi	XiYi	Xi²
0.11×10^{-5}	1.08×10^{7}	11.88	0.0121×10^{-10}
0.33×10^{-5}	0.52×10^{7}	17.16	0.1089×10^{-10}
0.31×10^{-5}	0.77×10^{7}	23.87	0.0961×10^{-10}
0.23×10^{-5}	1.22×10^{7}	28.06	0.0529×10^{-10}
0.30×10^{-5}	1.18×10^{7}	35.54	0.0900×10^{-10}
0.40×10^{-5}	1.03×10^{7}	41.20	0.16×10^{-10}
0.90×10^{-5}	0.51×10^{7}	45.90	0.8100×10^{-10}

Xi = 2.58×10^{-5}; $(\Sigma Xi)^2 = 6.6564 \times 10^{-10}$; $\Sigma Yi = 6.31 \times 10^{7}$; $\Sigma Xi^2 = 1.33 \times 10^{-10}$; $\Sigma XiYi = 203.61$. $\beta(Cu(OH)_2FA) = 1.18 \times 10^{7}$; lg $\beta = 7.07$.

For the calculation of average stability constant of copper dihydroxy fulvate at pH = 8.0, Leden function F(L) was used.[41]

$$F(L) = F(FA) = [Cu(OH)_2FA]/\{[Cu(II)_{free}][FA_{free}]\} = ([Cu(II)_{total}] - [Cu(II)_{free}])/([Cu(II)_{free}][FA_{free}]) = \beta_1 + \beta_2[FA_{free}] \quad (8.16)$$

where

$$[FA_{free}] = [FA_{total}] - [Cu(OH)_2FA] = [FA_{total}] - \{[Cu(II)_{total} - Cu(II)_{free}]\} \quad (8.17)$$

In solubility method, the concentration of $[Cu(II)_{free}]$ equals to the initial concentration of metal before adding of ligand in solution. It is worth mentioning that the usage of moderate molecular weights of "active associates" in given systems is both real and forced (as shown clearly in Table 8.1). Meanwhile, during the use of the solubility process, at any point, the molar concentration of the fulvate complex exceeded the molar concentration of the total associate (in which a moderate molecular weight at pH = 8 is reasonable enough). That's why it could not determine the free ligand. In the absence of this, it is not possible to calculate the stability constant of fulvate complex. And whenever $[FA_{free}]$ declines to zero, β could be found by the graphical procedure.

The section which is cut on the ordinate by the straight line built in coordinates F(FA)—$[FA_{free}]$ equivalents to the stability constant. The value of β is determined by the square method:

$$\beta = (\Sigma y_i - a\Sigma x_i)/n. \quad (8.18)$$

Where

$$a = (n\Sigma x_i y_i - \Sigma x_i \Sigma y_i)/[(n\Sigma x_i^2 - (\Sigma x_i)^2)] \quad (8.19)$$

$$x_i = [FA_{free}] \text{ and } y_i = F(FA) \quad (8.20)$$

The necessary data for the calculation of the stability constants of copper dihydroxy fulvate complex are given in Tables 8.5 and 8.6.

$$\beta(Cu(OH)_2FA) = 1.18 \times 10^7; \lg \beta = 7.07.$$

8.4 CONCLUSION

We have shown that in diluted water solutions, copper dihydroxy complex $Cu(OH)_2^0$ is the dominant form at pH = 8.0. It was shown that during the complex formation process, an associate of FA, with Mw at pH = 8 equals to 6260 divides and every 0.35 part of this associate inculcates into copper's inner coordination sphere as an integral ligand.

Such part of the associate of fulvic acids was conventionally considered an "active associate." The meaning of Mw of the "active associate" of FA (Mw = 2191) was used for determination of the concentration of free ligand ($[FA_{free}]$) and average stability constant. It is shown that using average molecular weights of "active associates" in given systems is both real and forced.

It is observed that in the $Cu(OH)_2$(solid)-$Cu(OH)_2^0$(solution)-FA-H_2O system at pH = 8.0, the copper dihydroxy fulvate complex dominates with the structure of 1:1, with average stability constant $\beta(Cu(OH)_2FA) = 1.18 \times 10^7$; $\lg \beta = 7.07$. Through the obtained results, it will be possible to calculate the migration forms of copper in natural waters, determine toxicity and bioavailability, and evaluate the ecological condition of water reservoirs.

KEYWORDS

- **fulvic acids**
- **copper**
- **stability constant**
- **active associate**
- **water reservoirs**

REFERENCES

1. Makharadze, G. A.; Supatashvili, G. D.; Varshal, G. M. The Research of the Forms of Copper in Surface Waters. *Hydrochem. Mat.* **1988,** *103*, 3–16.
2. Varshal, G. M. Migration Forms of Fulvic Acids and Metals in Natural Waters, Dissertation, Vernadsky Institute of Geochemistry and Analytical Chemistry of Russian Academy of Sciences, 1994.
3. Osadchyy V.; Nabyvanets, B.; Linnik, P.; Osadcha, N.; Nabyvanets, Y. *Processes Determining Surface Water Chemistry*. Springer International Publishing: Switzerland, 2016.
4. Olk, D. C.; Bloom, P. R.; De Nobili, M.; Chen, Y.; McKnight, D. M.; Wells, M. J. M.; Weber, J. Using Humic Fractions to Understand Natural Organic Matter Processes in Soil and Water: Selected Studies and Applications. *J. Environ. Qual.* **2019,** *48*, 1633–1643.

5. Nor, Y. M. Ecotoxicity of Copper to Aquatic Biota: A Review. *Environ. Res.* **1987,** *43*, 274–282.
6. Muna, M.; Heinlaan, M.; Blinova, I.; Vija, H.; Kahru, A. Evaluation of the Effect of Test Medium on Total Cu Body Burden of Nano CuO-exposed Daphnia Magna: A TXRF Spectroscopy Study. *Environ. Pollut.* **2017,** *231*, 1488–1496.
7. Traudt, E. M.; Ranville, J. F.; Smith, S. A.; Meyer, J.S. A Test of the Additivity of Acute Toxicity of Binary-metal Mixtures of ni with Cd, Cu, and Zn to Daphnia Magna, Using the Inflection Point of the Concentration-response Curves. *Environ. Toxicol. Chem.* **2016,** *35*, 1843–1851.
8. Dudal, Y.; Gerard, F. Accounting for Natural Organic Matter in Aqueous Chemical Equilibrium Models: A Review of the Theories and Applications. *Earth-Sci. Rev.* **2004,** *66*, 199–216.
9. Smithm K. S.; Balistrierib, L. S.; Todd, A. S. Using Biotic Ligand Models to Predict Metal Toxicity in Mineralized Systems. *App. Geochem.* **2015,** *57*, 55–72.
10. Dulaquas, G.; Waeles, M.; Gerringa, L. A.; Midag, R.; Rijkenberg, M.; Riso, R. G. The Biogeochemistry of Electroactive Humic Substances and Its Connection to Iron Chemistry in the North East Atlantic and the Western Mediterranean Sea. *J. Geophys. Res: Oceans* **2018,** *123*, 5481–5499.
11. Makharadze, G.; Goliadze, N.; Khaiauri, A.; Makharadze, T.; Supatashvili, G. Fulvic and Humin Acids in Surface Waters of Georgia. In *High-performas Polymers for Engineering-based Composites*; Apple Academic Press: Waretown, NJ, USA, 2016; 167–179.
12. Moiseenko, T. I.; Dinu, M. I.; Gashkina, N. A.; Kremlevaa, T. A. Occurrence Forms of Metals in Natural Waters Depending on Water Chemistry. *Water Res.* **2013,** *40*, 407–416.
13. Adusei-Gyamfi, J.; Ouddane, B.; Rietveld, L.; Cornard, J.; Criquet, J. Natural Organic Matter-cations Complexation and Its Impact on Water Treatment: A Critical Review. *Water Res.* **2019,** *160*, 130–147.
14. Whitby, H.; Planquette, H.; Cassar, N.; Bucciarelli, E.; Osburn, C. L.; Janssen, D. J.; Jay, T.; Cullen, J. T.; González, A. G.; Völker C.; Sarthou, G. A Call for Refining the Role of Humic-like Substances in the Oceanic Iron Cycle. *Sci. Rep.* **2020,** 106144–106156.
15. Boguta, P.; Sokolowska, Z. Zinc Binding to Fulvic Acids: Assessing the Impact of pH, Metal Concentrations and Chemical Properties of Fulvic Acids on the Mechanism and Stability of Formed Soluble Complexes. *Molekules* **2020,** *25*, 1297–1321.
16. Xu, H.; Xu, D. C.; Wang, Y. Natural Indices for the Chemical Hardness/Softness of Metal Cations and Ligands. *ACS Omega* **2017,** *2*, 7185–7193.
17. Makharadze, T.; Makharadze, G. The Calculation of the Stability Constant of Nickel Fulvate, Using an Average Molecular Weight of Active Associate of Fulvic Acids at pH=8. *Chall. Adv. Chem. Sci.* **2021,** 4, 27–33.
18. Maccarthy, P.; O'Cinneide, S. Fulvic Acids: Interactions with Metal Ions. *Eur. J. Soil Sci.* **2006,** *25*, 429–437.
19. Zhu, B.; Ryan, D. K. Characterizing the Interaction Between Uranul Ion and Fulvic Acid Using Regional Integration Analysis (RIA) and Fluorescence Quenching. *J. Environ. Radioact.* **2016,** *153*, 97–103.
20. Bertoli, A. C.; Garcia, J. S.; Trevisan, M. G.; Ramalho, T. C.; Matheus, P.; Freitas, M. P. Interactions Fulvate-metal (Zn^{2+}, Cu^{2+} and Fe^{2+}): Theoretical Investigation of Thermodynamic, Structural and Spectroscopic Properties. *Biometals* **2016,** *29*, 275–285.

21. Constantino, C.; Comber, S. D. W.; Scrimshaw, M. D. The Effect of Wastewater Effluent Derived Ligands on Copper and Zinc Complexation. *Environ. Sci. Pollut. Res.* **2017,** *24*, 8363–8374.
22. De Oliveira Vaz, D.; Fernandes, A. N.; Szpoganicz, B. Complexations of Divalent Metallic Ions with Fulvic Acids. *Eclética Química* **2018,** *43*, 54–58.
23. Chakraborty, P.; Chakraborty, Ch.L. Competition from Cu(II), Zn(II) and Cd(II) in Pb(II) Binding to Suwannee River Fulvic Acid. *Water Air Soil Pollut.* **2008,** *195*, 63–71.
24. Xue, H.; Sigg, L. Comparison of the Complexation of Cu and Cd by Humic or Fulvic Acids and by Ligands Observed in Lake Waters. *Aquatic Geochem.* **1999,** *5*, 313–335.
25. Buffle, J.; Greter F.; Haerdi, W. Measurement of Complexation Properties of Humic And fulvic Acids in Natural Waters with Lead and Copper Ion-selective Electrodes. *Anal. Chim. Acta.* **1977,** *118*, 29–44.
26. Cabaniss, S. E.; Shuman, M. S. Copper Binding by Dissolved Organic Matter: I. Suwanee River Fulvic Acid Equilibria. *Geochim. Cosmochim. Acta.* **1988,** *52*, 185–193.
27. Mantoura, R. F. C.; Dickson, A.; Riley, J. P. The Complexation of Metals with HUMIC materials in Natural Waters. *Estuarine Coastal Mar. Sci.* **1978,** *6*, 387–408.
28. McKnight, D. M.; Wershaw, R. L. Complexation of Copper by Fulvic Acid from the Suwannee River – Effect of Counter-ion Concentration. In *Humic Substances in the Suwannee River, Georgia: Interactions, Properties, and Proposed Structures*; 1989; 63–69.
29. Turner, D. R.; Varney, M. S.; Whitfield, M.; Mantoura, R. F. C.; Riley, J. P. Electrochemical Studies of Copper and Lead Complexation by Fulvic Acid. 1. Potentiometric Measurements and a Critical Comparison of Metal Binding Models. *Geochim. Cosmochim. Acta.* **1986,** *50*, 289–297.
30. Schnitzer, M.; Kerndorff, H. Reactions of Fulvic Acid with Metal Ions. *Water Air Soil Pollut.* **1981,** *15*, 97–108.
31. Hirata, S. Stability Constants for the Complexes of Transition-metal Ions with Fulvic and Humic Acids in Sediments Measured by Gel-filtration. *Talanta* **1981,** *28*, 809–815.
32. Taishi, K.; Hirotake, M.; Takayuki, S.; Ikuji, T.; Hatsumi, Y. Determination of Apparent Formation Constants of Eu(III) with Humic Substances by Ion Selective Liquid Membrane Electrode. *Am. J. Analy. Chem.* **2012,** *3*, 462–469.
33. Schnitzer, M.; Skinne, S. I. M. Organo-metallic Interactions in Soils: 5. Stability Constants of Cu++, Fe++, and Zn++fulvic Acid Complexes. *Soil Sci.* **1966,** *102*, 361–365.
34. Bertoli, A. C.; Garcia, J. S.; Trevisan, M. G.; Ramalho, T. C.; Freitas, M. P. Interactions Fulvate-metal (Zn^{2+}, Cu^{2+} and Fe^{2+}): Theoretical Investigation of Thermodynamic, Structural and Spectroscopic Properties. *Biometals* **2016,** *29*, 275–285.
35. Town, R. M.; Van Leeuwen, H. P.; Buffle, J. Chemodynamics of Soft Nanoparticulate Complexes: Cu(II) and Ni(II) Complexes with Fulvic acids and Aquatic Humic Acids. *Environ. Sci. Technol.* **2012,** *46*, 10487–10498.
36. Makharadze, G.; Makharadze, T. Method of Calculation of Stability Constants of Fulvic Complexes on the Example of Copper. *J. Chemist. Chem. Eng.* **2014,** *8*, 108–111.
37. Gamble, D. S.; Schnitzen, M.; Hoffman, I. Cu^{2+}-Fulvic Acid Chelation Equilibrium in 0.1 KCl at 25°C. Can. *J. Chem.* **1970,** *48*, 3197–3204.
38. Schnitzer, M.; Hansen, E. H. An Evaluation of Methods for the Determination of Stability Constants of Metal-Fulvic Acid Complexes. *Soil Sci.* **1970,** *109*, 333–340.

39. Makharadze, T.; Makharadze, G. Investigation of Complex Formation Process of Copper with Macromolecular Organic Substances, Isolated from Natural Waters. *Org. Chem. Plus* **2020,** *1*, 1–8.
40. Makharadze, G.; Cypatashvili, G.; Makharadze, T. New Version of Calculation of Stability Constant of Metal-fulvate Complexes on the Example of Zinc Fulvate. *Int. J. Environ. Sci. Tech.* **2018,** *15*, 2165–2168.
41. Beck, M. T; Nagypal, I. *Chemistry of Complex Equilibria*; Chichester: Horwood, New York, 1990.

CHAPTER 9

Newfangled Dimension of Scientists' Responsibility: Solutions in the Context of Sustainable Development and Evolution of Chemical Processes for Clean Environmental Management

MARINA AVALIANI[1], ELENA SHAPAKIDZE[2], and VAJA CHAGELISHVILI[1]

[1]*R. Agladze Institute of Inorganic Chemistry and Electrochemistry, I. Javakhishvili Tbilisi State University, Georgia*

[2]*A. Tvalchrelidze Caucasian Institute of Mineral Resources, Tbilisi, Georgia*

ABSTRACT

Scientists must face global warming and other challenges of the 21st century by analyzing together the role of chemical pollution due to human activity to think how to reduce the environmental impact of harmful products in order to adhere to the various concepts of green environmental protection.

Overall, research work in any field should be correlated with the global realities of economic development and should focus on scientific fields based on the existing progressive and valuable trends. Facing the current scientific challenges, based on the principles of clean environmental management our research team has been conducting research on the synthesis of new compounds, using condensation process which gives the vast scope of possibilities for synthesis of inorganic polymers containing phosphorus. Our activities were conducted in two directions: (1) domain of condensed inorganic polymer's chemistry; in other words, on the mysterious phenomenon of condensation that lead to the various classes of original compounds

Environmental Technology and Sustainability: Physical, Chemical, and Biological Technologies for Environmental Protection. Tamara Tatrishvili, Ann Rose Abraham, & A. K. Haghi (Eds.)

(simplified scheme below); (2) applied science by supporting the scientific potential of young people, developing the direction of the ecological field and inclusion of students in the research practice in order to stimulate their involvement in scientific environment.

Simplified Schema for obtaining of compounds by heating of components at diverse molar ratio- phosphoric acid + M_2O_3 +M_2O in the temperature range 100-500°C

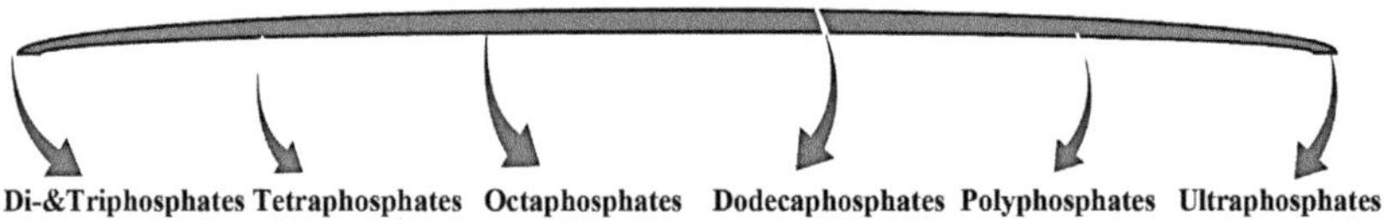

High thermal stability, elevated content of phosphorus of condensed phosphates—these preconditions have led to many applications such as raw and thermo-resistant materials, fertilizers, detergents, as fillers, also as laser materials, ion-exchangers, and catalytic agents, flame, etc. The luminescent and vibrational properties determine their large scale use in quantum electronics. Condensed compounds synthesized by us are not toxic and what is really important—do not pollute the environment.

9.1 INTRODUCTION

Sustainable development is a global process, inseparable from the general trend of the world, and to participate in this process means to inherit the spiritual, intellectual, and material values not only of one country but of all humanity. The reorientation of human activity toward sustainable development is also a global process, which involves various aspects, one of the main objectives of which is the reduction of harmful effects on the environment and of course taking into account the conservation of natural resources. According to a study in the years 2008–2010, the destruction of nature had cost humans approximately of $6.6 trillion euros per year and every year this loss in increasing at an alarming speed.

Generally, scientists in concerned fields have their own responsibility for the protection of nature, in searching of sustainable materials for green environmental protection by implementation of clean technologies. In fact, the concept of responsibility of scientists—this is the subject, the content of which is complex, inexhaustible, and wide-ranging and it is almost impossible to define it in words for giving a definitive conception. It is obvious, to destroy is much easier than to build and at this new stage of the world development, it is necessary to use a new approach to the notion of *Responsibility of the Scientists*, to the reorientation of common efforts to the subject of clean energy for changing the "consumer mentality"; it is crucial – a sine qua non – to realize what represents the notion of responsibility of scientists in the domain of clean technologies.

It is essential to know what is the modern ecological approach to nature, to scientific research, so that future generations live and survive not in ecological disasters, arid deserts, ruinous drought, or flood conditions, but in a balanced situation, taking into account Green Environmental Protection. Everything is connected on our world: the cataclysms of nature – mostly the results of "unreasonable consumption" approach of consumers: environmental pollution by chemicals, by polluting emissions, wastage of energy resources, and irrational use of energy resources. On that basis, scientists are obliged to face global warming and other challenges of the 21st century by analyzing together the role of chemical pollution due to human activity; how to reduce the environmental impact of harmful products to adhere to the various concepts of green environmental protection. The research work in any field should be correlated with the global realities of economic development and must develop scientific fields based on the existing progressive and valuable trends.[1–5]

As a simple illustration: one of the measurements of the progress of civilization is the kind of materials that are available and accessible to society.[6] It can be said that just as the Bronze and Iron Age marked the start of a new stage in human civilization at the end of the Stone Age, the discovery of organic polymers indicated a new era. Traditional predictable materials such as steel, various metals, wood, glass, or ceramics are now being supplemented and already very often replaced by polymeric advanced materials.[6] Starting from commonplace and banal implementations such as synthetic polymer with elastic properties, for example, rubber, fibers, thermoplastics to high-tech applications – in electronics, optical, biological, and other spheres, polymers have been more and more used in the production of advanced materials.

For organic polymers, two important factors are of significant importance: (1) the essential capacity of carbon to form bonds with itself and with other heteroatoms, which makes possible the synthesis of enormous number of polymeric compounds; (2) the consumption of cheap petroleum feedstocks has possible and accessible polymer synthesis widely available, which has favored the production of organic polymers.[6] But last decennia despite this extensive use and their omnipresence, carbon-based polymers do not have the required volume to meet all demands or needs of new applications as advanced materials.[6–15] For example, for many organic polymers is not appropriate application at high temperature (they are degraded by oxidation), besides, they become very brittle at low temperatures.

Thus, last 2–3 decades, the increased developments of the chemistry of inorganic polymers namely of the field of condensed crystal chemistry, based on the principles of Green Environmental Protection, was due to the quick development of progressive methods of analysis, as well as to the evolution and accomplishment in this field of condensed phosphates chemistry. In the search of sustainable materials by appropriate application of clean technologies, numerous inorganic polymers, namely condensed poly- and cyclo-phosphates with diverse formulas were obtained and described last year. Condensed phosphates of polyvalent metals, notably double phosphates containing alkali metals possess a number of more attractive, important, and appreciable properties, which clarify the prospects of their application. Actually, anions are known for $n = 3, 4, 5, 6, 8, 9, 10$, and 12.[17–28] High thermal stability and elevated content of phosphorus—these preconditions have caused their application as raw components for manufacture of phosphates glasses, the use of crystalline and non-crystalline ultraphosphates in quantum electronics are predetermined by specific properties. Some cyclophosphates were primarily synthesized; firstly, examined and determined in collaboration with Dr. N. Chudinova and Acad. Tananaev[29–31] using the experimental technique of research, named "Synthesis methods of Tananaev." Facing the current scientific challenges, based on the conceptual principles of clean environmental management, our research team is conducting research on the synthesis of new condensed compounds, using the condensation process which gives the vast scope of variants and possibilities for the synthesis of inorganic polymers containing phosphorus.

9.2 EXPERIMENTAL METHODS AND MATERIALS

9.2.1 OBJECTS OF THE EXPERIMENTAL STUDY

Contemporary technological development and real challenges require the rationalization of conceptual principles of technological schemes and educational approaches. Our group of scientists works not only in the field of research of new materials for advanced technical domains, but also to keep the succession of the educational system accomplished over decades.

To achieve this goal in order to promote Clean Technologies within our University/Institute, a group of scientists has organized Educational and Training Center "RATIO." Main goal of this Center is the encouragement of researchers to conduct their work for Green Environmental Protection. Several training courses have been organized, quite a few educational steps in the ecological field have already taken place. Then, Seminars & International Conferences were organized with the participation of scientists – researchers from Georgia and various foreign countries. It should be emphasized—with the involvement of students of educational establishments, including not only Institutes and Universities but also by the inclusion of learners of institutions for disabled people. Faced present-day defies and challenges due to the pandemic and the rather difficult situation in the educational field, our training course was called "Modern learning process and interactive forms of education by implementation of telework method." The basic point of our work was the implementation of continuity of education on the chemical and ecological domains, in the frame of clean environmental management.

In aims of search of sustainable materials for ecological protection of the environment using clean technologies, our research is conducted in the open systems, containing main components: $M^{III}{}_2O_3$-P_2O_5-H_2O and $M^I{}_2O$-$M^{III}{}_2O_3$-P_2O_5-H_2O at the temperature range 80–500°C, without using any toxic elements and w/o polluting the environment. Our method of synthesis describes the possibilities of synthesis normal and double condensed compounds, so-called inorganic polymers.[19,20,32–36] The interaction in the multi-component systems was studied, and novel phases formation and the solubility of obtained compounds were examined in the above-mentioned systems. The major task was also to examine the temperature range of areas of crystallization of new phosphates of mono- and trivalent metals in polyphosphoric acid melts.

Other important objectives were: (1) an analysis by different methods of newly obtained compounds; (2) the determination and evaluation of several properties of the synthesized condensed phosphates.

9.2.2 MATERIALS

In the present study, we review solely and exclusively the interaction, phase formation, and the solubility in the multi-component systems $M^I{}_2O$ - $M^{III}{}_2O_3$-P_2O_5-H_20, where M^I – is Rb and M^{III} are Ga and In and Sc at 50–450 (500)°C. The initial components are Rb_2CO_3, Ga_2O_3, In_2O_3, Sc_2O_3, and ortho-phosphoric acid (85%). The molar ratio of initial components P/M^{III} was constant n = P/M^{III} = 20 for each experiment and consequently, for each temperature range, but sometimes depending on the tasks, it was varied in the range from 15 to 18. Duration of the synthesis process up to 2–14 days, in correlation with the temperature regime.

9.2.3 CHARACTERISTICS OF THE METHODS

9.2.3.1 SYNTHESIS METHOD

The above-mentioned compounds, such as H_3PO_4 of percentage 85%, Rb_2CO_3, trivalent metal oxides Ga_2O_3, In_2O_3, or Sc_2O_3 were mixed in a special carbon-glass or platinum crucible during some munities. A mixture of these reagents was used for phase formation and heated at temperatures between 50°C and 350–360°C, sometimes until 500°C. Depending on the envisaged and desired condensed matters, in our experimental studies, the duration of the synthesis process was very variable, took 1 day for synthesis T = 400–500°C, or 2 days for T = 300–350°C and up to 10–12 days (synthesis T = 50–250°C). After mixing rubidium carbonate, gallium oxide, or indium oxide, or scandium oxide with ortho-phosphoric acid and after placing this mixture in the thermostatically controlled furnace at a predetermined temperature, the process of phase formation is started. The choice of the optimal temperature range, as well as the initial components and their preliminary ratio are the main factors that influencing the final results of the experiments, which allowed to obtain new numerous types of double condensed phosphates/so-called inorganic polymers with various programed and predetermined properties.

Afterward, the accomplishment of crystallization which has been checked by microscope or other methods, the crystalline phases, extracted by using half-litter of hot distilled water after melting of poly-phosphoric acid. Then the single crystals must be washed with ethanol and acetone, then dried in the thermostat or in the air.

In the course of the investigation of poly-component systems in the various temperature range, synthesized numerous inorganic polymeric compounds—double condensed phosphates were studied by various methods such as gravimetric analysis, method of paper chromatography, thermos-gravimetric analysis, roentgen phase's analysis, IR spectroscopy, scanning micro-spectroscopic analysis, and flame-spectrophotometric method.

9.2.3.2 GRAVIMETRIC ANALYSIS

The elemental composition of the newly synthesized condensed compounds was examined by gravimetric analysis. Phosphorus was determined by precipitation method using a molybdate mixture and a quinoline solution. Gallium, indium, and scandium were determined by the hydroxyquinoline precipitation method, appropriated by us for crystalline condensed matter forms.[16,20,32–36]

9.2.3.3 METHOD OF PAPER CHROMATOGRAPHY

The quite high stability of inorganic polymeric phosphates makes it able to classify them and determine the condensation degree by the method of paper chromatography. Synthesized di-, tri-, tetra-, octa-, dodeca-phosphates, and polymeric compounds were checked using paper chromatography in the following way: obtained crystalline samples were decomposed with H-form cation type ion-exchanger. In these experiments, 0.1 g of the substance was contacted for ca. 1 h with an aqueous suspension of 2 g (KU-2, Dowex) type sulfonated cation exchanger at ~0°C. After the decomposition of the main mass of the substance, according to the following scheme: cationite-H(solid) + M-phosphate(solid) → cationite-M(solid) + H-phosphate(liquid) the solution, containing phosphoric acids was neutralized with sodium bicarbonate and chromatographed on Filtrak FN-11 paper by acidic solvent CCl_3-COOH-CH_3COOH-CH_3OH. Chromatograms were sprayed with 5% solution of ammonium molybdate $(NH_4)_2MoO_4$ and irradiated with ultraviolet light (λ = 400 nm).[19,20,34–36]

9.2.3.4 THERMO-GRAVIMETRIC ANALYSIS

For thermos-gravimetric analysis (TGA), it was used a Derivatograph Q1500-D with a heating rate of 10 degree/min, in air atmosphere and maximum temperature of 1280°C (sometimes to 1600°C).

9.2.3.5 SCANNING MICRO-SPECTROSCOPIC ANALYSIS

Scanning electronic microscope measurements were accomplished on a JEOL scanning electronic microscope JSM–6510LV (well-appointed by energy-disperse X–Max N 20 micro-X-ray spectral analyzer produced by Oxford Instruments). SEM measurements were carried out by means of reflected (BES) as well as secondary (SEI) electrons at an accelerating voltage (at 20 kV). The employed distance was approximately 15 mm. Micrographs have been taken at the diverse enlargements. Micro-spectroscopic analysis was performed from the sampling point zones and their surface area.[19,20,34–36]

9.2.3.6 ROENTGEN PHASE'S ANALYSIS

The basic structural framework of synthesized double-condensed compounds is determined by the X-ray diffraction method and IR spectroscopy. The powder diffraction intensity data collections were performed on a DRON–3M diffractometer, with anodic Cu–K_α radiation in the range of 2θ = 10–60°, detector's speed 2° min^{-1}, lattice spacing d_α/n in Å, and I/I_0—is the relative intensity (used model/standard data—by ASTM—American Society for Testing and Materials).[35–36]

9.3 RESULTS AND DISCUSSION

9.3.1 STUDY OF POLY-COMPONENT SYSTEMS IN THE PRESENCE OF TRIVALENT METALS AND CONTAINING RB AS THE MONOVALENT CATION

9.3.1.1 STUDY OF THE INTERACTION IN THE SYSTEM RB2O - GA2O3 - P2O5 - H2O AT TEMPERATURE RANGE 80–500°C

This work continues our systematic research on the synthesis of monovalent metal-polyvalent metal double condensed phosphates, so-called inorganic polymers, which demonstrate various potential applications. The interaction in the multi-component systems, containing mono- and trivalent metals in the presence of phosphoric acid is studied. The temperature range of synthesis was various: from 80°C to 500°C. In the Schema, N1 is presented the correlation of the condensation degree of synthesized compounds on the temperature during interaction in the system $Rb_2O\_Ga_2O_3–P_2O_5–H_2O$ on the molar ratio P_2O_5/Ga_2O_3 = 15/1.5. Molar ratio for Rb_2O/Ga_2O_3 = n was following: 2.5; 3.5; 5.0; 6.0; 7.5; 8.0; 10.0. At the low temperature (from 80 to approximately 100°C), only amorphous phase or metastable phases (at 105–120°C) were obtained. The gradual increase of the temperature in the reaction area of phosphoric acids melts lead to the formation of oligo-phosphate: triphosphate of rubidium–gallium $RbGaHP_3O_{10}$ in the temperature range from 130–140°C to 280–300°C (for n = 2.5 and 3.5). In the case of n = 7.5, this compound is stable from 200°C to 500°C. For an initial ratio of components n = 8.0 and 10.0, this triphosphate is also stable between 200 and 350°C. Cyclo-octaphospate of rubidium–gallium was obtained at molar ratio n = 5.0 and 6.0, at synthesis temperature from 315–320°C to 380°C. In addition, other oligo-phosphates, such as double acidic diphosphate $RbGa(H_2P_2O_7)_2$ and double normal diphosphate $RbGaP_2O_7$ were synthesized also (schema 9.1).

Temperature (°C)	Molar ratio n = 2.5	Molar ratio n = 3.5	Molar ratio n = 5.0	Molar ratio n = 6.0	Molar ratio n = 7.5	Molar ratio n = 8.0	Molar ratio n =10.0
80	Amorphous phase	Amorphous phase	Amorphous phase	Amorphous phase	Amorphous phase	Amorphous phase	Amorphous phase
100	Amorphous phase	Amorphous phase	Amorphous phase	Amorphous phase	Amorphous phase	Amorphous phase	Amorphous phase
130	Δ	Δ	□	□	□	□	□
150	Δ	Δ	□	□	□	□	□
180	Δ	Δ	□	□	□	□	□
200	Δ	Δ	Δ	Δ	Δ	Δ	Δ
250	Δ	Δ	Δ	Δ	Δ	Δ	Δ
300	○	○	Δ	Δ	Δ	Δ	Δ
315	○	○	☼	☼	Δ	Δ	Δ
350	○	○	☼	☼	Δ	Δ	Δ
355	○	○	☼	☼	Δ	■	■
360	○	○	☼	☼	Δ	■	■
380	○	○	☼	☼	Δ	■	■
400	○	○	◊	◊	Δ	■	■
450	○	○	◊	◊	Δ	■	■
500	○	○	◊	◊	Δ	■	■

◊, $Ga(PO_3)_3$-C; ○, $Ga(PO_3)_3$-C^I; Δ, $RbGaHP_3O_{10}$; ☼, $Rb_2Ga_2P_8O_{24}$,
□, $RbGa(H_2P_2O_7)_2$; ■, RbGa P_2O_7.

SCHEMA 9.1 Dependence of composition of condensation degree of condensed compounds from synthesis temperature T and molar ratio $n = Rb_2O/Ga_2O_3$.

As it was already marked the powder diffraction intensity, data collections were performed on a DRON–3M diffractometer, with anodic Cu–K_{α} radiation in the range of $2\theta = 10–60°$, detector's speed 2° min^{-1}. It was determined that acidic double diphosphate $RbGa(H_2P_2O_7)_2$ was crystallized in two interesting modifications, one of the mentioned forms is isomorphic to obtained by us, double compound of potassium–gallium $KGa(H_2P_2O_7)_2$-form I, and other to analogous oligo-phosphate $KGa(H_2P_2O_7)_2$-form II. Similarly, double acidic triphosphate $RbGaHP_3O_{10}$ was crystallized in two forms, one of which is isomorphic to $KGaHP_3O_{10}$-form I (whose synthesis temperature range is quite more which is 150–260°C). After heating of components at a temperature higher than 265°C, the form II appears; this modification is not isomorphic to analogous triphosphates of Li-Ga, Li-In or K-Ga, K-In, or Li-Sc, K-Sc, obtained earlier.

The dependency of composition and stability of double condensed phosphates from ion radius of M^{III} is analyzed by comparison of various condensed compounds obtained last year. It is necessary to emphasize that by examining the range of structures of inorganic polymers $M^IM^{III}(PO_3)_4$ where M^I is an alkali metal and M^{III} is some trivalent metal such as Gallium, Indium, or Scandium. It can be concluded as the radius of M^{3+} decreases, the identical period of the polyphosphate chain increases, due to the complication of its form factors; the cycles appear slowly, the number of structural types increases due to the correlation of the average distances between (M^{III} - O) and (M^I - O). The lower the correlation between them, the higher the probability of big cycles' formation becomes. XRD data for synthesized inorganic polymers of rubidium–gallium are presented in Tables 9.1 and 9.2. It is necessary to observe, that the main results obtained during the investigation of systems containing In and Sc with Rb are essentially similar to those of the analogous system examined above.[19,20,34–36]

TABLE 9.1 XRD Data for Synthesized Acidic Oligo-Phosphates: $RbGa(H_2P_2O_7)_2$-Form 1, $RbGa(H_2P_2O_7)_2$-Form 1I, Acidic Triphosphate $RbGaHP_3O_{10}$-Form 1, and $RbGaHP_3O_{10}$-Form II.

RbGa(H2P2O7)2-Form 1		**RbGa(H2P2O7)2-Form 1I**		**RbGaHP3O10-Form 1**		**RbGaHP3O10-Form II**	
***I*, %**	**d, Å**	***I*, %**	**d, Å**	***I*, %**	**d, Å**	***I*,%**	**d, Å**
42	7.83	28	8.42	3	7.11	84	7.69
31	7.77	19	5.90	3	6.58	39	5.75
14	7.08	30	6.75	54	5.93	28	5.09
20	6.70	6	5.09	22	5.54	5	4.94
7	5.60	7	4.82	37	4.75	8	4.44
7	4.94	28	4.57	100	4.23	54	4.19

TABLE 9.1 *(Continued)*

RbGa(H2P2O7)2-Form 1		RbGa(H2P2O7)2-Form II		RbGaHP3O10-Form 1		RbGaHP3O10-Form II	
I, %	d, Å	*I*, %	d, Å	*I*, %	d, Å	*I*,%	d, Å
45	3.94	14	4.29	4	4.05	74	3.80
26	3.91	6	4.07	5	3.99	59	3.74
20	3.73	20	3.78	5	3.77	43	3.55
10	3.53	46	3.76	80	3.39	35	3.24
31	3.47	16	3.67	48	3.24	19	3.22
41	3.42	46	3.43	90	2.99	100	3.03
25	3.32	49	3.33	80	2.89	47	2.86
100	3.14	100	3.20	31	2.83	4	2.76
76	3.09	70	3.11	64	2.77	68	2.56
17	3.04	54	3.07	7	2.72	62	2.50
22	2.97	8	2.98	23	2.66	35	2.46
26	2.89	9	2.94	16	2.43	9	2.43
11	2.79	8	2.71	3	2.40	21	2.27
17	2.57	8	2.63	10	2.37	5	2.22
12	2.54	33	2.49	6	2.29	12	2.10
35	2.52	8	2.44	16	2.19	9	2.07
17	2.40	16	2.34	17	2.15	13	2.04
42	2.36	35	2.32	5	2.11	21	1.90
20	2.32	13	2.29	3	2.04	19	1.89
13	2.26	6	2.22	13	2.00	21	1.86
10	2.24	12	2.08	18	1.98	22	1.79
25	2.17	12	2.04	14	1.95	18	1.77
10	2.15	12	1.93	20	1.92	17	1.70
14	2.13	5	1.89	14	1.87	27	1.65
21	2.07	17	1.84	18	1.83	8	1.59

TABLE 9.2 XRD Data for Synthesized Inorganic Polymeric Phosphate – Double Cyclo-octaphosphate of Rubidium–Gallium $Rb_2Ga_2P_8O_{24}$ and Double Diphosphate $RbGaP_2O_7$.

Rb2Ga2P8O24		RbGaP2O7		Rb2Ga2P8O24		RbGaP2O7	
I, %	d, Å	*I*, %	d, Å	*I*, %	d, Å	*I*,%	d, Å
85	9.82	4	7.62	15	2.58	4	2.10
80	8.10	5	5.76	10	2.49	28	2.09
100	6.15	10	5.21	15	2.46	6	2.03
18	5.90	13	3.95	11	2.44	26	2.00
43	4.95	16	3.64	7	2.36	9	1.94

TABLE 9.2 *(Continued)*

Rb2Ga2P8O24		RbGaP2O7		Rb2Ga2P8O24		RbGaP2O7	
I, %	d, Å	*I*, %	d, Å	*I*, %	d, Å	*I*,%	d, Å
5	4.71	36	3.43	4	2.32	9	1.86
10	4.23	15	3.48	4	2.14	5	1.75
16	4.15	60	3.24	5	2.11	9	1.79
20	3.96	25	3.18	18	2.04	14	1.73
20	3.90	9	3.12	11	1.94	9	1.68
8	3.53	19	3.07	6	1.88	5	1.64
45	3.40	78	2.97	6	1.86	19	1.62
25	3.30	100	2.94	4	1.76	3	1.60
10	3.23	54	2.91	21	1.70	14	1.54
98	3.16	12	2.63	10	1.66	15	1.50
28	3.06	15	2.44	5	1.61		
23	2.88	9	2.40	10	1.59		
59	2.75	9	2.33	7	1.52		
25	2.62	5	2.19				

As it has been said above, condensed compounds of rubidium–gallium, according to their composition and structure, correspond with phosphates of potassium–gallium, potassium–indium, and sometimes with rubidium–scandium. Table 9.3 esented XRD data for the synthesized inorganic polymeric phosphate—double cyclo-octaphosphate of potassium–gallium $K_2Ga_2P_8O_{24}$ and acidic double triphosphate. Another additional conclusion—the obtained phosphates are not analogous to corresponding compounds of rare earth elements. Likewise, it has been revealed that at moderately low temperatures, it's more feasible to produce double acidic phosphates, at relatively high temperature appears double cyclo-octaphosphate of K-Ga, K-In, and Rb-Ga. It must be emphasized that the influence of molar ratio is nevertheless very important.

TABLE 9.3 XRD Data for Synthesized Inorganic Polymeric Phosphate—Double Cyclo-octaphosphate of Kalium-Gallium $K_2Ga_2P_8O_{24}$ and Acidic Double Triphosphate $KGaHP_3O_{10}$.

KGaHP3 O10		K2Ga2P8O24		KGaHP3 O10		K2Ga2P8O24	
I, %	d, Å	*I*, %	d, Å	*I*, %	d, Å	*I*,%	d, Å
19	7.08	22	9.76	55	2.86	61	3.17
21	6.55	95	8.22	60	2.74	16	3.05
100	5.89	100	6.14	15	2.68	7	2.86

TABLE 9.3 *(Continued)*

KGaHP3 O10		K2Ga2P8O24		KGaHP3 O10		K2Ga2P8O24	
I, %	d, Å	*I*, %	d, Å	*I*, %	d, Å	*I*,%	d, Å
81	5.47	75	4.97	16	2.61	42	2.73
28	4.69	8	4.25	8	4.44	33	2.61
98	4.24	5	4.09	14	2.40	8	2.56
30	4.01	7	3.94	20	2.35	21	2.49
18	3.87	29	3.87	8	2.26	6	2.45
70	3.32	8	3.52	19	2.17	7	2.41
43	3.24	18	3.40	7	2.01	3	2.36
18	3.10	13	3.30	21	1.90	14	2.04
100	2.96	10	3.26	20	1.79	6	1.89

X-Ray patterns of synthesized samples of inorganic polymeric compounds of potassium–gallium, isomorphs to rubidium–gallium are presented in Figures 9.1–9.4.

The decomposing of triphosphates of the above-mentioned alkali metals by heating for 1 h shows the appearance of cyclo-octaphosphates $M_2Ga_2P_8O_{24}$, diphosphate $MGaP_2O_7$, and polyphosphate of gallium $Ga(PO_3)_3$. On the other hand, heating triphosphates for 2 days shows the final decomposition and obtaining only of $Ga(PO_3)_3$ and $KGaP_2O_7$ or similarly $RbGaP_2O_7$ (Figs. 9.2 and 9.3).

In the Figure 9.5, it is presented the diffractogram of the pure polyphosphate of gallium $Ga(PO_3)_3$-form C obtained from solution-melts of phosphoric acids at the temperature range 355–500°C.

9.4 ABOUT POSSIBILITIES OF POTENT APPLICATIONS OF CONDENSED PHOSPATES

Referring to the well-known scientists A. Durif, I.V. Tananaev and A. S. Wagh[26,29,31] chemically bounded phosphates are the real materials of the 21st century. Nevertheless, the development of their processing technology still lacks the necessary attention. Advances in the domain of inorganic polymer chemistry, it shows the important application of phosphates materials in several technical domains, including nanotechnologies. The chemistry of inorganic compounds of phosphorous has advanced intensively in the last time also for the purpose that condensed compounds of phosphorus are greatest applicable, useful, and convenient to promote development of the

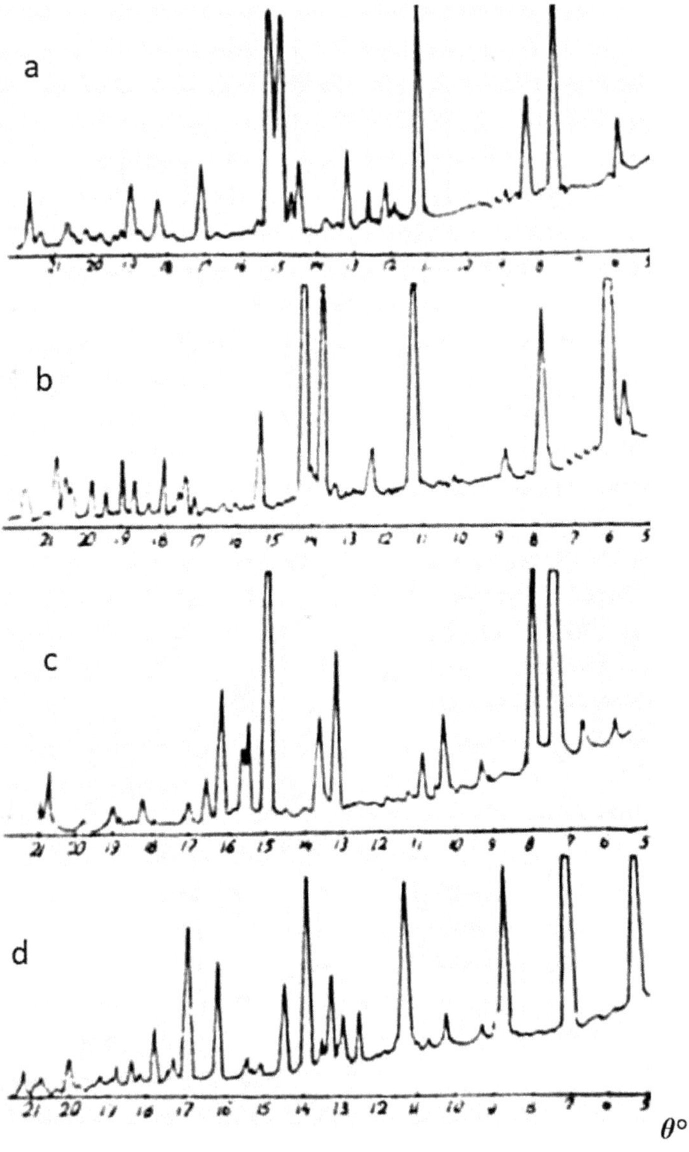

FIGURE 9.1 X-Ray patterns of synthesized samples of potassium–gallium (isomorphs to rubidium–gallium polymeric compounds): (a) $KGaP_2O_7$, (b) $KGa(H_2P_2O_7)_2$, (c) $KGaHP_3O_{10}$, and (d) $K_2Ga_2P_8O_{24}$.

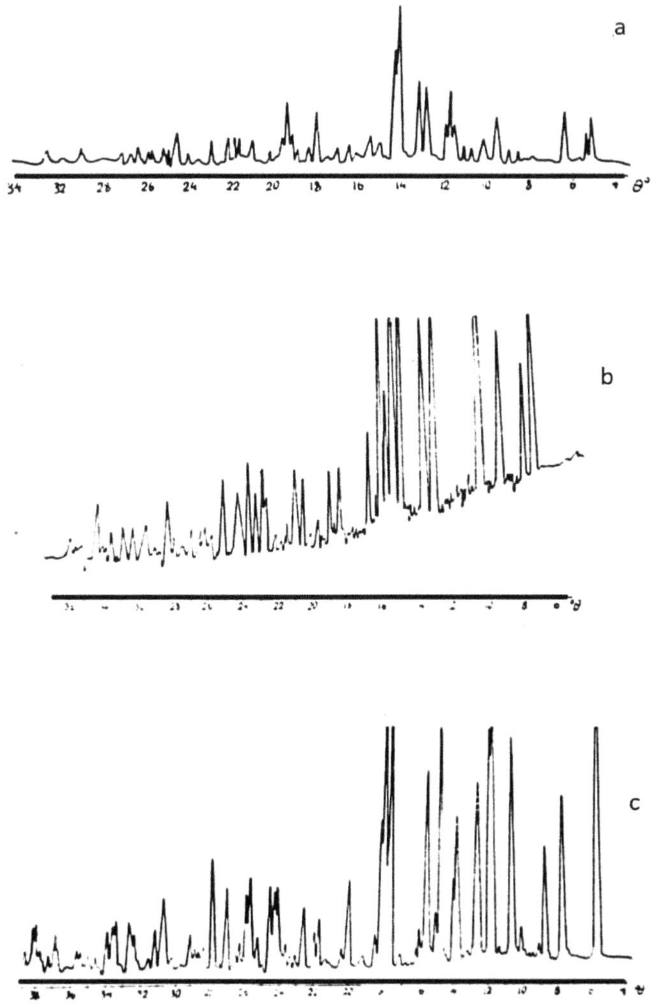

FIGURE 9.2 X-Ray patterns of synthesized samples of rubidium–gallium condensed phosphates (isomorphs to relevant potassium–gallium polymeric compounds): (a) $RGa(H_2P_2O_7)_2$-form II, (b) $RbGaHP_3O_{10}$-form I, and (c) $RbGaHP_3O_{10}$-form II.

chemistry of inorganic polymers, and last but not least—they are reasonably presumed as best fertilizers, detergents, and as materials used in engineering, construction, and other areas, such as raw materials, glasses, thermo-resistant constituents, effective applying nourishments, cleaners, cement substances, ion-exchange ingredients, and also catalytic agents, as

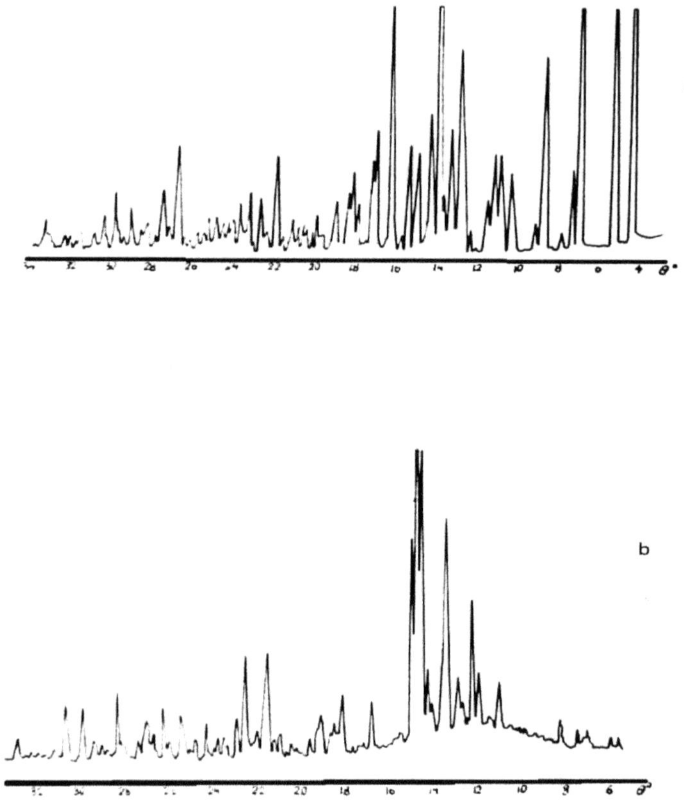

FIGURE 9.3 X-Ray patterns of synthesized samples of rubidium–gallium condensed phosphates (isomorphs to related potassium–gallium polymeric compounds): (a) $Rb_2Ga_2P_8O_{24}$ and (b) $RbGaP_2O_7$.

fire-retardants, etc.[20,23,24,28,32] Thermal behavior of the synthesized compounds shows that cyclo-octaphosphates of potassium–gallium, rubidium–gallium, and rubidium–indium demonstrated high thermal stability. They start to decompose at temperature of about 650°C. Final crystal products of thermal decomposition of the compounds are polyphosphate of appropriate metals (or, in any cases, diphosphate). Performed data allow to conclude that cyclo-octaphosphates having high chemical and thermal stability may be used as fire retardant additives to thermoplastics, similarly to the same compounds described in very interesting works.[37,41] Apropos polymers' degradation and stability—the synergy between predictable phosphorus fire retardants and

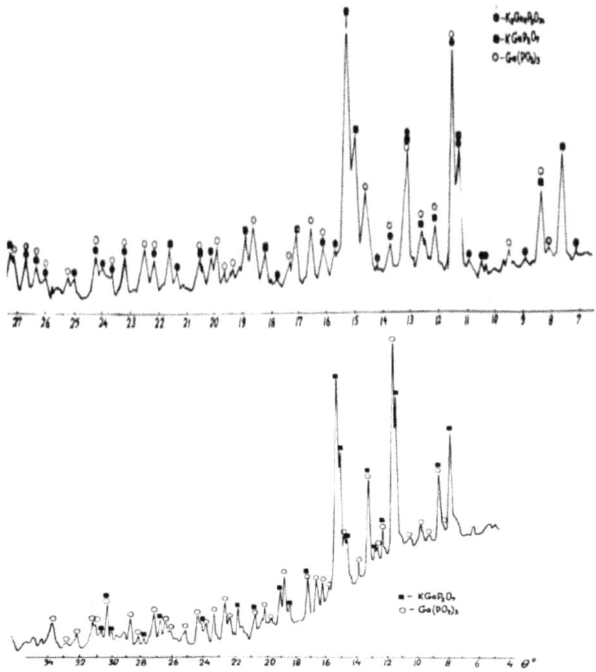

FIGURE 9.4 X-Ray patterns of potassium–gallium triphosphate's $KGaHP_3O_{10}$ heated to 600°C during 1 h (on the left) and his final decomposition by heating up to 650°C during 2 days (on the right).

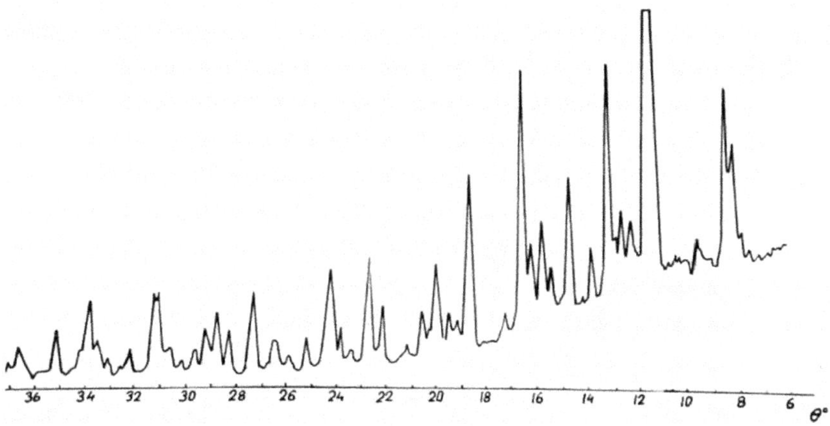

FIGURE 9.5 The diffractogram of the obtained long chain polyphosphate $Ga(PO_3)_3$.

organically modified clays can lead to fire retardancy of styrenics.[37–40] The structure and thermal properties, as well as the vibrational and luminescent properties of condensed compounds determine their use in quantum electronics. The biomaterials appear on the base of hydroxyl apatite and polyphosphates rapidly for the causes of the development of advanced methods of analysis and of the significant application of phosphates materials in several technical domains, including modern nano-science. The marketable requests for condensed phosphates applications are: in the production of fire-retardant and heat-resistant coatings and sealants, in medicine, in high-temperature ceramics, for the encapsulation of new types of fibrous or toxic, and radioactive waste components, and as an additive component for cement and concrete as raw ingredients/component for the production of phosphates glasses. By T. Gourley "...inorganic polymers are a family of materials that only in the last few decades, have been given their rightful place in the spectrum of inorganic materials." The production of crystalline and/or non-crystalline condensed phosphates has triggered their application in quantum electronics, which is predetermined by the specific and improved properties of these compounds. In this point of view, the achievements in the frame of clean energy technologies and synthesis' experimental technique, used by us "method of Tananaev" is very attractive. In fact, this method excludes the harmful impact on the ecology. Based on the above, the production of new environmentally friendly and cost-effective materials based on inorganic polymeric compounds is of great importance. As a final point, we would like to underline that it had been studied in detail the possibilities of application of condensed cyclic octa-phosphates of rubidium–gallium and potassium–gallium. According to the structural analysis, the diameter of the channel in the case of gallium cyclic octa-phosphate is 5.2 angstrom, which is objectively a very good sign. The cyclo-octaphosphates $K_2Ga_2P_8O_{24}$ and $Rg_2Ga_2P_8O_{24}$ synthesized by us were studied for the catalytic activity in the State University, the results showed that it has unique properties as an inorganic polymer and can be used as a best catalyst for organic synthesis reactions, specifically for getting low molecular weight dienic olefins. As we have already indicated the properties of this compound in the model dehydration reaction of n-butyl alcohol by the impulsion method are studied. Based on the piloted experiments, it was founded that the overall conversion was from 52 to 65%. The studied sample is comparable to the activity of the compound BPO_3 (50%) achieved by mixing the initial constituents and exceeded the activity of the zeolite catalyst sample NaZr-A (32%), tested under similar and comparable conditions. It has also been found that the

C_4 olefinic hydrocarbons were definitely and firmly retained by the catalyst under experimental conditions.[42]

9.5 CONCLUSIONS

In aims for searching of sustainable materials for ecological protection of the environment using clean technologies, our fundamental research was conducted in the open poly-component systems M^I_2O-$M^{III}_2O_3$-P_2O_5-H_2O for the purpose to obtain double, triple, polymeric, and substituted phosphates.

We have synthesized and investigated new inorganic polymers, between the temperature range 100 and 500 (600)°C (where M^I –alkali metals K and Rb and M^{III}–Ga, In and Sc). Various double acidic di- $M^IM^{III}(H_2P_2O_7)_2$ and triphosphates $M^IM^{III}HP_3O_{10}$, cyclo-octaphosphates $M^I_2M^{III}_2P_8O_{24}$ and long chain polyphosphates $Ga(PO_3)_3$-C and $Ga(PO_3)_3$-C^I were obtained. All inorganic polymers were studied by diverse methods such as chemical analysis, method of paper chromatography, thermos-gravimetric analysis, roentgen phase's analysis, IR spectroscopy, scanning micro-spectroscopic analysis, and flame-spectrophotometric method.

The dependence of the composition of condensation degree from synthesis temperature T and molar ratio $n = M^I_2O/M^{III}_2O_3$ was studied. Investigation of the thermal behavior of the compound has been demonstrated high thermal stability of obtained phosphates, especially for cyclic compounds. These properties have potential applications in the fire-retardant industry, as additives to thermoplastics, which were already successfully demonstrated for similar phosphates.

Synthesized cyclo-octaphosphates $K_2Ga_2P_8O_{24}$ and $Rg_2Ga_2P_8O_{24}$ have unique properties as inorganic polymers and can be used as a best catalyst for organic synthesis reactions, specifically for getting low molecular weight dienic olefins.

In order to promote Clean Technologies within University, we have organized the Educational and Training Center "RATIO," whose main objectives are: firstly, research and accomplishments in the frame of synthesis of new inorganic polymers-double condensed phosphates, and secondly, the encouragement of researchers conducted their work for Green Environmental Protection. Several training courses have been organized, quite a few educational steps in the ecological field have already taken place, various interesting workshops and international conferences were organized.

ACKNOWLEDGMENTS

We would like to dedicate this modest work to the memory of well-known scientist Professor O. MUKBANIANI, who was much beloved among all our colleagues and whose life's valued work deserves our sincere gratitude. We express our deepest gratitude to Professor Omari for his most important contribution in the field of macromolecular chemistry. We would also like to express our sincere gratefulness and appreciation for his wonderful support to our study for many years. The bright memory of his personality will always remain in our hearts.

KEYWORDS

- **inorganic polymers**
- **condensation**
- **triphosphates**
- **polyphosphates**
- **cyclooctaphosphates**
- **cyclododecaphosphates**
- **environment**

REFERENCES

1. Barroso, J. M. Europe 2020: A Strategy for Smart, Sustainable and Inclusive Growth, European Commission: Brussels, 2020; p 11.
2. Burmeister, M.; Rauch, F.; Eilks, I. Education for Sustainable Development (ESD) and Chemistry Education. *Chem. Educ. Res. Pract.* **2012,** *13* (2), 59– 69.
3. Avaliani, M. A.; Gvelesiani, M. K. Areas of Crystallization of Condensed Scandium and Cesium Phosphates and Regularities of their Formation. *J. Proc. Georgian Acad. Sci. Chem. Ser.* **2006,** *32*, 52–58.
4. Avaliani, M. A. Investigation and Thermal Behavior of Double Condensed Phosphates of Gallium, Scandium and Silver. *Bull. Intern. Nucl. Inform. Syst. INIS-IAEA* **2021,** *52*, 31.
5. Burden J. Twenty First Century Science: Developing a New Science; *Science in School*, 2007; vol *5* (74), p 77.
6. Chandrasekhar, V. Inorganic Polymers. A Review of Synthetic Strategies; Problems and Prospects. In *Inorganic and Organometallic Polymers*; Springer: Berlin, 2005; pp 2–26. https://doi.org/10.1007/3-540-26215-6_1

7. Lalena, J. N.; Cleary, D. A.; Carpenter, E. E.; Dean, N. F. In *Inorganic Materials Synthesis and Fabrication*; John Wiley & Sons, Inc., Springer: New Jersey, Canada, 2008; pp 5–314.
8. Kriven, W. M. Synthesis of LiFePO4 Powder by the Organic–Inorganic Steric Entrapment Method. *J. Mater. Res.* **2015,** *30,* 2133–2143.
9. Ponnusamy, S. K. Clean Technologies for Sustainable Environment. *IET Nanobiotechnol.* **2021,** *15* (2), 147–148. DOI: 10.1049/nbt2.12052.
10. Mukbaniani O.; Londaridze, L.; Markarashvili, E.; Tatrishvili, T.; Aneli, J. In *Triethoxysilylated Styrene as a New Coupling Agent in Wood Composites (Symposium speech)*, Proceedings of the 7th International Caucasian Symposium on Polymers and Advanced Materials - "ICSP&AM 7", 2021; p 73.
11. Averbouch-Pouchot, M. Th.; Durif, A. In *Topics in Phosphate Chemistry*; World Scientific, 1996; pp 174–209.
12. Kvinikadze, N.; Londaridze, L.; Zedgenidze, A.; Dzidziguri, D.; Mukbaniani, O. In *Wood Polymer Composites on the Basis of New Coupling Agent*, Proceedings of the 7th International Caucasian Symposium on Polymers and Advanced Materials - "ICSP&AM 7," 2021; p 60.
13. Grunze, I.; Cudinova, N. N.; Palkina, K. K.; Avaliani, M. A.; Guzeeva, L. S.; Maksimova, S. Structure and Thermal Rearrangements of Binary Cesium-Gallium Phosphates. *Energy Citations Database J. Inorg. Mater.* **2009,** *23* (4), 539–544.
14. Zanello. P. In *Chains, Clusters, Inclusion Compounds, Paramagnetic Labels*, 2nd ed.; Elsevier, 2012; pp 100–141.
15. Davidovits, J. Geopolymers: Ceramic-Like Inorganic Polymers. *J. Ceram. Sci. Technol.* **2017,** *08*, 335–350.
16. Avaliani, M. A.; Todradze, G.; Shapakidze, E.; Ukleba, M. H.; Chikovani, N.; Vibliani, M.; Magradze, G. In *Optimization of the Gravimetric Method for Determination of Trivalent Metals using Oxiquinoline*, Proceedings of the International Conference on Analytical Chemistry-Modern Trends 2020, Kyiv, Ukraune, 2020. http://kcacmt.univ.kiev.ua/en
17. Avaliani, M.; Gvelesiani, M.; Barnovi, N.; Purtseladze, B.; Dzanashvili, D. New Investigations of Poly-component Systems. *J. Proc. Georgian Acad. Sci. Chem. Ser.* **2016,** *42*, 308–311.
18. Avaliani, M.; Main Types of Condensed Phosphates Synthesized in Open Systems from Solution-Melts of Phosphoric Acids. *Nano Stud.* **2016,** *1*, 135–138.
19. Avaliani, M.; Chagelishvili, V.; Shapakidze, E.; Gvelesiani, M.; Barnovi, N.; Kveselava, V.; Esakia, N. Crystallization Fields of Condensed Scandium-Silver and Gallium-Silver Phosphates. *Eur. Chem. Bull.* **2019,** *5,* 164–170.
20. Avaliani, M.; Shapakidze, E.; Barnovi, N.; Dznashvili, D.; Todradze, G.; Kveselava,V.; Gongadze, N. Regio-controlled Synthesis of Double Condensed Oligo-, Poly- and Cyclo- Phosphates, Their Characterization and Possibilities of Applications, Due to their Solid-State Properties. *J. Nano Stud. Eur. Chem. Bull.* **2019,** *II* (19), 273–284.
21. Avaliani, M.; Shapakidze, E.; Barnovi, N.; Gvelesiani, M.; Dzanashvili, D. About New Inorganic Polymers-Double Condensed Phosphates of Silver and Trivalent Metals. *J. Chem. Chem. Eng.* **2017,** *11*, 60–64.
22. Avaliani, M.; Todradze, G.; Shapakidze, E.; Kveselava, V. Synthesis of Condensed Phosphates of Mono- and Polyvalent Metals, Development of the Optimal Methods for

the Study of their Properties and Composition. *J. Nano Studies Eur. Chem. Bull.* **2020,** *20,* 71–94.
23. Avaliani, M.; Chagelishvili, V.; Barnovi, N.; Shapakidze, E.; Esakia, N. Condensed Phosphates: New Inorganic Polymers with a Variety of Applications and Improvement of Their Gravimetric Determination Methods. In *Advanced Materials Polymers and Composites, II*, Chapter 18; Apple Acad. Press: USA, 2021; pp 255–276.
24. Avaliani, M. Some Innovative Results-Oriented Scientific Researches which Lead to the Development in the Field of Inorganic Polymer's Science. In *Materials Science/ Composite Materials Engineering: Modeling and Technology*; Apple Academic Press: USA, 2019; pp 61–72. DOI: 10.13140/RG.2.2.16514.32969
25. Avaliani, M.; Gvelesiani, M.; Barnovi, N.; Purtseladze, B.; Dzanashvili, D. New Investigations of Poly-component Systems. *J. Proc. Georgian Acad. Sci. Chem. Ser.* **2016,** *42*, 308–311.
26. Wagh A. S. Chemically Bonded Phosphate Ceramics – A Novel Class of Polymers. *Ceram. Trans.* **2005,** *165*, 107–116.
27. Avaliani, M.; Shapakidze, E.; Chagelishvili, V.; Todradze, G. In *Investigating the Influence of the Trivalent and Monovalent Metals Ionic Radius on the Structure of the Synthesized Double Condensed Compounds*, Proceedings of IX International Conference on Chemistry and Chemical Education-Sviridov Readings, Minsk, 2021; pp 109–110.
28. Avaliani, M.; Shapakidze, E.; Chagelishvili, V.; Gvelesiani, M.; Barnovi, N.; Vibliani, M.; Chikovani, K. Properties of the Synthesized Inorganic Polymeric Phosphate Materials and the Possibilities for Production of New Environmentally Friendly and Economically Viable Supplies from Local Industrial Waste. *Nano Stud. Eur. Chem. Bull.* **2020,** *20,* 63–70.
29. Tananaev, I. V. Some Aspects of the Chemistry of Phosphates and Their Practical Application. In *Problems of Chemistry and Chemical Technology*, Nauka, 1987 **;** vol *1,* pp 4–65.
30. Tananaev, I. V.; Grunze, I.; Chudinova, N. N. Prior Directions and Results in the Domain on Condensed Phosphates' Chemistry. *J. Neorg. Mater.* **1984,** *20* (6), 887–900.
31. Durif, A. In *Crystal Chemistry of Condensed Phosphates*, 3rd ed.; Plenum Press, 2014 ; pp 5–498.
32. Avaliani, M. A. In *About Development in Chemistry and Application of Condensed Phosphates*, Proceedings of International Conference; GEOHET, TSU : Tbilisi, Georgia, 2011; vol *1,* pp 162–163.
33. Avaliani, M. A.; Shapakidze, E. V.; Chagelishvili, V. A.; Todradze, G. A. In *Investigating the Influence of the Trivalent and Monovalent Metals Ionic Radius on the Structure of the Synthesized Double Condensed Compounds.* Proceedings of IX International Conference on Chemistry and Chemical Education Sviridov Readings, Minsk, 2021; p 109. DOI: 10.13140/RG.2.2.13652.94088
34. Avaliani, M.; Shapakidze, E.; Todradze, G.; Barnovi, N.; Vibliani, M.; Magradze, G. In *Double Condensed Phosphates – As Analogs of Inorganic Polymeric Compounds. Geopolymers-the Bilateral Materials, Reciprocals of Organic and/or Inorganic Polymers*, Proceedings of 7th International Caucasian Symposium on Polymers & Advanced Materials, Tbilisi, Georgia, 2021; p 7
35. Avaliani, M. In *Modern Approach to the Synthesis of Inorganic Polymers, Taking into Account the Prospects for their Application*, Proceedings of International Conference

on Natural and Synthetic Polymers for Medical and Technical Purposes, Minsk, 2022; pp 143–147; DOI: 10.13140/RG.2.2.28245.55528

36. Shapakidze, E. V.; Avaliani, M. A.; Nadirashvili, M.; et al. In *Variations on the Theme-Geopolymers as Inorganic Polymers and Possibilities of their Applications*, Proceedings of International Conference on Natural and Synthetic Polymers for Medical and Technical Purposes, Minsk, 2022; pp 221–224.
37. Abramovich, E. A.; Selevich, A. F. In *Synthesis and Characterization of Ammonium-Vanadium (III) Double Cyclo-Phosphates* $(NH_4)_2V_2P_8O_{24}$ *and* $(NH_4)_3V_3P_{12}O_{36}$, Proceedings of IX International Conference on Chemistry and Chemical Education Sviridov Readings, Minsk, 2021; p 13.
38. Weil, E. D.; Levchik, S. V. In *Flame Retardants for Plastics and Textiles. Practical Applications*; Carl Hansen Verlag: Munich, 2009; pp 2–39.
39. Jabishvili, N.; Klarjeishvili, N.; Karalashvili, I.; Urotadze, S. In *The New Phosphate Composition (antipyrene)- Increasing the Fire Resistance of Various Materials*, 3td Rafiael Agladze Conference RAC-3 on Applied Chemistry, Tbilisi, Georgia, 2011; pp 93–94.
40. Brostow, W.; Cañadas, I.; Fałtynowicz, H.; Levinskas, R.; Garcia, J. R. In *Fire Resistance of Materials*, Proceedings of the 7th International Caucasian Symposium on Polymers and Advanced Materials - "ICSP&AM 7", 2021; p 14.
41. Selevich, A. F.; Abramovich, E. A.; Ivashkevich, O. A. In *Synthesis of Ammonium-Containing Double Oligo-, Cyclo- and Polyphosphates of Trivalent Metals in the Ammonium Polyphosphate Flux*, Proceedings of International Conference on Natural and Synthetic Polymers for Medical and Technical Purposes, Minsk, 2022; pp 212–215.
42. Avaliani, M.; Dzanashvili, D.; Shapakidze, E.; Esakia, N. In *New Inorganic Polymers: Condensed Phosphates and Diverse Spheres of their Applications*, Proceedings POLYCHAR -26, World Forum on Advanced Materials, Tbilisi, Georgia, 2018; p 101. DOI: 10.13140/RG.2.2.16320.58885

CHAPTER 10

Synthesis and Quantum-Chemical Investigation of Some New Types of 1,2-Trans-Glycosides

NELI SIDAMONIDZE, RAMAZ GAKHOKIDZE, RUSUDAN VARDIASHVILI, and LEVAN SAMARGULIANI

Department of Chemistry, Iv.Javakhishvili Tbilisi State University, Tbilisi, Georgia

ABSTRACT

The condensation reaction of α-chloro-2,3,4,6-tetra-O-acetyl-a-D-glucopyranose and α-chloro-2,3,4,6-tetra-O-acetyl-a-D-galactopyranose with 4,4,8,8-tetramethyl-2,3,6,7-dibenzo-9-oxabicyclo-(3,3,1)-nonan-1-N-triptamin-5-ol, 4,4,8,8-tetramethyl-2,3,6,7-dibenzo-9-oxabicyclo-(3,3,1)-nonan-1-N-5-methoxytriptamin-5-ol, and 4,4,8,8-tetramethyl-2,3,6,7-dibenzo-9-oxabicyclo-(3,3,1)-nonan-1-N-4-methyltiazolylethylamino-5-ol in the presence of the silver carbonate catalyst has been studied at the first time. As the experimental results have been obtained corresponding dibenzooxabicyclo-aminocontaining 1,2-*trans*-glucosides, their structure was determined by physico-chemical methods of analysis. In order to justify theoretically the direction of the reactions, quantum-chemical calculations have been performed.

10.1 INTRODUCTION

In the series of derivatives of carbohydrates, a significant number of substances have been found that are used in medicine as medicines for various purposes. Chemical modification of known drugs based on carbohydrates is one of the

Environmental Technology and Sustainability: Physical, Chemical, and Biological Technologies for Environmental Protection. Tamara Tatrishvili, Ann Rose Abraham, & A. K. Haghi (Eds.)

promising areas in the search for new biologically active substances. Modified saccharide derivatives are currently widely used in medicine, for example, as effective antiviral and anticancer drugs. In this regard, glycosyl halides compare favorably among carbohydrates, which serve as starting compounds for the synthesis of various sugar derivatives at the glycosidic center.

Recent studies have shown that the synthesis of low-toxic compounds is becoming relevant in biological and pharmacological studies. Therefore, it is of interest to use carbohydrates to modify some heterocyclic compounds, which can lead to a significant change in the nature of the action of drugs.[1–4]

There is little information in the literature about carbohydrates containing heterocyclic compounds. Nevertheless, these compounds are of certain practical interest for the synthesis of new types of 1,2-trans-glucosides.

10.2 RESULTS AND DISCUSSION

In a number of previous works,[5–7] the condensation reactions of 1-chloro-2,3,4,6-tetra-O-acetyl-α-D-glucopyranose and 1-chloro-2,3,4,6-tetra-O-acetyl-α-D-galactopyranose with heterocyclic amines. The corresponding 1,2-trans-glycosides have been synthesized.

In continuation of these works, we carried out the synthesis of tryptamine-, methyltryptamine-, and 4-methylthiazolyl containing 1,2-trans-glycosides. By condensation of 1-chloro-2,3,4,6-tetra-O-acetyl-α-D-glucopyranose **(1)** and 1-chloro-2,3,4,6-tetra-O-acetyl-α-D -galactopyranose **(2)** with 4,4,8,8-tetramethyl-2,3,6,7-dibenzo-9-oxabicyclo-(3,3,1)-nonan-1-N-tryptamine-5-ol **(3)** 4,4,8,8-tetramethyl-2,3,6,7-dibenzo-9-oxabicyclo-(3,3,1)-nonan-1-N-(5-methoxytryptamine)-5-ol **(4)** and 4,4,8,8-tetramethyl-2,3,6,7-dibenzo-9-oxabicyclo-(3,3,1)-nonane-1-N-(4-methylthiazolylethylamino)-5-ol **(5)** synthesized 5-O-(2,3,4,6-tetra-O-acetyl-β-D-glucopyranosyl)-4,4,8,8-tetramethyl-2,3,6,7-dibenzo-9-oxabi-cyclo-(3,3,1)-nonan-1-N-tryp-tamine **(6),** 5-O-(2,3,4,6-tetra-O-acetyl-β-D-glucopyranosyl)-4,4,8,8-tetramethyl-2,3,6,7-diben-zo-9-oxabicyclo-(3,3,1)-nonan-1-N-5-methoxytryptamine **(7),** 5-O-(2,3,4,6-tetra-O-ace-tyl-β-D-glucopyr-nosyl)-4,4,8,8-tetramethyl-2,3,6,7-dibenzo-9-oxabicyclo-(3,3,1)-nonan-1-N-4-methylthiazolylethylamine **(8),** 5-O-(2,3,4,6-tetra-O-acetyl-β-D-galactopyranosyl)-4,4,8,8-tetramethyl-2,3,6,7-dibenzo-9-oxabicyclo-(3,3,1)-nonane-1-N-tryptamine **(9),** 5-O-(2,3,4,6-tetra-O-acetyl-β-D-galactopyranosyl)-4,4,8,8-tetramethyl-2,3,6,7-dibenzo-9-oxabicyclo-(3,3,1)-nonan-1-N-5-methoxytryptamine(10), 5-O-(2,3,4,6-te-tra-O-acetyl-β-D-galactopyranosyl)-4,4,8,8-tetramethyl-2,3,6,7-dibenzo-9-oxabicyclo-(3,3,1)-nonane-1-N-4-methylthiazolylethyl-amine **(11).**

The reaction was carried out at a temperature of 30–35°C in the presence of a catalyst— freshly prepared silver carbonate in ether solution. Ultimately, the reaction was monitored by thin layer chromatography. The reactions lasted 12–14 h. As a result, 1,2-trans glycosides **(6–11)** were mainly obtained, although the presence of the 1,2-cis isomer was also found in a small amount. The cis–trans isomers were separated on a column.

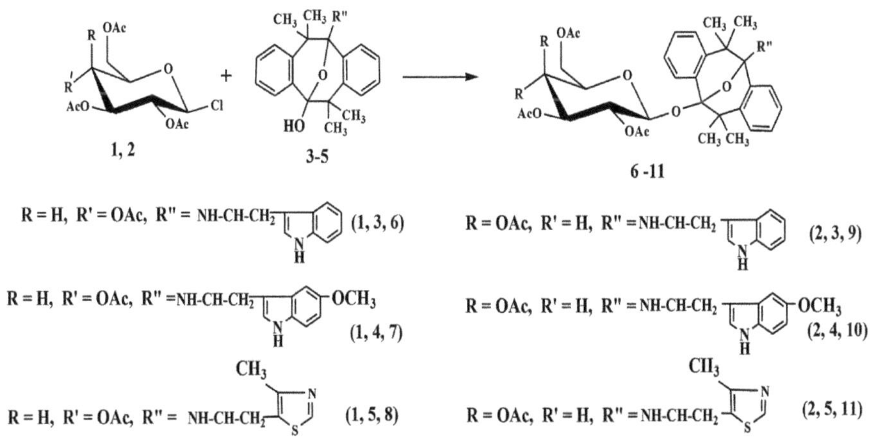

In order to theoretically substantiate the direction of these reactions, we carried out quantum-chemical calculations.

Calculations were performed using CS MOPAC 2000 Version 1.11. Before each calculation by the AM1 method (Austin Model 1), the connection was optimized - energy minimization, both by the molecular mechanics (MM) method and by the quantum-chemical method.[8]

The condensation reaction of 1-chloro-2,3,4,6-tetra-O-acetyl-a-D-glucopyranose **(1)** with 4,4,8,8-tetramethyl-2,3,6,7- dibenzo-9-oxabicyclo-(3,3,1)-nonan-1-N-(4-methylthiazolylethyl-amino)-5-ol **(5).** Two possible pathways of the reaction are considered; with the formation of 1,2-trans-glucoside (structure **I**) and with the formation of 1,2-cis-glucoside (structure **II**).

Determination of the heat of formation of the reaction products showed that the probability of the appearance of the product of the **I** structure, $\Delta H_{form} = -922.77$ kJ/mol (the product of the **II** structure $\Delta H_{form} = 844.88$ kJ/mol), prevails, which is confirmed by the data of PMR spectroscopy. In the PMR spectra of the obtained compounds **8** and **11,** the signals of the anomeric proton H-1 bound to C-1 are located in the region d = 4.45–5.60 and splitting as a result of interaction with hydrogen atoms at C-2 into two lines with a spin–spin constant interactions J1,2 = 8.0 Hz. This value is typical for the axial–axial arrangement of the interacting atoms (1,2-trans-glycosides).

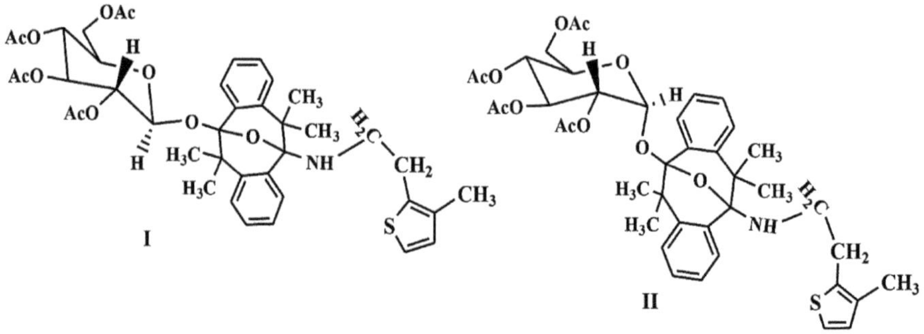

I

II

The mechanism for obtaining 1,2-trans-glycosides from α-chloro derivatives **(1,2)** of sugars can be explained as follows: the halogen atom in acyl halogenoses is easily subjected to nucleophilic substitution. Nucleophilic substitution at the glycosidic center can proceed both by the SN1 mechanism and by the SN2 mechanism. In accordance with the laws of stereochemistry of these reactions, partial or complete racemization at the glycosidic center or configuration reversal can occur simultaneously with substitution. Nucleophilic substitution in derivative cyclic forms of sugars proceeds with the participation of the neighboring acyloxy group. Reversal condensation at C1 results in 1,2-trans glycosides. The resulting glycosyl cation **(12)** is immediately stabilized by an intramolecular nucleophilic attack of the ester group at C2 to form a cyclic ion **(13).** Attack of the alcohol on the glycosidic center of this ion leads to 1,2-trans-glycosides **(14).** The reaction proceeds without isolation of anomeric mixtures (stereospecifically), due to stereochemical control due to the participation of the neighboring acyloxy group:

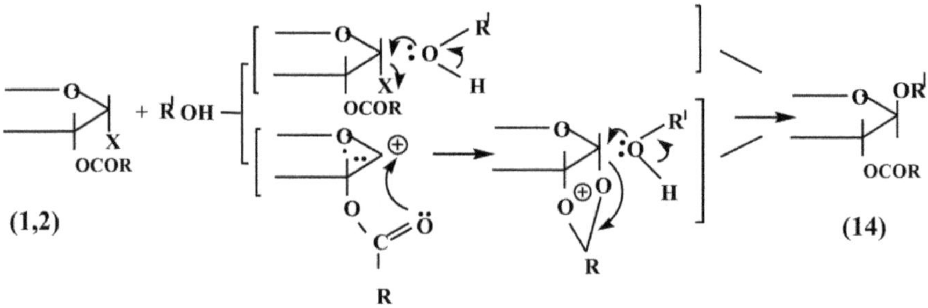

(1,2)

(12)

(13)

(14)

The structure of the synthesized compounds was also confirmed by the data of IR and 13C NMR and 1H NMR spectrometric analysis.

In the 13C NMR spectrum, signals of carbon atoms of gem-dimethyl groups are observed with a shift of 24.34 and 24.13 ppm ($2C(CH_3)_2$ and 46.52 and 44.74 ppm ($2C(CH_3)_2$; 45.22–156 ppm and 46.9–150.7 ppm (C1-8 cyclooctal ring), 125.2, 121.0, 127.2, 108.4 (C1'-4' benzene ring, ppm I), 118.8, 122.43, 116.05, ppm 120.22 (C1"-4" benzene ring II); 41.2 and 49.6 ppm (NH–CH_2-); 32.7 and 18.6 ppm (NH–CH_2–CH_2-), 159.7 and 160.3 ppm (-C=C-), and 70.02 and 66.8 ppm (-N=CH-), etc.

^{13}C NMR spectra of some synthesized compounds are presented in the table below.

Compound	Chemical shifts (CDCI3), δ, ppm
8	95.5 (C-1), 78.5 (C-2), 75.9 (C-3), 72.3 (C-4), 70.2 (C-5), 63.0 (C-6), 167-171.5 ($4\underline{C}OCH_3$), 12.2–21.4 ($4CO\underline{C}H_3$), 46.52 и 44.74 ($2\underline{C}(CH_3)_2$), 24.34 и 24.13 ($2C(\underline{C}H_3)_2$ (gem-dimethyl group), 156.8, 138.5, 59.8, 45.4, 153.2, 133.6, 64.2, 45.22 ($C_{1\text{-}8}$ cyclooctal ring), 124.39, 120.28, 133.93,113.87 ($C_{1'\text{-}4'}$ benzene ring I), 120.61, 129.69, 119.05, 125.73 ($C_{1''\text{-}4''}$ benzene ring II), 49.6 (NH–$\underline{C}H_2$-), 32.7 (NH–CH_2–$\underline{C}H_2$-), 161,6 и 163.63 (-$\underline{C}$=$\underline{C}$-), 70.2 (-N=$\underline{C}$H-).
9	104.8 (C-1); 77.4 (C-2); 79.2 (C-3); 74.0 (C-4); 69.6 (C-5); 61.2 (C-6); 169-173.5 ($4\underline{C}OCH_3$); 18.24.6 ($4CO\underline{C}H_3$); 46.52 и 44.74 ($2\underline{C}(CH_3)_2$); 24.34 и 24.13 ($2C(\underline{C}H_3)_2$ gem-dimethyl group)); 155.8; 140.2; 63.4; 46.5; 153.2; 132.5; 64.8; 43.54; ($C_{1\text{-}8}$ cyclooctal ring); 124.3; 120.2; 133.9; 112.8 ($C_{1'\text{-}4'}$ benzene ring I); 120.6; 129.9; 118.05; 125.7 ($C_{1''\text{-}4''}$ ring II), 40.6 (NH–$\underline{C}H_2$-); 23.4 (NH–CH_2–$\underline{C}H_2$-); 127.0; 119.7; 119.3; 122.1; 118.8; 135.5 (indole benzene ring).
10	95.5 (C-1), 78.5 (C-2), 75.9 (C-3), 72.3 (C-4), 70.2 (C-5), 63.0 (C-6), 167-171.5 ($4\underline{C}OCH_3$), 12.2-21.4 ($4CO\underline{C}H_3$), 46.52 и 44.74 ($2\underline{C}(CH_3)_2$), 24.34 и 24.13 ($2C(\underline{C}H_3)_2$ (gem-dimethyl group) 156.8, 138.5, 59.8, 45.4, 153.2, 133.6, 64.2, 45.22 ($C_{1\text{-}8}$ cyclooctal ring), 124.39, 120.28, 133.93, 113.87 ($C_{1'\text{-}4'}$ benzene ring I), 120.61, 129.69, 119.05, 125.73 ($C_{1''\text{-}4''}$ benzene ring II), 49.6 (NH–$\underline{C}H_2$-), 32.7 (NH–CH_2–$\underline{C}H_2$-), 161,6 и 163.63 (-$\underline{C}$=$\underline{C}$-), 70.2 (-N=$\underline{C}$H-).
11	$4\underline{C}OCH_3$), 16.5-22.6 ($4CO\underline{C}H_3$), 46.52 и 44.74 ($2\underline{C}(CH_3)_2$), 24.34 и 24.13 ($2C(\underline{C}H_3)_2$ 105.2 (C-1), 78.2 (C-2), 79.1 (C-3), 74.5 (C-4), 68.7 (C-5), 62.0 (C-6), 165.2–170.4 (gem-dimethyl group), 150.7, 124.2, 63.0, 49.4, 148.8, 130.2, 61.0, 46.9 ($C_{1\text{-}8}$ cyclooctal ring), 125.2, 121.0, 127.2, 108.4 ($C_{1'\text{-}4'}$ benzene ring I), 118.8, 122.4, 116.05, 120.22 ($C_{1''\text{-}4}$ benzene ring II "), 41.2 (NH–$\underline{C}H_2$-), 18.6 (NH–CH_2–$\underline{C}H_2$-), 159,7 и 160.3 (-$\underline{C}$=$\underline{C}$-), 66.8 (-N=$\underline{C}$H-).

EXPERIMENTAL PART

Optical rotation was studied on an SU-3 universal saccharimeter at 20 ± 2°C. The purity of the compounds obtained and the R_f values were determined on Silufol UV-254 using solvent systems: chloroform–methanol 19:1 (system a); chloroform–methanol 3:2 (system b); chloroform–methanol 3:1 (system c). Substances were detected by spraying developers: sulfuric acid–water 95:5; aniline phthalate. The IR spectra of the samples were obtained on a UR-20 spectrometer in tablets with KBr. ^{13}C NMR spectra were recorded on a Bruker AM-300 spectrometer (75.5 and 300 MHz) in, $CDCI_3$ and PMR spectra were recorded on a Bruker WM-250 spectrometer (250 MHz, $CDCI_3$), internal standard TMS.

The synthesis of 1-chloro-2,3,4,6-tetra-O-acetyl-α-D-gluco(galacto) pyranose **(1,2)** is described in,[13] while 4,4,8,8-tetramethyl-2,3,6,7-dibenzo-9-oxabicyclo-(3,3,1)-nonan-1-N-(4-methyl-thiazolylethylamino)-5-ol (3) – obtained according to the procedure.[14]

5-O-(2,3,4,6-tetra-O-acetyl-β-D-glucopyranosyl)-4,4,8,8-tetramethyl-2,3,6,7-dibenzo-9-oxabi-cyclo-(3,3,1)-nonane-1-N-tryptamine (6). Take 0.09 g of freshly prepared carbonate silver. The reaction was carried out in a nitrogen atmosphere with constant stirring for 12 h (30–35C). After filtering and evaporating the filtrate, the remaining syrup was dissolved in chloroform, treated with activated charcoal and again evaporated in vacuo. After separation on a column (system a, silicagel L 50/100), a chromatographically pure product was obtained with a yield of 0.87 g (55.7%), R_f 0.58 (system a). m.p. 163–164°C, $[\alpha]_D^{17}$ + 31.3° (c 0.6, $CHCI_3$). Found, %: C 67.28; H 5.95; N 3.22; $C_{44}H_{50}O_{11}N_2$. Calculated, %: C 67.51; H 6.39; N 3.58; **IR spectrum (ν, cm−1):** 948–1320 (gemm-dimethyl group), 1010, 1220 (C–O–C); 2920(NH); 1726 (C=O); 580 (C-Harom); 3315 (indole NH groups). **NMR spectrum (δ, ppm, J/Hz), TMS:** 10.7 (1H, s, N<u>H</u>-tryptamine); 7.5–6.8 (12H, m, aromatic protons); 3.0–2.5 (4H, m, NH–$C\underline{H}_2$–$C\underline{H}_2$); 1.32 and 1.41 (12H, s, gem-dimethyl groups, 4$C\underline{H}_3$); 4.43 (1H, H-1, d, $J_{1,2}$ = 8.0,); 5.05 (1H, H-2, dd, $J_{2,1}$ = 8.0, $J_{2,3}$ = 9.5,); 3.75 (1H, H-3, dd, $J_{3,2}$ = 9.5, $J_{3,4}$ = 3.0,); 5.20 (1H, H-4, dd, $J_{4,3}$ = 3.0, $J_{4,5}$ = 9.5); 3.84 (1H, H-5, dd, $J_{5.4}$ = 9.5, $J_{5.6'}$ = 5.0, $J_{5,6''}$ = 2.5); 4.11

(1H, H-6', dd, $J_{6',6''}$ = 12, $J_{6',5}$ = 2.5, C$\underline{H}_2$OCOCH_3); 4.22 (1H, H-6'', dd, $J_{6,'6''}$ = 12, $J_{6'',5}$ = 5.0, CH_2OCOC$\underline{H}_3$); 2.21, 2.10, 1.98, 1.93 (12H, m, 4CO–C$\underline{H}_3$); 3.55 (1H, s, N$\underline{H}$).

5-O-(2,3,4,6-tetra-O-acetyl-β-D-glucopyranosyl)-4,4,8,8-tetramethyl-2,3,6,7-dibenzo-9-oxabi-cyclo- (3,3,1)-nonane-1-N-5-methoxytryptamine (7). In a similar way, upon the condensation of 0.733 g (0.002 mol) of 1-chloro-2,3,4,6-tetra-O-acetyl-α-D-glucopyranose in 50 mL of dry ether and 1.17 g (0.0025 mol) of compound 4 in the presence of 0.09 g of freshly prepared silver carbonate, in a nitrogen atmosphere with constant stirring for 14 h (30–35°C), compound **7** was synthesized with a yield of 0.76 g (46.8%). R_f 0.57 (system c). m.p. 159–160°C, $[\alpha]_D^{17}$ + 37.4° (c 0.47, $CHCl_3$) Found, %: C 66.38; H 6.11; N 3.14. $C_{45}H_{52}O_{12}N_2$. Calculated, %: C 66.50; H 6.40; N 3.44.

IR spectrum (ν, cm−1): 1375–1390 (gemm-dimethyl group), 1030–1160 (C–O–C), 2915 (NH), 1710 (C=O), 1430 (CH_3), 590 (C-Harom). **NMR spectrum (δ, ppm, J/Hz), TMS:** 10.7 (1H, s, N$\underline{H}$-tryptamine); 7.5–6.8 (12H, m, aromatic protons); 3.0–2.5 (5H, m, NH–C$\underline{H}_2$–C$\underline{H}_2$); 1.32 and 1.41 (12H, s, gemm-dimethyl groups, 4C$\underline{H}_3$); 4.46 (1H, H-1, d, $J_{1,2}$ = 8.0); 5.09 (1H, H-2, dd, $J_{2,1}$ = 8.0, $J_{2,3}$ = 9.5); 3.70 (1H, H-3, dd, $J_{3,2}$ = 9.5, $J_{3,4}$ = 3.0); 5.13 (1H, H-4, dd, $J_{4,3}$ = 3.0, $J_{4.5}$ = 9.5); 3.84 (1H, H-5, dd, $J_{5,4}$ = 9.5, $J_{5,6'}$ = 5.0, $J_{5,6''}$ = 2.5); 4.11 (1H, H-6', dd, $J_{6',6''}$ = 12, $J_{6'5}$ = 2.5, C$\underline{H}_2$OCOCH_3); 4.20 (1H, H-6'', dd, $J_{6',6''}$ = 12, $J_{6'',5}$ = 5.0, CH_2OCOC$\underline{H}_3$); 2.08, 2.02, 1.98, 1.96 (12H, m, 4CO–C$\underline{H}_3$); 3.20 (1H, s, N$\underline{H}$).

5-O-(2,3,4,6-tetra-O-acetyl-β-D-glucopyranosyl)-4,4,8,8-tetramethyl-2,3,6,7-dibenzo-9-oxa-bicyclo- (3,3,1)-nonane-1-N-4-methylthiazolylethylamine (8). Take 0.11 g of freshly prepared carbonate silver. The reaction was carried out in a nitrogen atmosphere with constant stirring for 13 h (30–35C). After filtering and evaporating the filtrate, the remaining syrup was dissolved in chloroform, treated with activated charcoal and again evaporated in vacuo. After separation on a column (system c, silicagel L 50/100), a chromatographically pure product was obtained with a yield of 1.00 g (65.44%), R_f 0.64

(system a). m.p. 152–153.5°C, $[\alpha]_D^{18}$ + 31.6° (c 0.52, $CHCI_3$). Found, %: C 62.45; H 6.15; N 3.34; S 3.95. $C_{40}H_{48}O_{11}N_2S$. Calculated, %: C 62.82; H 6.28; N 3.66; S 4.18. **IR spectrum (ν, cm−1):** 1370–1390 (gemm-dimethyl group), 1050–1180 (C–O–C), 2910 (NH), 1725 (C=O), 680 (C–S–C), 2977, 1340 (CH_3), 580–620 (C-Harom). **NMR spectrum (δ, ppm, J/Hz), TMS:** 4.45 (1H, d, $J_{1,2}$ = 8.0, H-1); 5.0 (1H, dd, $J_{2,1}$ = 8.0, $J_{2,3}$ = 9.5, H-2); 3.72 (1H, dd, $J_{3,2}$ = 9.5, $J_{3,4}$ = 3.0, H-3); 5.20 (1H, dd, $J_{4,3}$ = 3.0, $J_{4,5}$ = 9.5, H-4); 3.88 (1H, ddd, $J_{5,4}$ = 9.5, $J_{5,6'}$ = 5.0, $J_{5,6''}$ = 2.5, H-5); 4.14 (1H,H-6', dd, $J_{6',6''}$ = 12, $J_{6',5}$ = 2.5, $C\underline{H}_2OCOCH_3$); 4.20 (1H, H-6", dd, $J_{6'.6''}$ = 12, $J_{6'',5}$ = 5.0, $CH_2OCOC\underline{H}_3$); 2.21, 2.10, 1.98, 1.93 (12H, m, 4CO–CH_3); 9.98 (1H, s, =C$\underline{H}$ of thiazoline); 7.15–7.6 (8H, m, aromatic protons); 3.2–2.2 (4H, m, NH–$C\underline{H}_2$–$C\underline{H}_2$); 3.35 (1H, s, NH); 1.44 and 1.47 (12H, s, gemm-dimethyl groups), 4$C\underline{H}_3$).

The derivatives of β-D-galactopyranose were synthesized in a similar way.

5-O-(2,3,4,6-tetra-O-acetyl-β-D-galactopyranosyl)-4,4,8,8-tetramethyl-2,3,6,7-dibenzo-9-oxabicyclo- (3,3,1)-nonane-1-N-tryptamine (9). Yield 0.79 g (50.5%). m.p.161–162.5°C, R_f 0.59 (system c), [α $_D^{16}$ + 37.4° (c 0.47, $CHCI_3$). Found, %: C 67.92; H 6.61; N. 3. 67. $C_{44}H_{50}O_{11}N_2$. Calculated,%: C 67.51; H 6.39; N 3.58; **IR spectrum (ν, cm−1):** 1365–1380 (gemm-dimethyl group), 1020–1230 (C-O-C); 2920(NH); 1720 (C=O); 2977, 1320 (CH_3); 1610(C=Carom); 3330 (NH indole).

5-O-(2,3,4,6-tetra-O-acetyl-β-D-galactopyranosyl)-4,4,8,8-tetramethyl-2,3,6,7-dibenzo-9-oxabicyclo- (3,3,1)-nonane-1-N-5-methoxytryptamine (10). Yield 0.72 g (44.5%). R_f 0.37 (system b). m.p. 149–150°C, $[\alpha]_D^{19}$ + 40° (c 0.37, $CHCI_3$) Found,%: C 66.21; H 6.78; N 3.83. $C_{45}H_{52}O_{12}N_2$. Calculated,% : C 66.50; H 6.40; N 3.44;

5-O-(2,3,4,6-tetra-O-acetyl-β-D-galactopyrano-zyl)-4,4,8,8-tetramethyl-2,3,6,7-dibenzo-9- oxabicyclo-(3,3,1)-nonan-1-N-4-methyl-thiazolylethylamine (11). In a similar way, with the condensation of 0.733 g (0.002 mol) of 1-chloro-2,3,4,6-tetra-O-acetyl-α-D-galactopyranose in 50 mL of dry ether and 1.08 g (0.0025 mol) of compound 3 in compound 5 was synthesized in the presence of 0.11 g of freshly prepared silver carbonate under nitrogen with constant stirring for 14 h (30–35°C) with a yield of 0.9 g (58.9%). R_f 0.52 (system b). m.p. 137–139°C, $[\alpha]_D^{16}$ + 39.5° (c 0.61, $CHCI_3$) Found, %: C 63.27; H 6.04 N 3.29; S 3.63. $C_{40}H_{48}O_{11}N_2S$. Calculated, %: C 62.82; H 6.28; N 3.66; S 4.18. **IR spectrum (ν, cm-1):** 1370–1380 (gemm-dimethyl group), 1040–1120 (C–O–C), 2923 (NH), 1715 (C=O), 720 (C–S–C), 2854, 1627 (CH_3), 580–600 (C-Harom).

NMR spectrum (δ, ppm, J/Hz), TMS: 5.60 (1H, d, $J_{1,2}$ = 8.0, H-1), 5.28 (1H, dd, $J_{2,1}$ = 8.0, $J_{2,3}$ = 9.5, H-2), 4.95 (1H, dd, $J_{3,2}$ = 9.5, $J_{3,4}$ = 3.0, H-3), 5.30 (1H, dd, $J_{4,3}$ = 3.0, $J_{4,5}$ = 9.5, H- 4), 3.94 (1H, ddd, $J_{5,4}$ = 9.5, $J_{5,6'}$ = 5.0, $J_{5,6''}$ – 2.5, H-5), 4.12 (1H, H-6', dd, $J_{6',6''}$ = 12, $J_{6',5}$ = 2.5, $C\underline{H}_2OCOCH_3$), 4.20 (1H, H-6", dd, $J_{6',6''}$ = 12, $J_{6'',5}$ = 5.0, $CH_2OCOC\underline{H}_3$), 2.08, 2.10, 1.98, 1.93 (12H, m , 4CO-CH_3), 8.90 (1H, s, =CH thiazoline), 7.1–7.4 (8H, m, aromatic protons), 3.0–2.4 (4H, m, NH-$C\underline{H}_2$-$C\underline{H}_2$), 3.30 (1H, s , NH), 1.43 and 1.45 (12H, s, gemm-dimethyl groups), 4CH3)

KEYWORDS

- **4-Methylthiazolylaminoderivatives**
- **triptamin**
- **5-methoxytriptamin**
- **silver carbonate**
- **1,2-trans-glucosides**

REFERENCES

1. Kren, V.; Martínková, L. *Curr. Med. Chem.* **2001,** *11*, 1303–1328.
2. Heinrich, M.; Barnes J.; Gibbons, S.; Williamson, E. M. In *Fundamentals of Pharmacognosy and Phytotherapy*, 1st ed.; Churchill Livingstone: Edinburgh, 2004.
3. Kochetkov, N. K.; Bochkov, A. F.; Dmitriev, B. A.; Usov, A. I.; Chizhov, O. S.; Shibaev, V. N. In *Carbohydrate Chemistry*; Moscow, 1967.
4. Lonas, G.; Stadler, R. *Acta Polym.* **1994,** *45*, 14.
5. Sidamonidze, N. N.; Gakhokidze, R. A.; Ramishvili, M. A.; Samsonia, T. G. *Georgian Chem. J.* **2007,** *7* (1), 44–48.

6. Sidamonidze, N. N.; Vardiashvili, R. O; Gverdtsiteli, M. I.; Gakhokidze, R. A. *Chem. Natl. Comp.* **2009,** *45* (4), 496–499.
7. Sidamonidze, N.; Vardiashvili, R.; Isakadze, M.; Djaniashvili, L. *Pharm. Chem. J.* **2007,** *41* (3), 131–134.
8. Dewar, M. J. S.; Zoebisch, E. G.; Healy, E. F.; Stewart, J. J. P. *J. Am. Chem. Soc.* **1985,** *107,* 3902–3909.
9. Rouvray, P. H. In *Chemical Application of Topology and Graph Theory*; Elsvier Science Publ.: Amsterdam, 1983.
10. Gamziani, G.; Gverdtsiteli, M. Isomery Phenomenon from Point of View of Mathematical Chemistry; Metsniereba, Tbilisi, 1992.
11. Gverdtsiteli, M.; Gamziani, G.; Gverdtsiteli, I. The Contiguity Matrices of Molecular Graphs and their Modifications; Tbilisi University Press: Tbilisi, 1996.
12. Kupatadze, K.; Lobzhanidze, T.; Gverdtsiteli, M. Algebric Chemical Investigation of some Organic Moleculs and their Transformations; Publishers Universal: Tbilisi, 2007.
13. Freudenberg, K.; Soff, K. *Chem. Ber.* **1936,** *69*, 1245–1247.
14. Lagidze, R. M.; Iremadze, N. G.; Kiriakidi, A. S. *Bull. Georg. Acad. Sci.* **2003,** *4*, 3–8.

CHAPTER 11

Preparation and Properties of Paper Containing Bactericidal Zeolite Adsorbents

VLADIMER TSITSISHVILI, NANULI DOLABERIDZE, NATO MIRDZVELI, MANANA NIJARADZE, ZURAB AMIRIDZE, and BELA KHUTSISHVILI

Petre Melikishvili Institute of Physical and Organic Chemistry, Tbilisi State University, Tbilisi, Georgia

ABSTRACT

In this chapter, silver-, copper-, and zinc-containing microporous materials have been composed on the basis of heulandite-clinoptilolite-containing tuff rock from the Rkoni plot of the Tedzami deposit (Eastern Georgia) using ion-exchange reactions between zeolite microcrystals (250 BBS mesh) and a salt of a comparable transition metal ($AgNO_3$, $CuCl_2$, or $ZnCl_2$) in the solid phase, accompanied by washing using distilled water. Obtained adsorbent-ion-exchangers reside in the zeolite crystal structure, contain 130 mg/g of silver, 63–72 mg/g of copper, and 30–58 mg/g of zinc, and show bacteriostatic activity against a wide range of microorganisms, while the mixtures of different forms are synergistic and have higher activities than individual materials. Filled papers containing up to 3.0, 1.6, and 1.4 mg/g of silver, copper, and zinc, respectively, were manufactured in a paper mill of GPM company, which produces paper and three-layer corrugated paperboard from recycled waste. It was concluded that the application of zeolite fillers leads to a change in the grammage, caliper, density, tensile, and stretching of the paper, and these characteristics of paper filled with metal-containing bactericidal zeolites are highly dependent on the nature of the filler and the content of metals in it. The introduction of zeolite fillers holding divalent metals brings

Environmental Technology and Sustainability: Physical, Chemical, and Biological Technologies for Environmental Protection. Tamara Tatrishvili, Ann Rose Abraham, & A. K. Haghi (Eds.)

about a notable change in the surface properties. This brings some remarkable changes in our experimental samples with a certain copper content as well. On the other hand, bacteriostatic activity testing by the colony-forming unit assay specifies that all metal-containing samples are active against gram-positive bacterium *Staphylococcus aureus*, and zinc-containing papers are more active against gram-negative bacterium *Escherichia coli* than those containing silver. These results are very decisive regarding the practical aspects, since they open up the potentiality of replacing expensive silver with cheaper copper and zinc. For the production of paper active against *E. coli*, a filler with a zinc content of 1.4 mg/g is recommended, and for the production of waterproof paper active against *Staphylococcus*, a filler with a copper content of about 1.6 mg/g is recommended.

11.1 INTRODUCTION

On the one hand, the current interest in bactericidal materials based on zeolites is due to the coronavirus pandemic, and on the other hand, due to the properties of zeolites (aluminosilicates of the general formula $M_x[Al_xSi_yO_{2(x+y)}]\cdot mH_2O$), in which ions of alkali (Na^+, K^+, etc.) or alkaline-earth (½Ca^{2+}, ½Mg^{2+}, etc.) metals are at least partially replaced by bioactive metal ions (Ag^+, Cu^{2+}, Zn^{2+}, etc.).[1] Studies that started at the beginning of 2001, continuing to this day and reflected in numerous publications, have shown that synthetic and natural zeolites enriched with bioactive metals exhibit antimicrobial activity against a wide range of microorganisms. Most of the research concerned the study of silver-containing zeolites, and based on the experimental result of recent works on silver-, copper-, and zinc-containing zeolites, it was concluded that the silver-containing zeolites are the most active. For example, silver-containing natural clinoptilolite and synthetic zeolite A are characterized by highest antibacterial activity against different isolates of gram-negative bacteria *Escherichia coli*, while copper-containing, and in particular, zinc-containing forms are inferior to them in activity.[2] Zeolite-rich tuff from the State of Chihuahua, Mexico modified with Ag^+ or Cu^{2+} ions showed microbicidal effect against *E. coli* and a yeast *Candida albicans*, but the action of Cu-form was 20–30 times weaker than of Ag-form.[3]

However, there are some disbenefits regarding the application of silver ions. These weaknesses are as follows:

Silver is not a cheap metal, and Ag^{+} is not stable in aqueous solutions; according to Ref. [4], it tends to be reduced to Ag^{0} and reacts with sulfate and other anions, forming insoluble salts. Therefore, the advantages of silver ions are not quite obvious in some cases. Thus, Zn^{2+} and Cu^{2+}-loaded samples of synthetic zeolite X proved desirable antimicrobial activities against three bacteria—gram-negative *E. coli* and *Pseudomonas aeruginosa*, gram-positive *S. aureus*, and a yeast *C. albicans* and a fungus *Aspergillus niger*.[5] According to the recent results,[6] Cu-X is more active against *S. aureus* and Zn-X against *E. coli*, although the zinc form of synthetic zeolite A is inactive.[2] For the preparation of bactericidal materials (while using natural zeolites), heulandite-clinoptilolite of various origins was utilized, and results for bioactive metal-enriched clinoptilolite were also inconsistent. On the one hand, it was found that the diameters of the zones of inhibition of the growth of *E. coli* by the Ag-, Cu-, and Zn-forms of clinoptilolite from Gördez, Turkey, are approximately the same; the maximum zone by Ag^{+} ions is 14 mm, by Cu^{2+} is 13 mm, and by Zn^{2+} is 12 mm.[7] On the other hand, for clinoptilolite from Mare Baia, Romania, "activated" by Cu^{2+} and Zn^{2+} ions and tested for antimicrobial activity against *E. coli*, *S. aureus*, and *C. albicans*, only the copper form and only against *E. coli* appeared to be active.[8]

Recently, we prepared and studied silver-, copper-, and zinc-containing phillipsites,[9] analcimes,[10] and heulandite-clinoptilolites.[11,12] Modified heulandite-clinoptilolites contain over 130 mg/g of silver, 70 mg/g of copper, and 55 mg/g of zinc. They present dynamic bacteriostatic activity against gram-negative bacterium *Escherichia coli*, gram-positive bacteria *Staphylococcus aureus* and *Bacillus subtilis*, fungal pathogenic yeast *Candida albicans*, and a fungus *Aspergilus niger*. Their mixtures reveal a synergistic effect—the silver form with the addition of copper and zinc forms is the most active against *Staphylococcus* and other microorganisms, and mixtures of copper and zinc forms are the most productive.

The purpose of our work was to prepare paper samples with heulandite-clinoptilolite bactericidal fillers and study their properties. The choice of heulandite-clinoptilolite of sedimentary origin from a wide range of zeolites of Georgia as a suitable raw material for the preparation of bactericidal fillers was made due to its properties—a high content of the zeolite phase and exchangeable ions of alkali and alkaline earth metals. In addition, it was taken into account that the legislation of the European Union recognizes clinoptilolite of sedimentary origin as a fodder additive for all animal species, which indicates its complete safety. The preliminary results of our research have been published.[13–15] This chapter summarizes and supplements these results.

In papermaking, zeolites have been in the application as fillers mainly to enhance optical properties, bulkiness, printability, and to augment retention at the wet end since the 1960s,[16] but according to Ref. [17,18], some studies on the introduction of germicidal zeolite fillers into paper have only started in recent years.

11.2 EXPERIMENTAL METHODS AND MATERIALS

11.2.1 ZEOLITE FILLERS

Heulandite-clinoptilolite-containing tuff rock from the Rkoni plot of the Tedzami deposit (Eastern Georgia) with zeolite phase content of 90% was used as a starting material.

Ion-exchange modification included the following steps: mixing of zeolite and salt ($AgNO_3$, $CuCl_2$, or $ZnCl_2$) powders (250 BBS mesh), grinding of mixtures in an agate mortar, and washing of mixtures transferred to a filter with distilled water until the complete disappearance of nitrate or chlorine anions, after which the modified samples were first dried in air and then at 100–105°C in a thermostat; details of the ion-exchange procedure are given in our recent publication.[11,15]

The chemical composition of native and modified samples described by averaged empirical formulas of dehydrated zeolites $(Na_aK_bCa_cMg_d)[AlSi_xO_{72}]$ calculated from the X-ray energy dispersive spectroscopy data obtained using an Oxford Instruments X-Max 20 analyzer, as well as calculated scientific ion-exchange capacity of samples are given in Table 11.1.

TABLE 11.1 Chemical Composition of Native and Modified Zeolites.

Sample	Empirical formula	Biometal content		Scientific ion-exchange capacity, meqv/g
		mg/g	mmol/g	
HCR	$(Na_{0.25}K_{0.06}Ca_{0.19}Mg_{0.15})Me_{0.08}[AlSi_{3.6}O_{72}]$			3.03
AgHCR	$Ag_{0.46}(Na_{0.11}K_{0.06}Ca_{0.10}Mg_{0.09})[AlSi_{3.6}O_{72}]$	130	1.2	3.03
$CuHCR_1$	$Cu_{0.36}(Na_{0.08}K_{0.06}Ca_{0.03}Mg_{0.04})[AlSi_{3.9}O_{72}]$	63	1.0	2.87
$CuHCR_2$	$Cu_{0.42}(Na_{0.03}K_{0.05}Ca_{0.02}Mg_{0.02})[AlSi_{4.1}O_{72}]$	72	1.14	2.78
$ZnHCR_1$	$Zn_{0.16}(Na_{0.10}K_{0.06}Ca_{0.13}Mg_{0.13})[AlSi_{3.65}O_{72}]$	30	0.46	3.00
$ZnHCR_2$	$Zn_{0.32}(Na_{0.07}K_{0.05}Ca_{0.05}Mg_{0.07})[AlSi_{3.9}O_{72}]$	58	0.89	2.87

Me—impurities (Fe, etc.).

Bacteriostatic properties of zeolite samples determined by the disk diffusion (Kirby–Bauer) method in standard conditions (details are given in Ref. [11–13]) using the cultures of different microorganisms are expressed in diameters of the zones of inhibition of their growth (see Table 11.2).

TABLE 11.2 Diameters (mm) of the Zones of Inhibition of the Growth of Microorganisms by Native and Modified Heulandites.

Microorganism	Sample					
	HCR	AgHCR	$CuHCR_1$	$CuHCR_2$	$ZnHCR_1$	$ZnHCR_2$
Escherichia coli (strain ATTC 8739)	0	21	15	36	0	35
Staphylococcus aureus (ATTC 6538)	0	19	19	53	0	52
Bacillus subtilis (ATTC 6633)	0	30	21–36[a]	41	19	42
Candida albicans (ATTC 10231)	0	20.5	15	54	15[a]	61
Aspergillus niger (ATTC 16404 – *A. brasiliensis*)	0	25	14[a]	65	17	69

[a]Secondary growth.

11.2.2 PREPARATION OF FILLED PAPER

The filled papers were fabricated on the production line of the paper mill of the GPM company (2 Akhvlediani str., line 10, Tbilisi, Georgia), which produces paper and three-layer corrugated paperboard from recycled waste paper. Usually fillers are added to wet paper to distribute them evenly throughout the volume, but to impart bactericidal properties to the paper surface, introduction of silver-, copper-, and zinc-containing zeolite fillers was carried out on almost dry paper, throughout coating, by including a zeolite suspension in a binder of cooked starch. It has been concluded that it is ideal to add 140–160 g (in terms of the initial heulandite-clinoptilolite matrix $H^+(H_2O)_3[AlSi_{3.6}O_{9.2}]$) of zeolite filler and extend it over 40–60 square meters of paper web entering the reel drum (Fig. 11.1); designation and description of prepared paper samples are given in Table 11.3.

11.2.3 CHARACTERIZATION OF FILLED PAPER

The digital trinocular compound microscope T690C-PL-10M (AmScope, the USA) and scanning electron microscope JSM6510LV (Jeol, Japan) were used to study the morphology of the paper samples. To determine the chemical composition of paper samples, X-ray energy-dispersive (XRED)

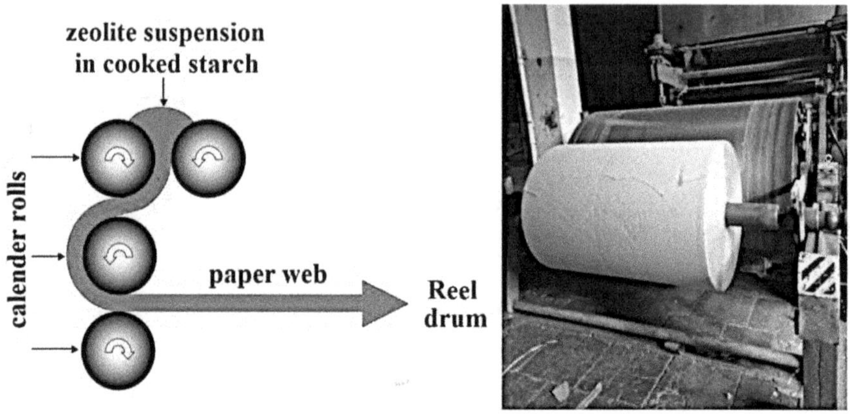

FIGURE 11.1 Scheme of entering zeolite fillers into paper.

Source: Reprinted from Ref. [15]. https://doi.org/10.51582/interconf.7-8.12.2021.050 © 2021 Copyright belongs to the authors of this chapter.

TABLE 11.3 Designation and Description of Paper Samples.

Designation	Description
P	Ordinary GPM paper
P-HCR	Paper with heulandite-clinoptilolite from Rkoni HCR filler
P-AgHCR	Paper with AgHCR filler containing 130 mg/g of silver
P-$CuHCR_1$	Paper with $CuHCR_1$ filler containing 63 mg/g of copper
P-$CuHCR_2$	Paper with $CuHCR_2$ filler containing 72 mg/g of copper
P-$ZnHCR_1$	Paper with $ZnHCR_1$ filler containing 30 mg/g of zinc
P-$ZnHCR_2$	Paper with $ZnHCR_2$ filler containing 58 mg/g of zinc
P-(½Zn + ½Cu)HCR	Paper with mixture of $ZnHCR_1$ and $CuHCR_1$ fillers in weight ratio 1:1
P-(¾Zn + ¼Cu)HCR	Paper with mixture of $ZnHCR_1$ and $CuHCR_1$ fillers in weight ratio 3:1

spectra and mappings of different atoms were recorded using X-Max 20 analyzer (Oxford Instruments, the UK). X-ray diffraction (XRD) patterns were obtained from a modernized Dron-4 X-ray diffractometer (the USSR) employing the Cu-K_α line (λ = 0.154056 nm); the paper samples were scanned in the 2Θ range from 5° to 70° with a 0.02° step at a scanning speed of 1°/min.

Accelerated aging of paper was done in an oven for 72 h (according to Ref. [19]) at a temperature of T = 90–100°C and a comparative humidity of 25% and 65%. Meanwhile, the relative weight loss was 1.8–2.2% and 2.7–3.3% for paper samples aged at modest and high relative humidity, respectively.

It is worth mentioning, before the experiments, samples of main and aged paper were conditioned at a temperature of 20–25°C and a relative humidity of 25–30%. Then, our samples were kept until they reached an equilibrium moisture content. It should be noted that this can only be achieved if, after two successive weighings of the sample (carried out after 1 h), the last mass differs from the previous one by no more than 0.25%.

Nevertheless, the basis weight (BW, grammage) of the paper was obtained on an electronic analytical balance (Type: FA 2204N, Made by JOAN Lab, China). Then the caliper (thickness) was measured with a micrometer as the perpendicular distance between two circular plane parallel surfaces under a pressure of 1 kg/cm^2. The maximum tensile force F developed in a test specimen (the paper strips 25 mm × 150 mm) and the increase in length of the test specimen ΔL before rupture were determined using a Universal Testing Machine (UTM) (made in Hongtuo, China). Meanwhile, we have done the following two measurements:

Measurements have been carried out in machine direction (MD) (this is in fact the direction of the paper web which is running on the machine) and measurement in a cross direction (CD) (which is in fact the direction that is perpendicular to the paper sheet that is running on the machine during paper production).

Evaluation of the bactericidal activity of paper samples was carried out by the disk diffusion test using the same cultures and strains as for characterization of zeolite fillers; colony-forming unit (CFU) counting was carried out against bacteria interacting with the paper surfaces, *E. coli* (ATTC 25922) and *S. aureus* (ATTC 25923); conditions are described in Ref. [14].

11.3 RESULTS AND DISCUSSION

11.3.1 STRUCTURE, CHEMICAL COMPOSITION, AND GENERAL PROPERTIES

11.3.1.1 STRUCTURE AND CHEMICAL COMPOSITION

Generally, paper is a heterogeneous mixture of plant material such as cellulose ($[C_6H_{10}O_5]_n$), hemicellulose, lignin, etc., filling materials such as calcium carbonate ($CaCO_3$), china clay (kaolinite $Al_2Si_2O_5(OH)_4$), etc., and chemical additives such as rosin, alum, starch, etc. depending on the grade of the paper. Plant material in the form of fibers and filling material in the form of particles are clearly visible in SEM images of an ordinary paper (Fig. 11.2).

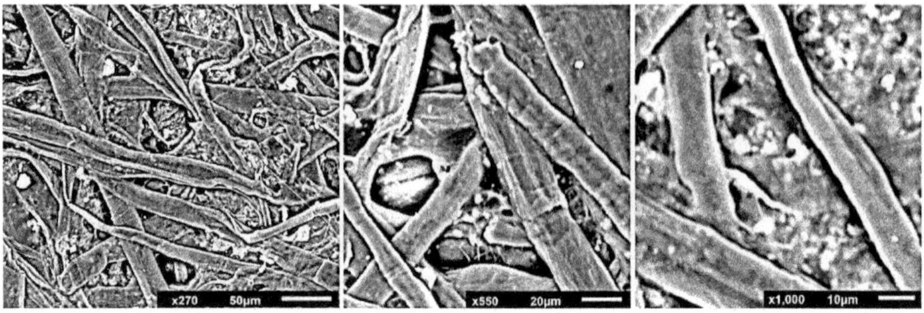

FIGURE 11.2 SEM images of ordinary paper.

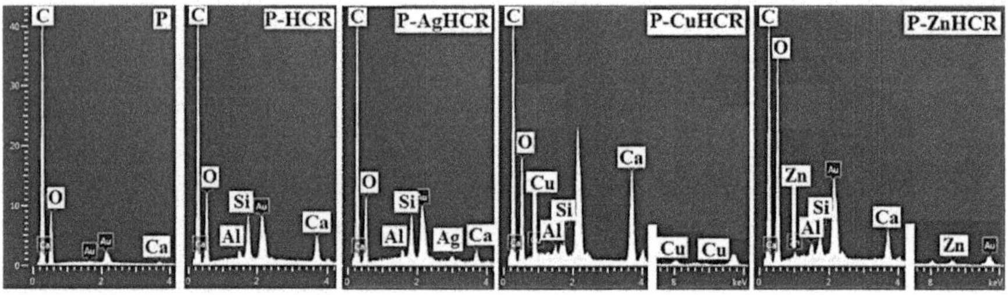

FIGURE 11.3 XRED spectra of ordinary (P) and fillered paper samples.

***Source*:** Reprinted from Ref. [15]. https://doi.org/10.51582/interconf.7-8.12.2021.050 © 2021 Copyright belongs to the authors of this chapter.

The large-scale XRED spectrum (taken from an area of 400 μm × 500 μm) of filler-free paper (Fig. 11.3, P) shows intense lines of carbon C and oxygen O atoms, as well as weak lines of calcium Ca atoms (all spectra also contain lines of gold used as a sputter coating material); spectra of filled paper show lines of silicon Si and aluminum Al (Fig. 11.3, P-HCR), as well as lines of bioactive metals Ag (Fig. 11.3, P-AgHCR), Cu (Fig. 11.3, P-CuHCR), and Zn (Fig. 11.3, P-ZnHCR).

The XRD patterns of paper samples (Fig. 11.4) show broad peaks of cellulose in the region $5° < 2\Theta < 30°$ and narrow peaks of calcite $CaCO_3$ at large 2Θ angles.

The XRD patterns of the fillered paper do not differ from the pattern of the original paper, since the cellulose broad peaks overlap the intense low-angle peaks of heulandite-clinoptilolite (at 2Θ = 9, 11, and 22°), and the XRD method turns out to be useless for the study of paper with zeolite fillers, but the XRED spectra confirm the presence of silicon and aluminum atoms in an approximate ratio of Si/Al = 4 in the fillered samples (Fig. 11.3, P-HCR, P-AgHCR, P-CuHCR, and P-ZnHCR).

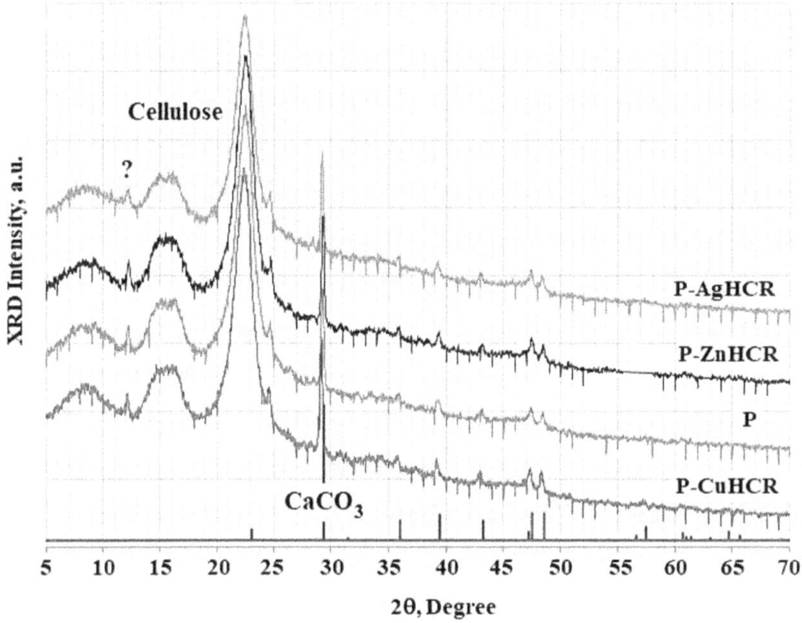

FIGURE 11.4 Powder XRD spectra of ordinary (P) and fillered paper samples.

***Source*:** Reprinted from Ref. [14]. https://doi.org/10.51582/interconf.19-20.07.2021.038 © 2021 Copyright belongs to the authors of this chapter.

The main inorganic constituent of GMP paper is calcium carbonate, but the paper also contains different impurities, and SEM images (Fig. 11.5) show micrometric and nanosized particles of various shapes and sizes.

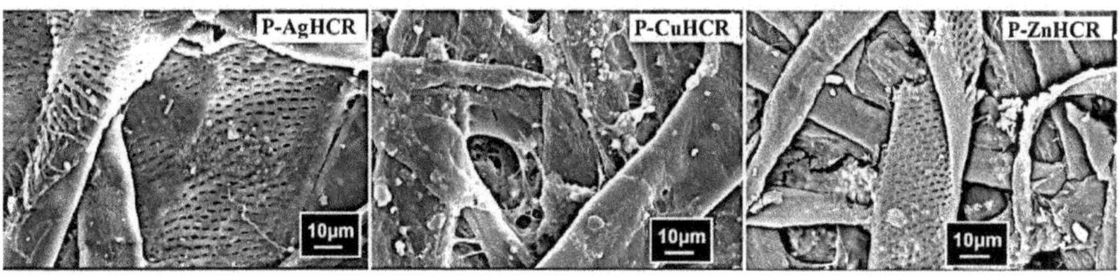

FIGURE 11.5 SEM images of paper with zeolite fillers.

Source: Reprinted from Ref. [14]. https://doi.org/10.51582/interconf.19-20.07.2021.038 © 2021 2021 Copyright belongs to the authors of this chapter.

Calcite ($CaCO_3$) particles were identified using XRED mapping of Ca atoms (Ca-XRED, Fig. 11.6a, right), and very common micrometric glass ($SiO_2 \cdot nNa_2O \cdot mCaO$) particles were highlighted by XRED spectra (Fig. 11.6a, left).

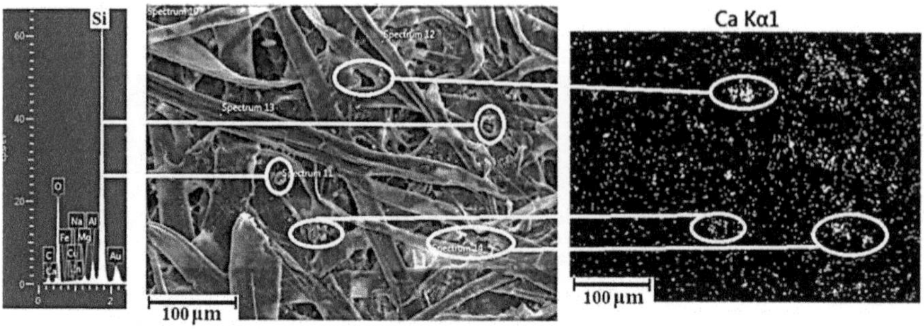

FIGURE 11.6A Part of XRED spectrum, SEM image, and XRED Ca-mapping of filler-free paper.

Source: Reprinted from Ref. [14]. https://doi.org/10.51582/interconf.19-20.07.2021.038 © 2021 2021 Copyright belongs to the authors of this chapter.

In paper, there is a small amount of zinc oxide in the form of nanoparticles and the silicate of magnesia (talc, $(MgO)_n(SiO_2)_m$) in the form of large particles (Fig. 11.7b).

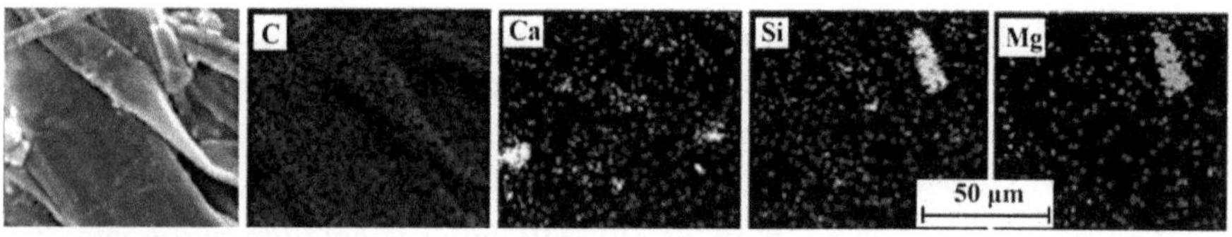

FIGURE 11.6B SEM image of a talc particle in the upper right corner, transparent in the XRED mappings of carbon (C) and calcium (Ca, showing two large calcite particles) and visible in the mappings of silicon (Si) and magnesium (Mg).

Source: Reprinted from Ref. [14]. https://doi.org/10.51582/interconf.19-20.07.2021.038 © 2021 Copyright belongs to the authors of this chapter.

In the XRED spectra of paper with a silver-containing filler (Fig. 11.3, P-AgHCR), the intensity of the silver signal at 2.5 keV increases with an increase in the amount of introduced filler, and on the XRED Ag-mappings, the density of silver-containing particles increases. Thus, when the AgHCR filler is introduced in an amount of 1 g/m, the average distance between the particles is 25 μm (Fig. 11.7a), and when the filler is introduced in an amount of 3 g/m, the average distance between the particles decreases to 5 μm (Fig. 11.7b). According to the XRED Ag-, Cu-, and Zn-mappings of filled paper (Figs. 11.8 and 11.9), bioactive metals are evenly distributed over the surface of the paper, and the distance between particles depends on the type and amount of zeolite filler introduced into the paper.

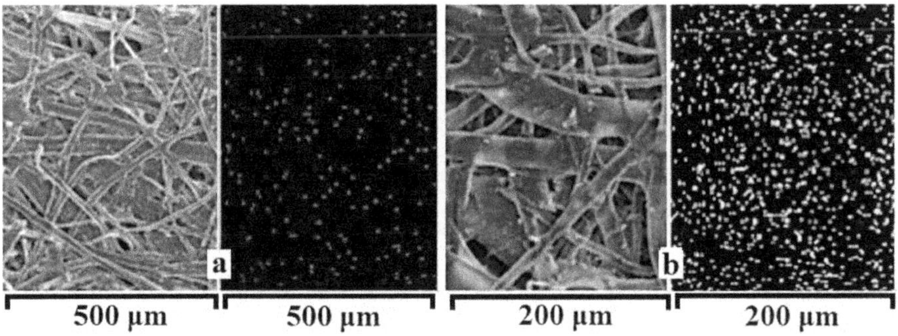

FIGURE 11.7 SEM images and XRED Ag-mapping of paper with a silver-containing filler AgHCR introduced in an amount of 1 g/m (a) and 3 g/m (b).

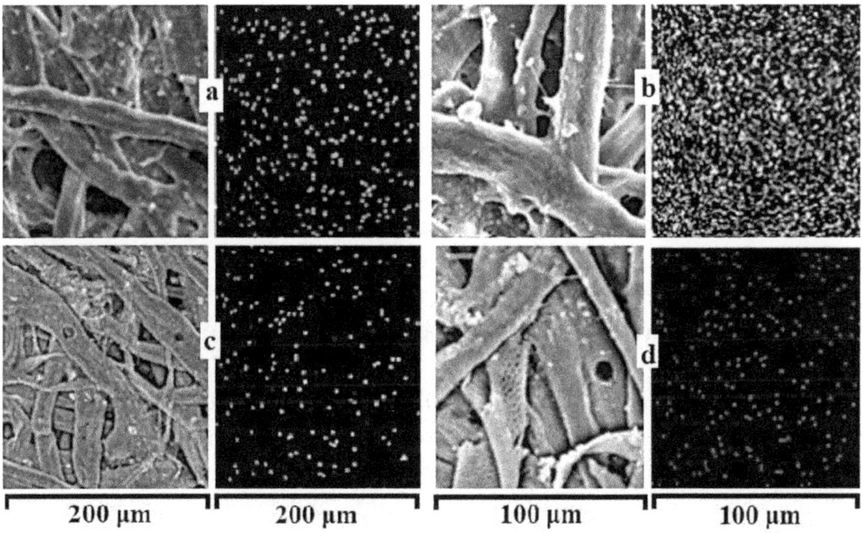

FIGURE 11.8 SEM images and XRED mappings of paper with copper-containing fillers $CuHCR_1$ (a) and $CuHCR_2$ (b), and zinc-containing fillers $ZnHCR_1$ (c) and $ZnHCR_2$ (d), introduced in an amount of 3 g/m.

The content of active metals was estimated from the XRED spectra in comparison with the carbon peak. The chemical composition of paper samples according to the data of XRED spectroscopy is given in Table 11.4. The content of zeolite filler (4 g/m^2 $\approx$ 2.3 wt.%) and silver (3 mg/g) in sample P-AgHCR approximately corresponds to that indicated in Ref. [17] for zeolites immobilized on cellulose nanofiber mats and in Ref. [18] for zeolite powders delivering antibacterial performance to cellulosic paper; the silver content is slightly lower, and the zinc content is slightly higher than in a FAU-type zeolite membranes with antimicrobial activity.[20]

TABLE 11.4 Bioactive Metals and Filler Contents in Filled Paper Samples.

Sample	Bioactive metal content, mg/g			Filler content,
	Ag	Cu	Zn	g/m2
P-HCR	0	0	0	3.5 ± 0.5
P-AgHCR	3.0 ± 0.5	0	0	4.0 ± 0.6
P-CuHCR$_1$	0	1.5 ± 0.25	0	3.7 ± 0.6
P-CuHCR$_2$	0	1.6 ± 0.25	0	3.8 ± 0.6
P-ZnHCR$_1$	0	0	0.75 ± 0.11	3.6 ± 0.6
P-ZnHCR$_2$	0	0	1.4 ± 0.2	3.7 ± 0.6
P-(½Cu + ½Zn)HCR	0	0.8 ± 0.12	0.7 ± 0.1	3.7 ± 0.6
P-(¾Zn + ¼Cu)HCR	0	0.4 ± 0.07	1.0 ± 0.1	3.8 ± 0.6

11.3.1.2 PHYSICAL AND MECHANICAL PROPERTIES

The values of basis weight (grammage), caliper (thickness), and density are shown in Figure 11.9 and the data on tensile strength in Figure 11.10.

It appears that fillers affect the grammage and thickness of the paper, but these figures are mainly determined by the production conditions. The density of ordinary GMP paper (0.64 g/cm^3) corresponds to the density of unbleached kraft paper (0.58–0.69 g/cm^3) and is slightly larger than the value for corrugated medium paperboard (0.61 g/cm^3), but the basis weight (~150 g/m^2) and thickness (235 μm) of the ordinary paper are quite low. The introduction of pure heulandite-clinoptilolite and metal-containing zeolite fillers generally leads to an increase in grammage and density of filled paper.

The ordinary GMP paper P has approximately doubled tensile in the MD compared to the CD, which is typical for cylinder machines. It is very important to note that the application of pure HCR zeolite filler can eventually lower the tensile strength by something around 20% in both directions while maintaining the same TS_{MD}/TS_{CD} ratio in the corresponding P-HCR paper. On the other hand, the metal-containing fillers affect tensile strength in a different manner as follows:

- For papers containing silver and copper, P-AgHCR and P-CuHCR$_1$, respectively, and for papers with mixed filler, P-(½Cu+½Zn)HCR have a higher "tensile strength."

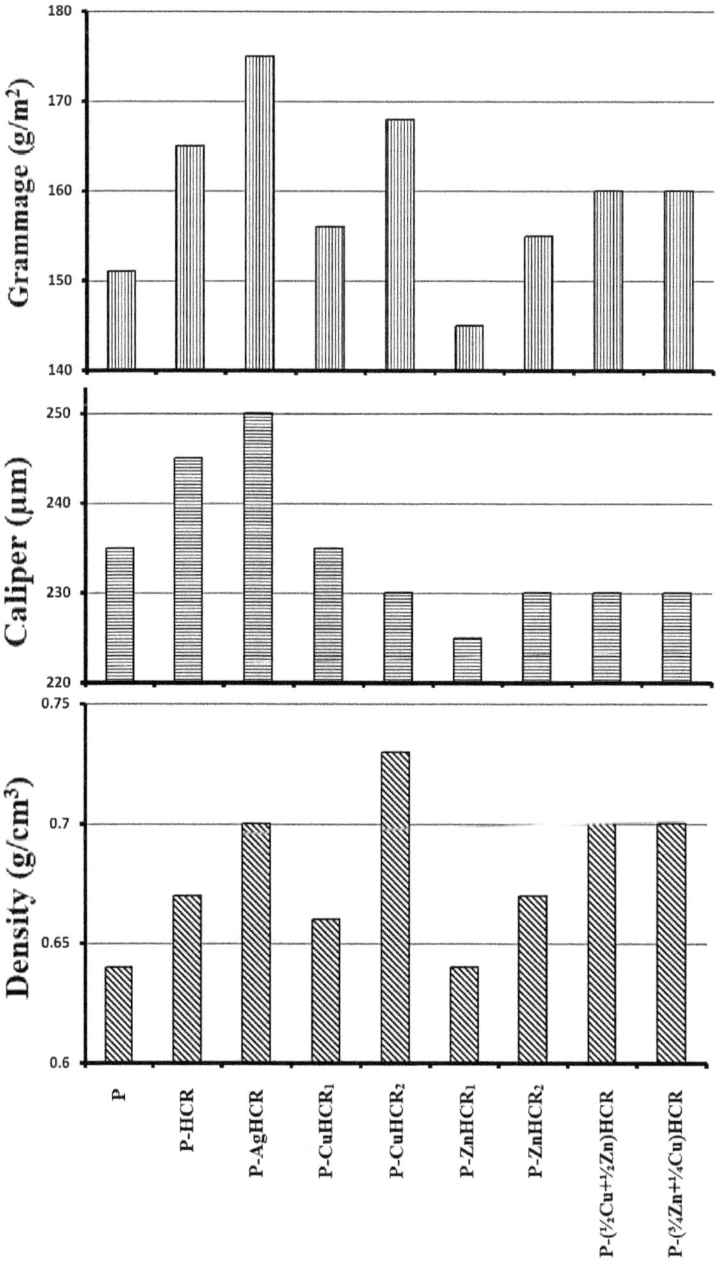

FIGURE 11.9 Grammage, caliper, and density of paper samples.

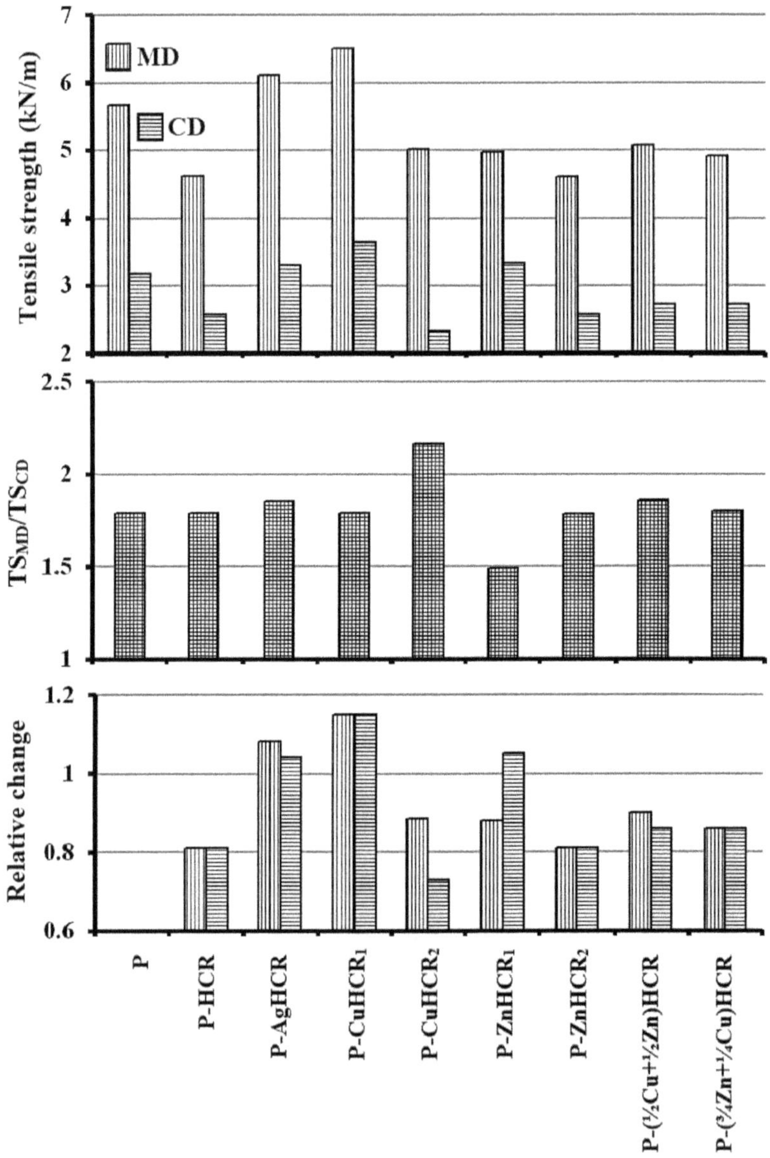

FIGURE 11.10 Tensile strength of paper samples.

- For papers with a high copper content P-$CuHCR_2$ have a low "tensile strength," especially in the cross direction.
- For papers with zinc or those papers containing P-$ZnHCR_1$ have a lower "tensile strength" in the machine direction and a raised "tensile strength" in the cross direction.

Tensile strength is determined by: (1) fiber strength, (2) fiber length, (3) fiber bonding, and (4) the structure and formation of the sheet in combination. It can be assumed that the strength of the fiber itself does not change with the addition of fillers. According to the data of optical and scanning electron microscopy, we have observed the following:

- the average fiber diameter is 7–12 μm,
- the maximum fiber diameter is 60 μm,
- the maximum fiber length is 2 mm, and
- these strength indicators also do not change when fillers are added to the paper.

It should be noted that the formation of the paper sheet depends on:

1. the parameters of the cylinder machine and
2. the conditions of papermaking that are the same for ordinary paper and filled paper.

Therefore, the structure of the paper along with fiber bonding are the main reasons for the changes in tensile strength in filled paper.

Based on the data of the Society of Mechanical Engineering (SEM), the microscopic structure of the paper does not rely on the application of zeolite fillers. Although, this cannot keep out the chance of some disorientation of the fibers in reference to the machine direction (since some of the fibers lie in the cross direction).

Accelerated aging of ordinary paper P leads to a reduction in TS, especially in the machine direction. And it should be noted that the TS_{MD}/TS_{CD} ratio decreases as well (please refer to Fig. 11.11). The same condition is true with most filled papers. For instance, for aged papers (P-$CuHCR_1$) expecting to become "square" ($TS_{MD}/TS_{CD} = 1$), but for silver papers and papers with a mixture of fillers P-(¾Zn+ ¼Cu)HCR, the opposite effect is expected (i.e., TS mainly reduces in the cross direction, and the TS_{MD}/TS_{CD} ratio rises).

The breaking length was calculated as BL(km) = 102.TS(kN/m)/BW(g/m^2),

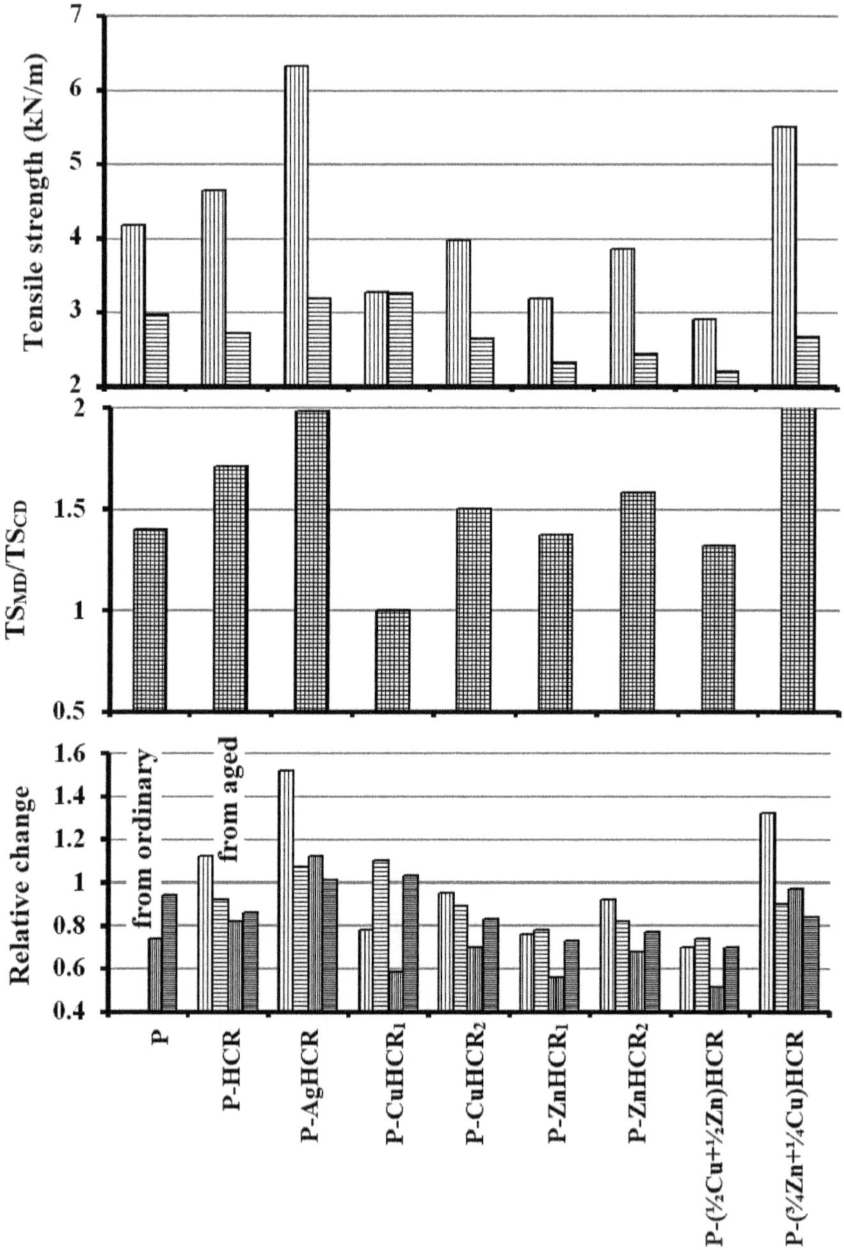

FIGURE 11.11 Tensile strength of aged paper samples.

where

(BL) = Limiting length of a strip of uniform width (calculated)**

(** beyond this value, if such a strip were suspended by one end, it is expected to break, due to its own weight).

Observed results are shown in Figure 11.12.

With the exception of copper-filled paper P-$CuHCR_1$, the breaking length in the filled paper samples decreases in both machine and cross directions.

Figure 11.13 shows the tensile strength limit or the value of σ (which is the ratio of the force F causing rupture of the sample to the cross-sectional area WC of test specimen, calculated as $\sigma(MN/m^2) = 10^{-6}.F(N)/W(m)C(m))$ and Figure 11.14 shows the tensile index TI (calculated as TI(Nm/g) = 1000. $TS(kN/m)/BW(g/m^2))$ for native and aged paper.

The tensile strength limit graph (Fig. 11.13), the calculated breaking length (Fig. 11.12), and the tensile index (Fig. 11.14) show the maximum rates in both directions for copper-containing paper P-$CuHCR_1$.

Stretch represents the capability of paper to accord to a desired contour or to survive nonuniform tensile stress. This is a very important factor in all papers, but is of particular importance in papers where stress–strain properties are being modified or controlled.[21] Stretch data in terms of percentage elongation (i.e., the ratio of the increase in length of the test specimen ΔL to the original test span L) are given as:

$$S(\%) = 100.\Delta L(m)/L(m)$$

These are shown in Figure 11.15.

It also should be noted that the elongation (%) of ordinary paper in the CD is greater than in the MD ($S_{MD}/S_{CD}<1$), and this pattern holds for filled papers as well.

Application of zeolite fillers leads to a stretch reduction in both directions; this effect is observed for a paper sample with a high copper content P-$CuHCR_2$. The value of stretch in the cross direction in most paper samples decreases to the greatest extent. But for the silver-containing paper P-AgHCR, the stretch in the machine direction changes more strongly, and in the zinc-containing paper P-$ZnHCR_1$, the stretch in the cross direction practically does not change.

Accelerated aging of ordinary paper P leads to a decrease in stretching mainly in the MD. Now, if we compare stretching in aged filled paper with the indicators of ordinary paper, then in most paper samples the elongation

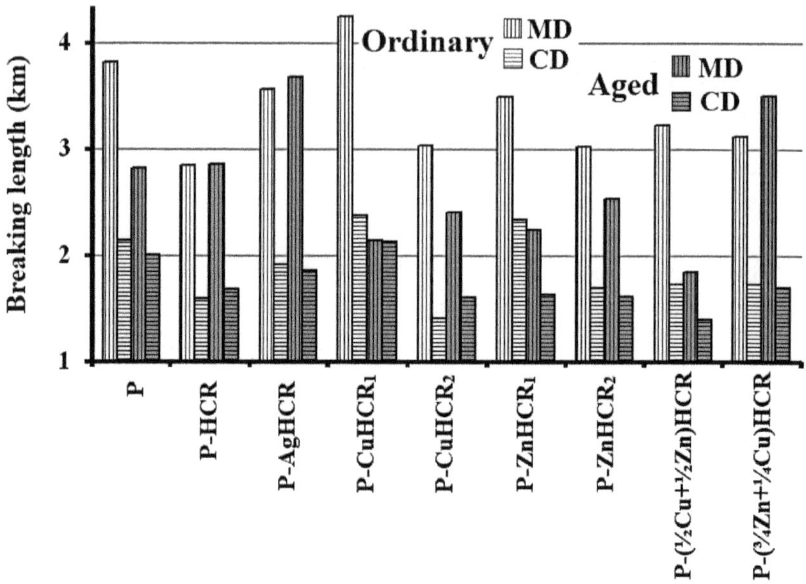

FIGURE 11.12 Breaking length of paper samples.

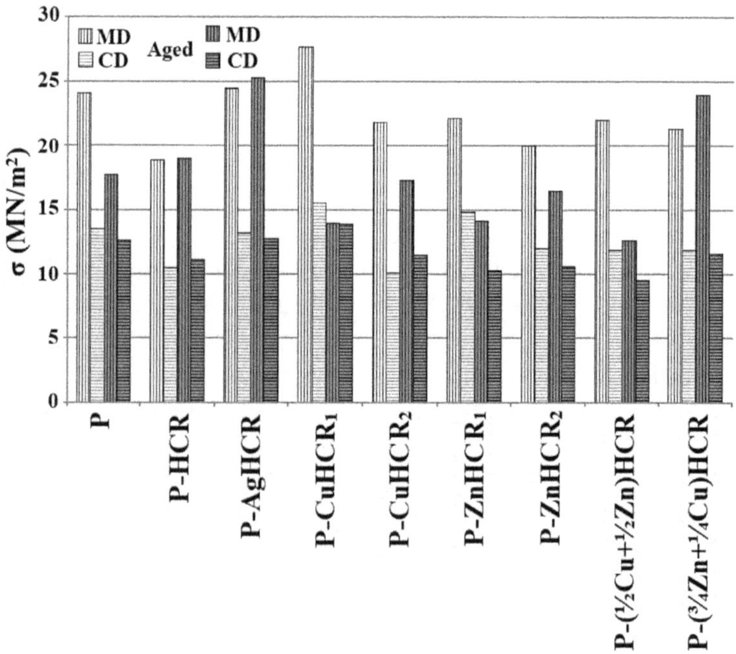

FIGURE 11.13 Tensile strength limit of paper samples.

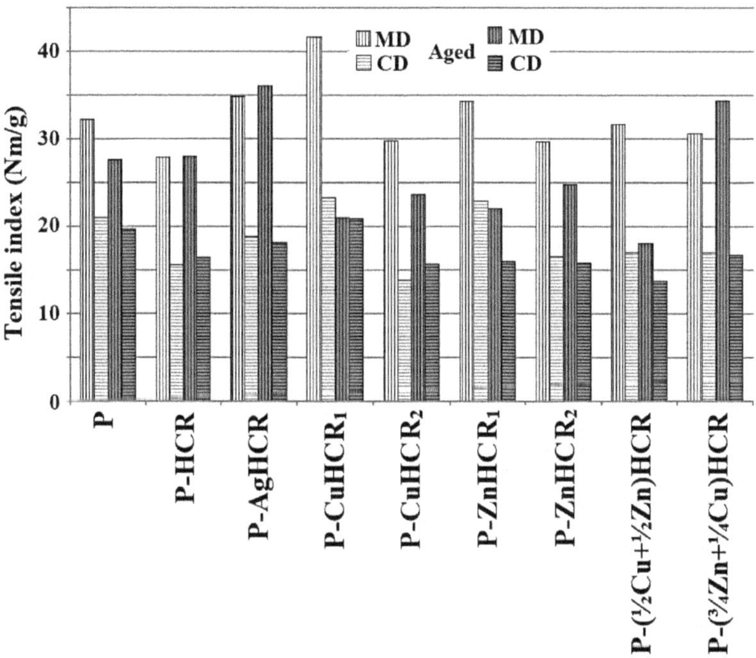

FIGURE 11.14 Tensile index of paper samples.

(%) reduces. This effect is, of course, the most pronounced for a paper sample with a mixture of fillers P-(½Cu + ½Zn)HCR. Nevertheless, in a number of samples, such as P-ZnHCR$_1$ and P-(½Cu + ½Zn)HCR, the elongation changes more strongly in the machine direction. But according to our experimental observations, in other samples, such as P-HCR, P-CuHCR$_2$, P-ZnHCR$_2$, and P-(¾Zn+ ¼Cu)HCR, the elongation (%) varies more strongly in the cross direction. It is worth mentioning that our particular interest is the sample of paper with a high copper content (P-CuHCR$_2$) where $S_{MD}/S_{CD}>1$. However, in the silver-containing sample of (P-AgHCR) paper, the elongation (%) augments in both directions. And also we have concluded that in copper containing sample of (P-CuHCR$_1$) paper, the elongation (%) augments in the cross direction and reduces slightly in the machine direction.

A comparison of stretching in aged filled paper with that of aged ordinary paper shows that the percentage elongation increases for most of the paper samples; this effect is most pronounced for the silver-containing paper sample. This means that we should only expect a significant reduction in elongation for a paper sample with a mixture of fillers P-(½Cu + ½Zn)HCR.

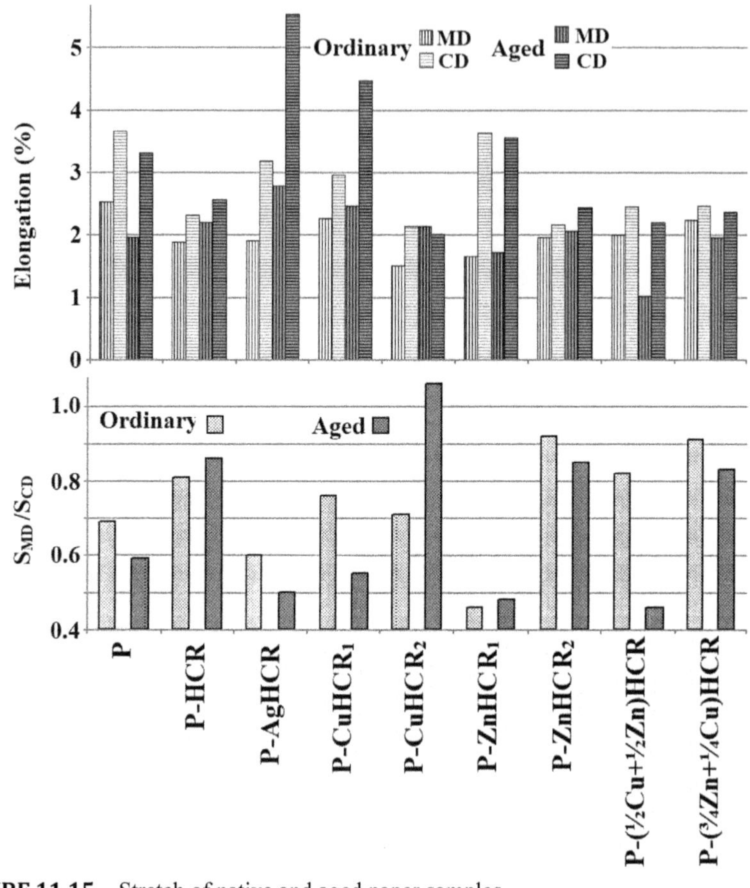

FIGURE 11.15 Stretch of native and aged paper samples.

11.3.1.3 WETTABILITY

The wettability of a paper or paperboard is determined based on the coatings and treatments done to it and is very important in a number of paper-related applications, including packaging. A drop of water on the surface of conventional GMP paper instantly spreads and is absorbed into the paper (Fig. 11.16); the same happens on the surface of paper with pure zeolite and silver-containing fillers.

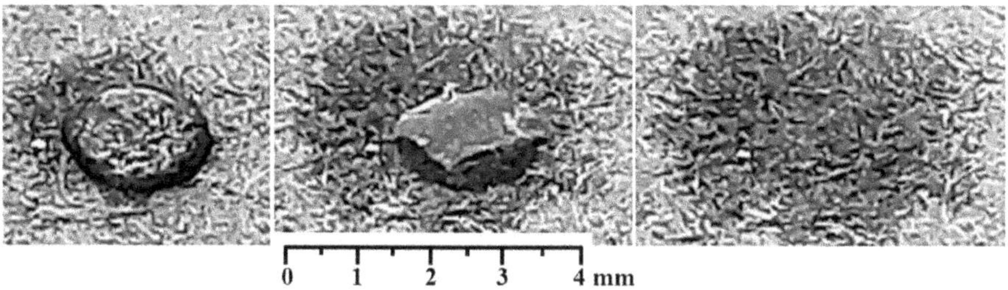

FIGURE 11.16 Spreading of a drop of water on the surface of the paper and its absorption, the interval between shots—0.3 s.

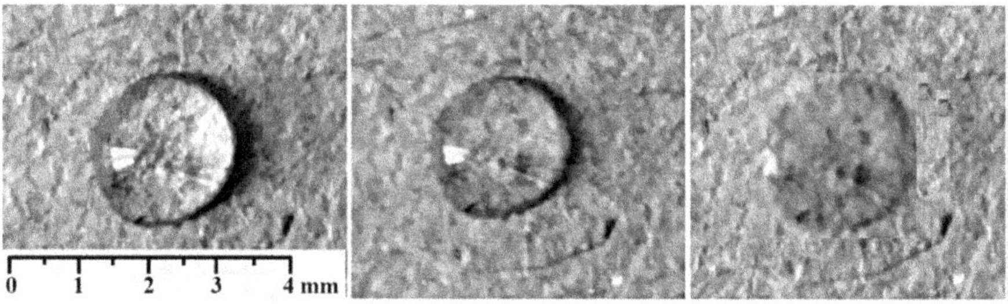

FIGURE 11.17 Drying of a drop of water on the surface of the paper P-$CuHCR_1$, the interval between shots—3 h.

On the surface of paper with fillers containing divalent metals, a drop of water does not spread and is absorbed into the paper much more slowly; a drop on the surface of paper with copper-containing filler does not absorb at all and dries out (Fig. 11.17), and this effect occurs both on the smoother top (felt) side of the paper and on the rougher wire side.

To obtain quantitative indicators, the specific water absorption of samples pre-dried, kept in water for 10 s, and then dried in air at room temperature for 1 h was determined. The results are shown in Table 11.5. After weighing the moisture-containing samples, they were dried for 20 h and weighed again for comparison with the weight of the original samples; weight changes do not exceed ± 0.65%. The P-$CuHCR_1$ sample exhibits the highest hydrophobicity; an increase in the copper content (P-$CuHCR_2$ sample) reduces the effect.

In the second series of experiments, the specific absorption of water vapor was determined for the dried samples placed in a desiccator for 95 h at a unit relative pressure of water vapor; the water absorption was calculated relative to the total surface and mass of the samples. The results are shown in Table 11.5. The introduction of fillers with divalent metals insignificantly increases the adsorption of water vapor, but this has no practical value.

TABLE 11.5 Water Absorption Indicators of Paper Samples.

Sample	Weight gain after wetting, %	Water vapor adsorption	
		g/g	mg/cm^2
P	73	0.215	1.08
P-HCR	91	0.206	1.03
P-AgHCR	72	0.206	1.03
P-CuHCR$_1$	2	0.227	1.14
P-CuHCR$_2$	7	0.230	1.15
P-ZnHCR$_1$	5	0.224	1.12
P-ZnHCR$_2$	6	0.244	1.22
P-(½Zn + ½Cu)HCR	10	0.240	1.21
P-(¾Zn + ¼Cu)HCR	12	0.223	1.12

11.3.2 BACTERICIDAL ACTIVITY

According to the disk diffusion tests, paper with a silver-containing filler P-AgHCR shows a bacteriostatic effect on *E. coli* and *S. aureus* (diameter of the inhibition zone 11 and 12 mm, respectively), paper with a zinc-containing filler P-ZnHCR$_1$ shows less activity against *B. subtilis* (inhibition zone of 18 mm, secondary growth), and all other paper samples do not prevent the growth of microorganisms on their surfaces.

On the contrary, the colony-forming unit (CFU) assay performed using gram-negative (*E. coli*) and gram-positive (*S. aureus*) bacteria confirmed the bacteriostatic activity of all paper samples with metal-containing fillers (Table 11.6).

TABLE 11.6 CFUs Counting in Relation to Control for Bacteria in Contact with the Surfaces of the Paper Samples.

Paper sample	Bacteria	
	E. coli	*S. aureus*
P	1.0	1.0
P-HCR	1.0	1.0
P-AgHCR	0.50	0.53
P-CuHCR$_1$	1.0	0.67
P-CuHCR$_2$	1.0	0.47
P-ZnHCR$_1$	0.52	0.70
P-ZnHCR$_2$	0.24	0.73
P-(½Zn + ½Cu)HCR	0.48	0.65
P-(¾Zn + ¼Cu)HCR	0.50	0.62

A similar situation is described in the study of the antibacterial activity of a synthetic FAU-type zeolite membrane: membranes exchanged with Ag^+ showed an agar-diffusive bactericidal activity against *E. coli* and no inhibition halo around the initial Na^+ and Zn^{2+} exchanged membranes, but a significant decrease in CFU for the bacterial solutions incubated with Zn^{2+} membranes. Daou et al.[20] suggested that "Zn^{2+} exchanged zeolite membrane presented a bacteriostatic activity that is less diffusive in agar." In the case of paper, the activity of zinc-containing samples against *E. coli* is comparable and even exceeds that of silver-containing ones—for P-$ZnHCR_2$ containing about 1.4 mg/g of zinc, the number of CFUs decreases by more than four times, while the silver-containing sample P-AgHCR kills about half of the *E. coli* CFUs. The CFU assay has shown that pure copper-containing forms P-$CuHCR_1$ and P-$CuHCR_2$ are not active against *E. coli*, and all metal-containing paper samples are active against *Staphylococcus*, the most active is paper P-$CuHCR_2$ with a copper content of 1.6 mg/g.

By this, zinc-containing fillers can be recommended for production of paper with bactericidal activity against *E. coli*, and copper-zinc-containing fillers for production of waterproof paper with bactericidal activity against *Staphylococcus*.

11.4 CONCLUSION

Application of zeolite silver-, copper-, and zinc-containing fillers prepared on the basis of natural heulandite-clinoptilolite into paper made by waste recycling leads to a change in the grammage, caliper, density, tensile, and stretching of the paper.

The "tensile strength" of paper filled with metal-containing bactericidal zeolites is related to the nature of the filler and its metal content. The application of pure zeolite filler (HCR) and metal-containing fillers into the structure of the paper evenly decreases the strength of the fibers binding in both directions. However, the application of the silver filler (AgHCR), and in particular, the filler with a low copper content ($CuHCR_1$) will augment the "tensile strength" in both directions. This shows that these fillers enhance the bonding between the paper fibers.

Accelerated aging of most papers will reduce "tensile strength." This is mainly in the machine direction, but Ag and Cu + Zn papers indicate the opposite effect.

The elongation (%) in the CD is higher than in the MD. However, for filled paper, it is strongly dependent on the nature and metal content of the filler. This emphasizes the impact of metal-containing zeolites on the binding forces (between paper fibers) and the effects of the nature and content of metals in the filler. Thus, changing indices of elongation are enhanced in the aged paper, making some of them more resistant to aging.

The introduction of zeolite fillers containing divalent metals copper and zinc causes a significant change in surface properties: papers with zinc (P-$ZnHCR_1$), and in particular, with copper filling (P-$CuHCR_1$) become waterproof, but with an increase in the content of metals (P-$CuHCR_2$ and P-$ZnHCR_2$) and for mixtures, this property is lost to a certain extent.

Paper samples contain up to 3.0, 1.6, and 1.4 mg/g of silver, copper, and zinc, respectively. According to the colony-forming unit assay, all metal-containing papers are active against *Staphylococcus*, and zinc-containing papers are more active against *E.coli* than those containing silver. From an industrial point of view, these results are very useful as they open up the possibility of replacing expensive silver with cheaper copper and zinc. For the production of paper active against *E. coli*, a filler with a zinc content of 1.4 mg/g is recommended, and for the production of waterproof paper active against *Staphylococcus*, a filler with a copper content of about 1.6 mg/g is recommended.

KEYWORDS

- **ion exchange**
- **paper filler**
- **silver**
- **copper**
- **zinc**

REFERENCES

1. Díez-Pascual, A. M. Antibacterial Activity of Nanomaterials. *Nanomaterials* **2018,** *8*, 359–365. https://doi.org/10.3390/books978-3-03897-050-7
2. Milenkovic, J.; Hrenovic, J.; Matijasevic, D.; Niksic, M.; Rajic, N. Bactericidal Activity of Cu-, Zn-, and Ag-containing Zeolites Toward *Escherichia coli* Isolates. *Environ. Sci. Pollut. Res. Int.* **2017,** *24*, 20273–20281. https://doi.org/10.1007/s11356-017-9643-8

3. Rossainz-Castro, L. G.; De la Rosa-Gomez, I.; Olguín, M. T.; Alcantara-Díaz, D. Comparison Between Silver- and Copper-Modified Zeolite Rich Tuffs as Microbicidal Agents for Escherichia coli and Candida Albicans. *J. Environ. Manag.* **2016,** *183*, 763–770. https://doi.org/10.1016/j.jenvman.2016.09.034
4. Vergara-Figueroa, J.; Alejandro-Martín, S.; Pesenti, H.; Cerda, F.; Fernández-Pérez, A.; Gacitúa, W. Obtaining Nanoparticles of Chilean Natural Zeolite and its Ion Exchange with Copper Salt (Cu^{2+}) for Antibacterial Applications. *Materials* **2019,** *12* (13), 2202–2220. https://doi.org/10.3390/ma12132202
5. Tekin, R.; Bac, N. Antimicrobial Behavior of Ion-Exchanged Zeolite X Containing Fragrance. *Micropor. Mesopor. Mat.* **2016,** *234*, 55–60. https://doi.org/10.1016/j.micromeso.2016.07.006
6. Yao, G.; Lei, J.; Zhang, W.; Yu, C.; Sun, Z.; Zheng, S.; Komarneni, S. Antimicrobial Activity of X Zeolite Exchanged with Cu^{2+} and Zn^{2+} on Escherichia coli and Staphylococcus aureus. *Environ. Sci. Pollut. Res. Int.* **2019,** *26* (3), 2782–2793. https://doi.org/10.1007/s11356-018-3750-z
7. Top, A.; Ülkü, S. Silver, Zinc, and Copper Exchange in Na-clinoptilolite and Resulting Effect on Antibacterial Activity. *Appl. Clay Sci.* **2004,** *27*, 13–19. https://doi.org/10.1016/j.clay.2003.12.002
8. Dolic, M. B.; Rajakovic-Ognjanovic, V. N.; Strbac, S. B.; Dimitrijevic, S. I.; Mitric, M. N.; Onjia, A. E.; Rajakovic, L. V. Natural Sorbents Modified by Divalent Cu^{2+}- and Zn^{2+}-ions and their Corresponding Antimicrobial Activity. *New Biotechnol.* **2017,** *39*, 150–159. https://doi.org/10.1016/j.nbt.2017.03.001
9. Tsitsishvili, V. G.; Dolaberidze, N. M.; Nijaradze, M. O.; Mirdzveli, N. A.; Amiridze, Z. S. Bactericidal Adsorbents Obtained by Ion Exchange Modification of Natural Phillipsite. *Chem. Phys. Technol. Surf.* **2019,** *10* (4), 327–333. http://dx.doi.org/10.15407/hftp10.04.327
10. Tsitsishvili, V.; Dolaberidze, N.; Mirdzveli, N.; Nijaradze, M.; Amiridze, Z. Properties of Bactericidal Adsorbents Prepared from Georgian Natural Analcime and Phillipsite. *Bull. Georg. Natl. Acad. Sci.* **2020,** *14* (4), 25–33. http://science.org.ge/bnas/t14-n4/04_Tsitsishvili_Physical-Chemistry.pdf
11. Tsitsishvili, V.; Dolaberidze, N.; Mirdzveli, N.; Nijaradze, M.; Amiridze, Z. Properties of Georgian Natural Heulandite-Clinoptilolite and its Silver, Copper, and Zinc-Containing Forms. *Bull. Georg. Natl. Acad. Sci.* **2021,** *15* (2), 60–67. http://science.org.ge/bnas/t15-n2/08_Tsitsishvili_Physical-Chemistry.pdf
12. Tsitsishvili, V.; Dolaberidze, N.; Nijaradze, M.; Mirdzveli, N.; Amiridze, Z.; Khutsishvili, B.; Virsaladze, K.; Kapanadze, T. Properties of Georgian Heulandite-Clinoptilolite and its Application for Production of Bactericidal Adsorbents. *InterConf.* **2021,** *59*, 633–642. https://ojs.ukrlogos.in.ua/index.php/interconf/article/view/13241
13. Tsitsishvili, V.; Dolaberidze, N.; Mirdzveli, N.; Nijaradze, M.; Amiridze, Z. Preparation of Bactericidal Fillers from Georgian Heulandite-Clinoptilolite and their Application for Paper Production. I. Bactericidal Fillers. *InterConf.* **2021,** *67*, 340–358. https://doi.org/10.51582/interconf.19-20.07.2021.037
14. Tsitsishvili, V.; Dolaberidze, N.; Mirdzveli, N.; Nijaradze, M.; Amiridze, Z. Preparation of Bactericidal Fillers from Georgian Heulandite-Clinoptilolite and their Application for Paper Production. II. Bactericidal Papers. *InterConf.* **2021,** *67*, 359–374. https://doi.org/10.51582/interconf.19-20.07.2021.038

15. Tsitsishvili, V.; Dolaberidze, N.; Mirdzveli, N.; Nijaradze, M.; Amiridze, Z.; Chubinishvili, Z. Tensile and Stretch of Paper Filled with Bactericidal Zeolites. *InterConf* **2021,** *90*, 437–451. https://doi.org/10.51582/interconf.7-8.12.2021.050
16. Narayanan, S.; Batchelor, W.; Webley, P. A. A Review on the use of Zeolites to Create Valuable Paper Products and Paper-like Adsorbent Materials. *Appita J.* **2013,** *66* (3), 235–245. https://search.informit.org/doi/10.3316/informit.426324938402492
17. Rieger, K. A.; Cho, H. J.; Yeung, H. F.; Fan, W.; Schiffman, J. D. Antimicrobial Activity of Silver Ions Released from Zeolites Immobilized on Cellulose Nanofiber Mats. *ACS Appl. Mater. Interfaces* **2016,** *8* (5), 3032–3040. https://doi.org/10.1021/acsami.5b10130
18. Lin, Y.; Xueren, Q. Functionally Modified Zeolite Powders for Delivering Antibacterial Performance to Cellulosic Paper. *Key Eng. Mater.* **2019,** *807*, 141–150. https://doi.org/10.4028/www.scientific.net/KEM.807.141
19. Rusch, R. H. Accelerated Aging Test for Paper. *Bur. Stand. J. Res.* **1931,** *7*, 465–475. https://doi.org/10.6028/JRES.007.026
20. Daou, T. J.; Dos Santos, T.; Nouali, H.; Josien, L.; Michelin, L.; Pieuchot, L.; Dutournie, P. Synthesis of FAU-Type Zeolite Membranes with Antimicrobial Activity. *Molecules* **2020,** *25* (15), 3414–3423. https://doi.org/10.3390/molecules25153414
21. Technical Association of the Pulp and Paper Industry. Tensile properties of Paper and Paperboard (Using Constant Rate of Elongation Apparatus; Tappi, 1998; pp T499–om06. https://www.tappi.org/content/sarg/t494.pdf; http://grayhall.co.uk/BeloitResearch/tappi/t494.pdf

CHAPTER 12

Investigation of Complex Formation Process of Lead (II) with Fulvic Acids at pH = 7: An Ecosystem Approach

TAMAR MAKHARADZE

R. Agladze Institute of Inorganic Chemistry and Electrochemistry, Ivane Javakhishvili Tbilisi State University, Tbilisi, Georgia

ABSTRACT

Lead is a highly toxic element, which has unfavorable outcomes on the biological, neurological, and cognitive functions of the body. Complexes of lead with macromolecular organic substances are characterized with less bioavailability and accordingly with less toxicity. Fulvic acids are macromolecular natural organic substances. Through the functional groups, they take part actively in complex formation and sorption processes that move along in natural waters, bottom sediments, and soils. Fulvic acids form stable complexes with heavy metals and radioactive elements, and stipulate their migration forms in natural objects.

Already existing data about stability constants of complex compounds of fulvic acids along with heavy metals are heterogeneous. Even for the same metals (among them lead), at the same value of pH, the values of stability constants of fulvate complexes differ in several lines from one another, which make it impossible to calculate the migration forms of metals in natural waters and to determine the toxicity and bioavailability through the obtained stability constants. Such a condition is largely designated by disregarding the dependence of the ionization degree of fulvic acids and average molecular weights of the associates of fulvic acids at the value of pH in the complex formation process, which ends up by getting the wrong results.

Environmental Technology and Sustainability: Physical, Chemical, and Biological Technologies for Environmental Protection. Tamara Tatrishvili, Ann Rose Abraham, & A. K. Haghi (Eds.)

In this work, the complex formation process of fulvate complexes of lead (II) was studied by the gel chromatographic method at pH = 7.0.

Pure samples of fulvic acids were isolated from natural waters. Their elemental compositions and dissociation constants were established. The average molecular weight of oligomer of fulvic acids (Mw = 1350) was utilized to observe the composition of lead fulvate complex (1:1) and stability constant $\beta = 3.14 \times 10^4$; lg $\beta = 4.50$. Through the obtained results, it will be possible to calculate the migration forms of lead in natural waters, determine the toxicity and bioavailability, and evaluate the ecological condition of water reservoirs.

12.1 INTRODUCTION

Lead contamination of natural water systems has adverse behavioral, physiological, and biochemical effects on humans. The most toxic forms of lead are Pb^{2+}, $[PbOH^+]$, $[Pb(OH)_2{}^0]$, and different inorganic complexes. They react with the biomolecules of the body to form extremely stable biotoxic compounds.[1–3] The complexes of heavy metals with macromolecular organic substances are characterized with less bioavailability and accordingly with less toxicity.[4,5]

The macromolecular natural organic substances, fulvic acids (FA), are one of the most important ligands governing the geochemical cycling of metals in the environment. They are the considerable organic matter of natural waters (please refer to Ref. [6–11]). It is very important to note that fulvic acids form stable complexes with heavy metals and radionuclides,[6,8,12–22] and stipulate migration forms related to natural waters and soils.[6,8,23–28] Consequently, fulvic acids could possibly have some effect on the toxicity and bioavailability of metals in the living environment. Even though many reliable studies as well as empirical details on stability constants (β) of complex compounds of FA with lead (as in the case of other metals) are heterogeneous, but they differ in several lines from each other.[20,29–41] This condition is normally designated by neglecting the average molecular weight (Mw) of the associates of fulvic acids, whose value in its turn relies on the value of pH, and in the end, produce the wrong results. For this reason, it is not very easy to study the complex formation processes;

- To take place in natural waters,
- To identify migration forms of Pb (II),
- To classify and estimate chemical–ecological condition of natural waters.

The main aim of this chapter is to study the complex formation process among the pure samples of fulvic acids, isolated from natural water, and Pb (II) to calculate the moderate stability constant of lead fulvate complex. It is worth mentioning that the complex formation process of fulvate complexes of lead was investigated using the gel chromatographic process at pH = 7.0. While studying the complex formation process, we have considered the moderate molecular weights of the associates of fulvic acids at pH = 7. Meanwhile, it should be noted that for the optimal determination, we have used Sephadex G-25 (the limits of fractionating 100–5000).

12.2 EXPERIMENTAL METHODS AND MATERIALS

In order to be able to provide the pure samples of fulvic acids, after filtration through membrane filters (with 0.45 μm pore size), water of lake Paravani was distilled by the frozen approach. These samples were then acidified with 6 M HCl to pH 2 and kept on water bath at 60°C (for 2 h) for coagulation of humic acids. After we have completed this stage of experiments, the solution was centrifuged for 10 min at 8000 rpm (Centrifuge T-23). Then, for the isolation of fulvic acids from the centrifugate, we have used the adsorption chromatographic method with charcoal as a sorbent. Afterward, we have carried out "desorption" of amino acids and carbohydrates with 0.1 M HCl. It is very important to note that for desorption of polyphenols, we have considered 90% acetone–water solution. Right after, the elution of the fraction of fulvic acids was performed using 0.1 N NaOH solution.[6,14,42] Hereafter, we obtained alkalic solution of fulvic acids for the purification process (this was done through a cation exchanger KU-2-8). In order to be able to study the concentration of fulvic acids in our experimental solutions, we have decided to use the gravimetric method. It is important to note that the part of the solution was dried under vacuum until we have reached the constant weight. Meanwhile, the elemental composition of standard samples of fulvic acids isolated from natural waters and the average meaning of dissociation constants, respectively equal to

$$C = 53.75\%,\ H = 4.29\%;\ O = 40.56\%,\ N = 0.89\%,\ S = 0.51\%;$$
$$pK_{H,COOH} = 4.37,\ pK_{H,Ph-OH} = 10.4;\ \text{ash} = 0.35\%.$$

The parameters of Sephadex G-25 are as follows:

- mass of dry gel equals to 17 g,
- height of swelled layer of gel equals to 42 cm,

- inner diameter of column equals to 1.6 cm.

It should be noted that for the calibration of Sephadex G-25, blue dextran, polyethylene glycol along with molecular weights 300, 600, 1000, and glucose were used. The titer of standard substances was 1 mg/mL, transmission speed 3 mL/min, and applied volume of solution 2 mL.

Increasing quantity of standard solutions of FA and lead (II) solutions with one and the same concentration, prepared from the standard solutions, considered for atomic absorption method, were placed in fluoroplastic cylinders. On the other hand, the constant ionic strength was made by adding 1 mL of 0.1 M potassium nitrate. At the end, the final volume of solutions was 10 mL. And the concentration of hydrogen ions was regulated by 0.01 M potassium hydroxide and 0.01 M nitric acid, in model solutions at pH = 7.

The aliquots of varied solutions (2–2 mL) were located at the top part of the column. The elution process was completed using 0.1 M potassium nitrate, which had obviously the same pH as the aliquots of solution. We obtained the quantity of metals connected with fulvic acids in fractions, which releases substances with molecular weights 300≤ Mw>5000. The concentration of lead was determined by atomic absorption spectrophotometer (Model: Agilent 280Z AA).

12.3 RESULTS AND DISCUSSION

Table 12.1 shows that the content of lead in high weight molecular fraction (300≤Mw>5000) increases with the augmentation of the concentration of fulvic acids (this may only be explained by the formation of fulvate complexes). During the determination of molar concentrations of fulvic acids, the fact that they form associates in water solutions was taking into account as well. Our experimental technique was established on water solutions of fulvic acids by the gel chromatographic method. This was done in the interval of pH 4–11, considering line dependence between the average molecular weight (Mw) of fulvic acids and the value of pH. And then this is expressed in the following manner: Mw = 1350 pH 4540,[1] considering pH = 7, Mw (FA) = 4910.

However, in diluted water solutions, the forms of lead (stability constants of lead (II) hydroxo complexes are $\beta_{1,1}$ = 1,84.106 and $\beta_{1,2}$ = 3,17.10^{12})[43] were calculated through the following eqs 12.1–12.3:

$$Pb^{2+}\% = 100/(1 + \beta_{1,1} [OH^-] + \beta_{1,2} [OH^-]^2) \tag{12.1}$$

$$PbOH^{+}\% = \beta_{1,1}[OH^{-}]100/(1 + \beta_{1,1}[OH^{-}] + \beta_{1,2}[OH^{-}] \quad (12.2)$$

$$Pb(OH)_2^{0}\% = \beta_{1,2}[OH^{-}]^2\,100/(1 + \beta_{1,1}[OH^{-}] + \beta_{1,2}[OH^{-}]^2 \quad (12.3)$$

As the results show, Pb^{2+} is the dominant form at pH = 7.0.

For the cases in which we cannot consider the charges of ion, the reactions of formation of lead fulvate complexes could be written in the following way:

$$Pb(II)_{free} + mFA_{total} = PbFA_m \quad (12.4)$$

$$\beta = [PbFA_m]/([Pb(II)_{free}]\,[FA_{total}]^m) \quad (12.5)$$

In homogeneous systems, where $[Pb(II)_{free}]$ is not a constant value, the numeral value of the stoichiometric coefficient (m) is equivalent to the tangent of the tilt angle of straight built-in coordinates lg([PbFAm]/$[Pb(II)_{free}]$) and lg[FA]. During the experimental approach by the gel chromatographic method, [PbFAm] is considered equivalent to the concentracion of metal in high weight molecular fraction (300≤Mw>5000) and

$$[Pb(II)_{free}] = [Pb\ (II)_{total}] - [PbFAm] \quad (12.6)$$

To observe the exact value of the tangent tilt angle of straight line, the least square method was used:

$$m = tg\alpha = (n\Sigma x_i y_i - \Sigma x_i \Sigma y_i)/(n\Sigma x_i^2 - (\Sigma x_i)^2) \quad (12.7)$$

where $x_1 = \lg\,[FA_{total}]$ and $y_i = \lg([PbFAm]/[Pb\ (II)_{free}])$ (12.8).

Consequently, the numeral value of "m" equals to 1.28 (please refer to Table 12.1 and 12.2).

TABLE 12.1 Data for Identification of the Composition of Lead Fulvate Complexes, pH = 7; Mw(FA) = 4910; 300≤ Mw(PbFA)>5000; $[Pb(II)_{total}] = 14.89 \times 10^{-5}$.

mol/L			[PbFAm]: [Pb(II)free]	lg [FAtotal]	lg([PbFAm]: [Pb(II)free])
[FAtotal]	[PbFAm]	[Pb(II)free]			
1.72×10^{-5}	4.95×10^{-5}	9.54×10^{-5}	0.52	–4.7645	–0.2840
2.58×10^{-5}	6.00×10^{-5}	8.48×10^{-5}	0.70	–4.5884	–0.1549
3.44×10^{-5}	7.74×10^{-5}	6.75×10^{-5}	1.15	–4.4634	0.0607
4.30×10^{-5}	8.82×10^{-5}	5.67×10^{-5}	1.55	–4.3665	0.1903
5.16×10^{-5}	10.05×10^{-5}	4.44×10^{-5}	2.26	–4.2873	0.3541
6.02×10^{-5}	10.23×10^{-5}	4.26×10^{-5}	2.40	–4.2204	0.3802
6.88×10^{-5}	10.55×10^{-5}	3.94×10^{-5}	2.68	–4.1624	0.4281

TABLE 12.2 Determination of Composition of Lead Fulvate Complexes Using Least Square Method, Mw (FA) = 4910; pH = 7; Xi = lg [FA_{total}], Yi = lg([Pb FA_m]:[Pb (II)free]).

Xi	Yi	XiYi	Xi2
–4.7645	–0.2840	1.3531	22.7005
–4.5884	–0.1549	0.7107	21.0534
–4.4634	0.0607	–0.2709	19.9219
–4.3665	0.1903	–0.8309	19.0663
–4.2873	0.3541	–1.5181	18.3809
–4.2204	0.3802	–1.6046	17.8112
–4.1624	0.4281	–1.7819	17.3255

∑Xi = –30.8529; $(∑Xi)^2$ = 951.9014; ∑Yi = 0.9745; $∑Xi^2$ = 136.2597; ∑XiYi = –3.9426; m = 1.28.

TABLE 12.3 Data for the Identification of the Composition of Lead Fulvate Complexes, pH = 7; Mw (FA) = 1350; 300≤ Mw(PbFA)>5000; $[Pb(II)_{total}] = 14.89 \times 10^{-5}$.

mol/L			[PbFAm]:	lg [FAtotal]	lg([PbFAm]:
[FAtotal]	[PbFAm]	[Pb(II)free]	[Pb(II)free]		[Pb(II)free])
6.25×10^{-5}	4.95×10^{-5}	9.54×10^{-5}	0.52	–4.2041	–0.2840
9.37×10^{-5}	6.00×10^{-5}	8.48×10^{-5}	0.70	–4.0283	–0.1549
12.50×10^{-5}	7.74×10^{-5}	6.75×10^{-5}	1.15	–3.9031	0.0607
15.62×10^{-5}	8.82×10^{-5}	5.67×10^{-5}	1.55	–3.8063	0.1903
18.75×10^{-5}	10.05×10^{-5}	4.44×10^{-5}	2.26	–3.7270	0.3541
21.87×10^{-5}	10.23×10^{-5}	4.26×10^{-5}	2.40	–3.6601	0.3802
25.00×10^{-5}	10.55×10^{-5}	3.94×10^{-5}	2.68	–3.6020	0.4281

TABLE 12.4 Determination of the Composition of Lead Fulvate Complexes by the Least Square Method, Mw(FA) = 4910; pH = 7; Xi = lg [FA_{total}], Yi = lg([Pb FA_m]:[Pb (II)free]).

Xi	Yi	XiYi	Xi2
–4.2041	–0.2840	1.1939	17.6744
–4.0283	–0.1549	0.6240	16.2272
–3.9031	0.0607	–0.2369	15.2342
–3.8063	0.1903	–0.7243	14.4879
–3.7270	0.3541	–1.3197	13.8905
–3.6601	0.3802	–1.3916	13.3963
–3.6020	0.4281	–1.5420	12.9744

∑Xi = –26.9309; $(∑Xi)^2$ = 725.2773; ∑Yi = 0.9745; $∑Xi^2$ = 103.8849; ∑XiYi = –3.3966, m = 1.28.

TABLE 12.5 Data of Stability Constants of Lead Fulvate Complexes by the Leden Method.

Mol/L				F(FA)
[FAtotal]	**[PbFA]**	**[Pb(II)free]**	**[FAfree]**	
6.25×10^{-5}	4.95×10^{-5}	9.54×10^{-5}	1.30×10^{-5}	3.99×10^{4}
9.37×10^{-5}	6.00×10^{-5}	8.48×10^{-5}	3.37×10^{-5}	2.10×10^{4}
12.50×10^{-5}	7.74×10^{-5}	6.75×10^{-5}	4.76×10^{-5}	2.41×10^{4}
15.62×10^{-5}	8.82×10^{-5}	5.67×10^{-5}	6.80×10^{-5}	2.29×10^{4}
18.75×10^{-5}	10.05×10^{-5}	4.44×10^{-5}	8.70×10^{-5}	2.60×10^{4}
21.87×10^{-5}	10.23×10^{-5}	4.26×10^{-5}	11.64×10^{-5}	2.06×10^{4}
25.00×10^{-5}	10.55×10^{-5}	3.94×10^{-5}	14.45×10^{-5}	1.85×10^{4}

pH = 7; $[Pb(II)_{total}] = 14.89 \times 10^{-5}$ M; Mw(FA) = 1350; 300≤Mw(PbFA)>5000; F(FA) = $[PbFA]/([Pb_{free}][FA_{free}])$; $[FA_{free}] = [FA_{total}] - [PbFA]$; $[Pb(II)_{free}] - [Pb(II)_{total}] - [PbFA]$.

TABLE 12.6 Conditional Stability Constants of Lead Fulvate Complexes by the Least Square Method, Mw (FA) = 1350; pH = 7; Xi = $[FA_{free}]$; Yi = F (FA).

Xi	Yi	XiYi	Xi2
1.30×10^{-5}	3.99×10^{4}	0.5187	1.69×10^{-10}
3.37×10^{-5}	2.10×10^{4}	0.7077	11.3569×10^{-10}
4.76×10^{-5}	2.41×10^{4}	1.1472	22.6576×10^{-10}
6.80×10^{-5}	2.29×10^{4}	1.5572	46.24×10^{-10}
8.70×10^{-5}	2.60×10^{4}	2.262	75.69×10^{-10}
11.64×10^{-5}	2.06×10^{4}	2.3978	135.4896×10^{-10}
14.45×10^{-5}	1.85×10^{4}	2.6732	223.5025×10^{-10}

$\sum Xi = 51.02 \times 10^{-5}$; $(\sum Xi)^2 - 2603.0404 \times 10^{-10}$; $\sum Yi = 17.3 \times 10^{4}$; $\sum Xi^2 = 516.6266 \times 10^{-10}$; $\sum XiYi = 11.2638$; $\beta = 3.14 \times 10^{4}$; $\lg \beta = 4.50$.

Hence, the system Pb(II)(solution)-FA-H_2O at pH = 7.0 dominates lead fulvate complex, with the structure 1:1. So the complex formation reaction (pH = 7.0) may be written in such a way:

$$Pb(II) + FA = PbFA \tag{12.9}$$

For the calculation of average stability constant of lead fulvate at pH = 7, Leden function F(L) was used.[44] The calculation of characteristics of Leden's function,

$$F(L) = F(FA) = [PbFA]/([Pb(II)_{free}][FA_{free}]) = \beta_1 + \beta_2[FA_{free}] \tag{12.10}$$

is not complicated during the study of fulvate complexes by the gel chromatographic method. The concentration of metal connected to fulvic acids

[PbFA] quantitatively equals the quantity of metals (mol/l) determined in high weight molecular fraction (300≤Mw>5000),

$$[FA_{free}] = [FA_{total}] - [PbFA] \quad (12.11)$$

and

$$[Pb(II)_{free}] = [Pb(II)_{total}]\ [PbFA] \quad (12.12)$$

whereas:

* $[FA_{total}]$—the total amount of fulvic acids in solution (mol/L),
* $[FA_{free}]$—the concentration of free ligand (mol/L),
* $[Pb(II)_{total}]$—the total amount (mol/L) of metal in the sample, which is a constant value according to the condition of the experiment.

The complexation of fulvic acids to metal ions may not be explained in strict frames due to the poorly defined nature of fulvic acids in contrast with the complexation of single ligands. So, in order to be able to obtain the stability constants of fulvate complexes, it is required to consider just a few assumptions. The obtained results show (Table 12.1) that in case of using the average molecular weight of FA associate at pH = 7 (Mw = 4910), it is impossible to calculate the concentration of free ligand. Without it, it's impossible to calculate the stability constant of lead fulvate complex.

Therefore, the meaning of Mw of oligomer of FA (Mw = 1350) was used for the determination of the concentration of free ligand ([FAfree]) and stability constant. The changes of molecular weights do not cause the changing of the numeral values of stoichiometric coefficient (m) (please refer to Tables 12.3 and 12.4). When $[FA_{free}]$ aspires to zero, β could be approached by the graphical method.

The section which is cut on the ordinate by the straight line built-in coordinates F(FA)—$[FA_{free}]$ is considered equivalent to the stability constant. The value of β determined by the "square method":

$$\beta = (\Sigma y_i - a\Sigma x_i)/n \quad (12.13)$$

where $x_i = [FA_{free}]$,

Y_i = F(FA) and

$$a = (n\Sigma x_i y_i - \Sigma x_i \Sigma y_i)/(n\Sigma x_i^2 - (\Sigma x_i)^2) \quad (12.14)$$

The necessary data for the calculation of stability constants of lead fulvate complex are given in Tables 12.5 and 12.6; $\beta = 3.14 \times 10^4$; lg $\beta = 4.50$.

12.4 CONCLUSION

We have observed that Pb(II)(solution)-FA-H_2O system at pH = 7.0 dominates the lead fulvate complex [PbFA] with the structure 1:1, $\beta = 3.14 \times 10^4$; lg $\beta = 4.50$. Through the obtained results, it will be possible to calculate the migration forms of lead in natural waters, determine the toxicity and bioavailability and evaluate the ecological condition of water reservoirs.

ACKNOWLEDGMENT

This research was supported by Shota Rustaveli National Science Foundation of Georgia (SRNSFG) [grant number №YS-21-3728].

KEYWORDS

- **lead fulvate**
- **stability constant**
- **gel chromatographic method**

REFERENCES

1. Balali-Mood, M.; Naseri, K.; Tahergorabi, Z.; Khazdair, M. R.; Sadeghi, M. Toxic Mechanisms of Five Heavy Metals: Mercury, Lead, Chromium, Cadmium, and Arsenic. *Front. Pharmacol.* **2021,** *12*, 643972.
2. Moiseenko, T. I. Bioavailability and Ecotoxicity of Metals in Aquatic Systems: Critical Levels ofPollution. *Geokhimia* **2019,** *64* (7), 675–688.
3. Duruibe, J. O.; Ogwuegbu, M. O. C.; Egwurugwu, J. N. Heavy Metal Pollution and Human Biotoxic Effects. *Int. J. Phys. Sci.* **2007,** *2*, 112–118.
4. Dudal, Y.; Gerard, F. Accounting for Natural Organic Matter in Aqueous Chemicalequilibrium Models: A Review of the Theories and Applications. *Earth Sci. Rev.* **2004,** *66*, 199–216.
5. Smith, K. S.; Balistrierib, L. S.; Todd, A. S. Using Biotic Ligand Models to Predict Metal Toxicity in Mineralized Systems. *Appl. Geochem.* **2015,** *57*, 55–72.

6. Varshal, G. M. Migration Forms of Fulvic Acids and Metals in Natural Waters, Dissertation, Vernadsky Institute of Geochemistry and Analytical Chemistry of Russian Academy of Sciences, 1994.
7. Dulaquas, G.; Waeles, M.; Gerringa, L. A.; Midag, R.; Rijkenberg, M.; Riso, R. G. The Biogeochemistry of Electroactive Humic Substances and Its Connection to Iron Chemistry in the North East Atlantic and the Western Mediterranean Sea. *J. Geophys. Res. Oceans*. **2018**, *123,* 5481–5499.
8. Osadchyy, V.; Nabyvanets, B.; Linnik, P.; Osadcha, N.; Nabyvanets, Y. In *Processes Determiningn Surface Water Chemistry*; Springer International Publishing: Switzerland, 2016.
9. Makharadze, G. A.; Supatashvili, G. D.; Varshal, G. M.; Humic Acids in Surface Waters of Georgia. *Hydrochem. Mater.* **1989,** *106*, 22–30.
10. Makharadze, G.; Goliadze, N.; Khaiauri, A.; Makharadze, T.; Supatashvili, G. Fulvic And Humin Acids in Surface Waters of Georgia. In *High-performas Polymers for Engineering-based Composites*; Apple Academic Press: Waretown, NJ USA, 2016; 167–179.
11. Pisarek, I.; Glowacki, M. Quality of Groundwater and Aquatic Humic Substances from Main Reservoire of Ground Water No. 333. *J. Ecol. Eng*. **2015**, *16*, 46–53.
12. Bertoli, A. C.; Garcia, J. S.; Trevisan, M. G.; Ramalho, T. C.; Matheus, P.; Freitas, M. P. Interactions Fulvate-Metal (Zn^{2+}, Cu^{2+} and Fe^{2+}): Theoretical Investigation of Thermodynamic, Structural and Spectroscopic Properties. *Biometals* **2016,** *29*, 275–285.
13. Xu, H.; Xu, D. C.; Wang, Y. Natural Indices for the Chemical Hardness/Softness of Metal Cations and Ligands. *ACS Omega* **2017,** *2*, 7185–7193.
14. Makharadze, T.; Makharadze, G. Investigation of Complex Formation Process of Copper with Macromolecular Organic Substances, Isolated from Natural Water. *Org. Chem. Plus* **2020,** *1*, 1–5.
15. Maccarthy, P.; O'Cinneide, S. Fulvic Acids: Interactions with Metal Ions. *Eur. J. Soil Sci.* **2006,** *25,* 429–437.
16. De Oliveira Vaz, D.; Fernandes, A. N.; Szpoganicz, B. Complexations of Divalent Metallic Ions with Fulvic Acids. *Eclética Química* **2018,** *43*, 54–58.
17. Boguta, P.; Sokolowska, Z. Zinc Binding to Fulvic acids: Assessing the Impact of pH, Metal Concentrations and Chemical Properties of Fulvic Acids on the Mechanism and Stability of Formed Soluble Complexes. *Molekules* **2020,** *25*, 1297–1321.
18. Zhu, B.; Ryan, D. K. Characterizing the Interaction between Uranul Ion and Fulvic Acid using Regional Integration Analysis (RIA) and Fluorescence Quenching. *J. Environ. Radioact.* **2016,** *153,* 97–103.
19. Wang, J.; Lü, C.; He, J.; Zhao, B. Binding Characteristics of Pb^{2+} to Natural Fulvic Acids Extracted from the Sediments in ake Wuliangsuhai, Inner Mongolia Plateau, P.R. China. *Environ. Earth Sci.* **2016,** *75*, 768–779.
20. Schnitzer, M.; Skinner, S. I. M. Stability Constants of Pb, Ni, Mn, Co, Ca and Mg Fulvic Acid Complexes. *Soil Sci.* **1967,** *103,* 247–252.
21. Makharadze, T; Makharadze, G. Measurement of Complex Formation Process of Nickel (II) with Freshwater Fulvic Acids Using the Solubility Method. *Fine Chem. Eng.* **2021,** *2*, 54–61.
22. Dinu, M.; Shkinev, V. M. Complexation of Metal Ions with Organic Substances of Humus Nature: Methods of Study and Structural Features of Ligands, and Distribution of Elements between Species. *Geochem. Int.* **2020,** *58*, 200–211.

23. Moiseenko, T. I.; Dinu, M. I.; Gashkina, N. A.; Kremleva, T.A. Occurrence Forms of Metals in Natural Waters Depending on Water Chemistry. *Water Resour.* **2013**, *40*, 407–416.
24. Adusei-Gyamfi, J.; Ouddane, B.; Rietveld, L.; Cornard, J.; Criquet, J. Natural Organic Matter-cations Complexation and its Impact on Water Treatment: A Critical Review. *Water Res.* **2019,** *160*, 130–147.
25. Dinh, Q. T.; Li, Z.; Tran, T. A.; Wang, D.; Liang, D. Role of Organic Acids on the Bioavailability of Selenium in Soil: A Review. *Chemosphere* **2017,** *184*, 618–635.
26. Makharadze, G. A.; Supatashvili, G. D.; Varshal, G. M. The Research of the Forms of Copper in Surface Waters. *Hydrochem. Mater.* **1988,** *103*, 3–16.
27. Whitby, H.; Planquette, H.; Cassar, N.; Bucciarelli, E.; Osburn, C. L.; Janssen, D. J.; Jay, T.; Cullen, J. T.; González, A. G.; Völker, C.; Sarthou, G. A Call for Refining the Role of Humic-like Substances in the Oceanic Iron Cycle. *Sci. Rep.* **2020,** *10*, 6144–6156.
28. Linnik, P.; Zhezheva, V.; Linnik, R.; Ivanchenko, Y. Influence of the Component Composition of Organic Matter on Relationship between Dissolved Forms of Metals in the Surface Waters. *Hidrobiol. J.* **2013,** *49*, 91–108.
29. Orsetti, S.; Marco-Brown, J. L.; Andrade, E. M.; Molina, F. V. Pb (II) Binding to Humic Substances: An Equilibrium and Spectroscopic Study. *Environ. Sci. Technol.* **2013,** *47*, 8325–8333.
30. Saar, R. A.; Weber, J. H. Lead(II)-Fulvic Acid Complexes. Conditional Stability Constants, Solubility, and Implications for Lead(II) Mobility. *Environ. Sci. Technol.* **1980,** *14*, 877–880.
31. Sahu, S.; Banerjee, D. K. Complexation of Copper (II), Cadmum (II) and Lead (II) with Humic and Fulvic Acids of Yamuna River Sediments. In *Chemistry for the Protection of the Environment*; NY, USA, 1996; pp 375–388.
32. Turner, D. R.; Varney, M. S.; Whitfield, M.; Mantoura, R. F. C.; Riley, J. P. Electrochemical Studies of Copper and Lead Complexation by Fulvic Acid. *Geochim. Cosmochim. Acta* **1986,** *50*, 289–297.
33. Chakraborty, P.; Chakraborty, C. L. Competition from Cu(II), Zn(II) and Cd(II) in Pb(II) Binding to Suwannee River Fulvic Acid. *Water Air Soil Pollut.* **2008,** 195, 63–71.
34. Pinheiro, J. P.; Mota, A. M.; Benedetti, M. F. Lead and Calcium Binding to Fulvic Acids: Salt Effect and Competition. *Environ. Sci. Technol.* **1999,** *33*, 3398–3404.
35. Christl, I.; Metzger, A.; Heidmann, I.; Kretzschmar, R. Effect of Humic and Fulvic Acid Concentrations and Ionic Strength on Copper and Lead Binding. *Environ. Sci. Technol.* **2005,** *39,* 5319–5326.
36. Xiong, J.; Koopal, L. K.; Tan, W. F.; Fang, L. C.; Wang, M. X.; Zhao, W.; Liu, F.; Zhang, J.; Weng, L. P. Lead Binding to Soil Fulvic and Humic Acids: NICA-Donnan Modeling and XAFS Spectroscopy. *Environ. Sci. Technol.* **2013,** *47*, 11634–11642.
37. Gondar, D.; López, R.; Fiol, S.; Antelo, J. M.; Arce, F. Cadmium, Lead, and Copper Binding to Humic Acid and Fulvic Acid Extracted from an Ombrotrophic Peat Bog. *Geoderma* **2006**, *135*, 196–203.
38. Orsetti, S.; Marco-Brown, J. L.; Andrade, E. M.; Molina, F.V. Pb(II) Binding to Humic Substances: An Equilibrium and Spectroscopic Study. *Environ. Sci. Technol.* **2013**, *47*, 8325–8333 .
39. Quan, G.; Yan, J. Binding Constants of Lead by Humic and Fulvic Acids Studied by Anodic Stripping Square Wave Voltammetry. *Russian J. Electrochem.* **2010,** *46***,** 90–94.

40. Gurjia, Z.; Supatashvili, G. D.; Makharadze, G. A.; Varshal, G. M.; Chitiashvili, Z. D. Complex Formation of Lead Ions with Fulvic Acids, Derived from Natural Waters.
41. Makharadze, T. Measurement of Complex Formation Process of Lead (II) with Fulvic Acids at pH=8.Current Topics on Chemistry and Biochemistry. *Heliyon* **2022**, *1*, 148–154.
42. Revia, R.; Makharadze, G. Cloud-Point Preconcentration of Fulvic and Humic Acids. *Talanta* **1999,** *48*, 409–413.
43. Gurjia, Z.; Supatashvili, G. D.; Varshal, G. M.; Makharadze, G. A.; Chitiashvili, Z. D. Monohydrolysis of lead ions in dilute solutions. *Proc. Georgian Natl. Acad. Sci.* **1992,** *18*, 22–26.
44. Beck, M. T.; Nagypal, I. In *Chemistry of Complex Equilibria*; Chichester, Horwood, New York, 1990.

PART II

Case Studies and Analysis

CHAPTER 13

Conditioned Feasibility of the Application of Georgian Natural Zeolite–Laumontite for Sustainable Development of Agriculture

OSIPOVA N., KVERNADZE T., and BURKIASHVILI N.

P. Melikishvili Institute of Physical and Organic Chemistry, Tbilisi, Georgia.

ABSTRACT

Natural zeolites are characterized by well-developed, ordered crystalline microporous structures. Zeolites represent unique natural minerals by their physical–chemical properties, and simultaneously, they are characterized by selective, ion-exchange, and adsorptive properties that condition the feasibility of their practical application.

Especially, an effective application of natural zeolites in agriculture for increasing the yield of crops and maintaining an ecologically healthy environment should be noted.

The ion-exchange property and adsorptive ability of zeolites against microelements and other ions in the soil positively affect the soil structure and the processes that condition germination, growth, development, and chemical indices of plants, as well as their yields.

13.1 INTRODUCTION

The 21st century puts forward a very real problem of increasing demand for food supplies. As a matter of fact, about 50% more food generation is needed to meet the demand of people by 2050. In modern days, climatic

Environmental Technology and Sustainability: Physical, Chemical, and Biological Technologies for Environmental Protection. Tamara Tatrishvili, Ann Rose Abraham, & A. K. Haghi (Eds.)

abnormalities and water scarcity necessitate more efficient agricultural production systems to be able to feed the booming population.[1,2]

The soil is a main source for meeting the demand for food products, but nowadays, the areas usable for agriculture are drastically decreasing because of natural and technical erosions that are often caused by the unforeseen actions of people. The outcomes of demanding practices are the degradation of soil and water qualities (i.e., heavy metal contamination that reduces crop productivity), consumption of soil organic carbon, and eventually, per capita food grain accessibility.[3] Long-term in-depth farming activities make the agricultural land useless, resulting in low soil retention capacity. Therefore, increasing the productivity of agricultural crops using both intensive and rational application of mineral fertilizers and recultivation of arid, contaminated areas are the main problems that the scientists of the world face today. In this regard, farming with zeolites, which are environment-friendly materials has attracted an attention owing to their benefits in agricultural projects. The use of these minerals as soil improver simplifies the improvement of soil's physical and chemical properties. Zeolites reduce the rate of nutrient release from both organic and inorganic fertilizers and allow better nutrient opportunities throughout the crop growth stages.[4]

The improvement of the wide range of agronomic crops concerning growth, yield, and quality traits with the application of zeolites has been well studied by many scientists.[5–11]

Because of poor application of mineral fertilizers, about 50% of them are not consumed by plants and are washed away by underground waters that pollute the environment. To neutralize the negative sides of mineral fertilizers and increase the yield of crops, the application of natural zeolites of sedimentary origin seemed encouraging. Natural zeolites have exclusivity for most indispensable nutrients, including "ammonium" (NH_4^+), "phosphate" (PO_4^{2-}), "nitrate" (NO_3^-), and "potassium" (K^+) in their distinctive porous structure that decreases nutrient leaching. It is very important to note that the unreasonable application of nitrogenous fertilizers promotes their easy discharge from soil to groundwater. This, of course, will have very negative anthropogenic impacts on the groundwater quality. Therefore, in modern agricultural systems, soil nutrient retention is a very important factor to account for maximum nutrient use and productivity. This will surely improve the soil nutrient status and prevent groundwater contamination.[12–14]

It should be noted that the slow-release nature of zeolites is very advantageous to avail the nutrients optimally throughout the crop growth. These unique features of zeolites improve fertilizer, water use, and productivity. This will eventually diminish the environmental pollution by reducing nitrate leaching. The foregoing features greatly improve the productivity, growth, and of course, the quality of crops.

In modern days, cereal crops play a dominant role in the global agriculture. They are main food products for people and forage for agricultural animals. Out of different sorts of cereal crops, wheat (both spring and winter) is the main and the most valuable food product. In its turn, winter wheat exceeds all the other cereals with quality of grain. On the whole, wheat can be considered to be the basic livelihood for mankind. Therefore, all the possible agrotechnical methods are used today to raise the yield of this crop and achieve high harvests. But in many respects, this depends not only on the weather conditions and correct choice of cultivation technology but also on weed vegetation, the agricultural land plot, and various crop diseases.

All the above are the limiting factors for raising the level of crop yield under the conditions of intensive farming.

Weeds, taking a lot of water and nutrients out of soil, oppress the growth and development of cultivated plant and reduce the crop capacity. Another reason causing sharp decrease in productivity of grain crops, sometimes up to 75%, is the possibility of diseases such as mildew, rust, and root rot.[15,16]

Destruction of weeds is one of the basic agrotechnical actions for maintaining steady high crop yield of agricultural crops. Usually, actions for the struggle against weeds are divided into two groups: precautionary and destructive, in turn subdivided into agrochemical and chemical ways of struggling. These methods are rather labor consuming, expensive, and in some cases, poorly effective. A very small dose of herbicide can poorly affect the weeds, and a larger dose causes destruction not only of weeds but also of cultivated plants.

In connection with an intensive use of natural sedimentary zeolites in plant growing,[17–19] the data given in a number of studies show that adding of natural zeolites into protected and open ground, appreciably hampers germination of weeds, and in some cases, completely excludes their spreading on the field plot.[20–22]

In most of the above studies, clinoptilolite-bearing rocks (natural zeolite) were more commonly used than chabazite- and philipsite-bearing rocks.[23]

The presented research shows the possibility of positive use of laumontite-bearing rocks in plant growing (for winter wheat as an example), and compares the obtained results with the results of the influence of traditional mineral fertilizers on this plant.

13.2 EXPERIMENTAL METHOD

Experiments were carried out in the laboratory and field conditions. All the experiments were performed in three series and 11 variants, each with four replications.

The first variant (absolute background) was pure soil without fertilizers.

The second variant (control) was the soil containing mineral fertilizers at a ratio of $N_{90}P_{100}K_{80}$ kg/ha (this norm of mineral fertilizers is recommended for wheat growing). Both variants served as objects of comparison.

AISI species (Production Association "Tavtavi" of Tserovani, Georgia) of winter wheat served as a test.

Laumontite-bearing rocks of the deposit in the vicinity of Tbilisi, near the botanical garden were used in the present work. The laumontite content in the rock amounts up to 60–75% calcium dominating the cation composition. The rock, alongside with the zeolites, contains montmorillonite, chlorine, pyroxene, and other minerals.[24] These zeolites are characterized by high cation-exchange capacity CEC (4.25 meq/g), which practically is twice as high as that of clinoptilolite.[25]

Laboratory investigations were carried out to show the influence of laumontite applied into the soil on germination, seed vigor of wheat, and shoot sprouting of the plant. The experiments were performed as follows: Special plates with corresponding holes for water drainage were used. A layer of gray cinnamonic soil with 6 cm height was placed on a filter paper; the plate area was 0.05 m^2; the number of sown seeds in all cases equaled 100. Laumontite was added to the soil in amount of 200, 400, and 600 g/m^2 accordingly in the first series of experiments. In the second series of experiments, mineral fertilizers, according to the agrochemical norms of NPK used in Georgia, were added to laumontite placed on special plates. In the third experiment, the same doses of zeolites as in the second one were used but at halved NPK norms.

For control, the pure soil (S) and the soil containing mineral fertilizers in accordance with the agrochemical norms (S-NPK) were used. Calculation of germination, seedling emergence, and height of the plant was evaluated by arithmetic means according to the eqs 13.1 and 13.2.

$$\text{RSE\%} = 100[(\text{s.e.(tp)}-\text{s.e.(rp)})]/\text{s.e.(rp)} \tag{13.1}$$

where RSE is relative seedling emergence; s.e.(tp) is the number of seedlings on the treated plots; s.e.(rp) is the number of seedlings in the control experiments (pure soil) or the soil with NPK addition.

$$RPH\% = 100[p.h.(tp)–p.h.(rp)]/p.h.(rp) \quad (13.2),$$

where RPH% is relative plant height; p.h.(tp) plant height on treated plot; p.h.(rp) seedling height in control experiment.

The field experiment was carried out at 18–30°C in the vicinity of a village Kvemo-Chala, Kaspi region (Georgia).

I series of experiments: In the third, fourth, and fifth variants, only laumontite-bearing rocks were introduced into the soil in amounts of 200, 400, and 600 g/m^2, correspondingly.

II series of experiments: In the sixth, seventh, and eighth variants, combinations of laumontite rocks and mineral fertilizers according to the agrochemical norms fixed for Georgia, N_{90}, P_{100}, and K_{90} kg/ha, correspondingly 200, 400, and 600 g/m^2 of laumontite-bearing tuff and full dose of mineral fertilizers, were used.

III series of experiments: In the ninth, 10th, and 11th variants, a combination of laumontite-bearing rocks and half the doses of mineral fertilizers in amount of 200, 400, and 600 g/m^2, correspondingly was applied.

The soil used for the experiments was gray-cinnamonic and weakly alkaline. Table 13.1 gives some indices of the original as well as the laumontite-enriched soil.

TABLE 13.1 Some Indices of Original Soil and the Soil Enriched in Laumontite-bearing Rock, 600 g/m^2.

Samples	pH H_2O	Humus, %	Content of uptake per mg/100g soil				
			Ca	Mg	H	Ca + Mg	% ratio of Ca and Mg
Soil	8.3	1.83	19.34	13.02	-	32.36	Ca–60% Mg–40%
Soil + laumontite	8.3	1.53	24.92	12.28	-	37.20	Ca–67% Mg–33%

Table 13.2 shows that the fraction of granulation <0.01 mm predominates in the soil; therefore, it may be attributed to the clayed, so-called heavy soil.[27]

TABLE 13.2 Granulometric (Fractional) Composition of Original Soil and Soil Enriched in Laumontite-bearing Rock, %.

Sample	Size of particles in mm and their percentage						
	1–0.25 mm	0.026–0.05 mm	0.05–0.01 mm	0.01–0.005 mm	0.005–0.001 mm	<0.001	<0.01
Soil	0	29	20	13	18	20	51
Soil + laumontite	0	32	22	8	15	23	46

The experiments were carried out on separate trial plots; each plot equaled 1 m^2; 44 plots were used in all. The distance between trial plots was 50–70 cm (Scheme).

SCHEME OF EXPERIMENTS

Replications	Variants										
I	I	2	3	4	5	6	7	8	9	10	11
II	II	10	9	8	7	6	5	4	3	2	1
III	I	2	3	4	5	6	7	8	9	10	11
IV	II	10	9	8	7	6	5	4	3	2	1

Absolute background (soil without fertilizers)—the first variant; Control (soil with fertilizers)—the second variant; I series of the experiments include the third, fourth, and fifth variants; II series of the experiments include the sixth, seventh, and eighth variants; III series of the experiments include the ninth, 10th, and 11th variants.

Prior to sowing, the soil was subjected to plug treatment. The mineral fertilizers of about 100 and 50 g/m^2 and the corresponding amounts of laumontite-containing rocks were applied.

Wheat was sown by manual embedding of seeds into the soil (600 grains/m^2). Germination of seeds proceeded in 10 days after sowing.

The yield of wheat and straw mass output from the four replications were calculated for q/ha, together with the relative indices of wheat grain yield and straw mass output and mass of 1000 wheat grains, RCY%, ROM%, and RMG%, against absolute background and control with NPK.[26]

The most important indices such as mass of 1000 wheat seeds characterizing technological properties of the grain and the plant root weight were determined according to the influence of laumontite as well as their mixes with the mineral fertilizers introduced into soil.

In all the variants of the experiments, some biomaterial indices of wheat were determined. Part of the given field experiments were subjected to mathematical statistical treatment with determination of the least significant difference (LSD) to establish the level of probability and relative accuracy of the experiment, Sx%.

13.3 RESULTS AND DISCUSSION

The main results of the laboratory experiments are given in Table 13.3. From the obtained data, it is seen that the same dependence of $E_{(seed)}$ on the laumontite amount introduced into the soil is observed for all the three variants.

TABLE 13.3 The Values of Relative Seedling Emergence $E_{(seedl)}$ of Wheat on the Treated Plots Compared with 1) Absolute Background (S) and 2) Control (S)-NPK.

Object of study	Control-(absolute background-soil)	Control-S-NPK
	$E_{(seed)}$%	$E_{(seed)}$%
Series I		
1) S+200gL	300	128
2) S+400gL	575	285
3) S+600gL	450	214
Series II		
1) S+200gL+NPK	200	71
2) S+400gL+NPK	325	142
3) S+600gL+NPK	225	85
Series III		
1) S+200gL+1/2 NPK	250	100
2) S+400gL +1/NPK	400	185
3) S+600gL+1/2 NPK	300	128

The maximum values of energy $E_{(seed)}$ are observed in the case of laumontite introduced in the amount of 400 g/m^2. However, in series II, we can observe the decrease in $E_{(seed)}$ value as compared with series I. While decreasing the value of NPK (series III), the increase in these values compared with series II is observed.

We have also found that the increase of laumontite content in soil enhances the growth of seedling height calculated by eq 13.2, which makes comparable with absolute background while using laumontite in amount of 200, 400, and 600 g/m^2. Similar dependence is observed while comparing the obtained results with those of the control experiment soil-NPK 25, 50, and 90%.

Thus, the results of the present study testify that in extreme conditions, the introduction of laumontite into the soil provides an increase in seedling emergence of wheat, while in series II and III, where the mixture of laumontite with mineral fertilizers is used, it appears to be effective, especially in the series where mineral fertilizers are used according to agrotechnical norms.

Phenological observation of the process of wheat growing in field conditions showed that the introduction of laumontite into the soil oppressed the growth of weeds and positively affected the immunity of this plant to diseases. A visual calculation of the weediness of the experimental field was carried out in May and July. According to the Maltsev[16] modified scale, weediness was less than 1%; weediness within the limits of 1–5%, 5–25%, 25–50%, and over 50%. The obtained results are given in Table 13.4.

The data given in this table show that the weediness of the field where zeolites have been used changes within the limits of 25–50%. On the field where laumontite-bearing rocks and their combination with mineral fertilizers were applied in the soil, the weediness did not exceed 5%, except for the sixth and seventh variant, where it exceeded 5%. That was caused by a little concentration of laumontite in the soil and a high dose of mineral fertilizers. It has been noticed that the increase in amount of zeolites applied in the soil caused decrease of the degree of germination of weeds.

In the same period, the heights of weeds (averaged from 20 plants from each plot) were determined; the higher they are, the stronger they oppress cultivated plants, in our case, wheat.

The comparative heights of weeds in relation to heights of wheat seedlings are given in Table 13.4 as well.

On the field where laumontite-containing rocks were not used, the heights of weeds and wheat seedlings were equal at the end of the experiment, while the heights of weeds evolved on the soil enriched in laumontite were much lower than those of wheat seedlings.

The results have confirmed the earlier assumption that applying of natural zeolites, in particular laumontite, into the soil oppresses the growth and development of weeds; at the same time, does not render negative influence on the growth of the cultivated plant, wheat in our case.

TABLE 13.4 Estimation of Amounts and Comparative Heights of Weeds on the Experimental Field.

Variants	Amount of weed										Weeds are higher than seedlings		Weeds and seedlings have equal heights		Weeds are 75% higher than seedlings		Weeds are 50% higher than seedlings		Weeds are very low	
	Less than 1%		1–5%		5–25%		25–50%		More than 50%											
	I	II	I	II	I	II	I	II	I	II	I	II	I	II	I	II	I	II	I	II
First					X			X					X	X						
Second					X			X					X	X						
Third			X			X										X	X			
Fourth			X	X														X	X	
Fifth			X	X														X	X	
Sixth					X	X										X	X			
Seventh					X	X										X	X			
Eighth			X	X														X	X	
Ninth			X	X												X	X			
Tenth			X	X														X	X	
Eleventh			X	X														X	X	

Applying of zeolite into soil protects the leaves against various diseases as well.[28] By phytopathological control, the disease of wheat, leaf rust was revealed on some fields, especially on the fields without laumontite-bearing rock where percentage of the disease was 10–20%; on the fields where laumontite-bearing rocks were used separately and in combination with mineral fertilizers, it did not exceed 10%.

The experiments of wheat growing under field conditions with laumontite-containing rocks introduced into the soil led us to very interesting results, not only for raising the yield of the crop but also for improving the technological indices of wheat grain.

The obtained results dealing with the influence of laumontite-containing rocks and their combination with mineral fertilizers on the productivity of wheat grain and straw mass outputs as well as the important indices of the estimation of technological properties of seed, such as a mass of 1000 wheat grains are summarized in Table 13.5.

As evidenced from the data given in Table 13.5, application of mineral fertilizers favorably affects the productivity of wheat grain and straw mass output, but markedly decreases the mass of 1000 wheat grains. All these data are the average values of four replications in separate variants. The mass of 1000 wheat grains determines the productivity of agricultural crops. The heavier the weight of seeds, the higher their productivity.[15] In our case, mineral fertilizers do not exert a favorable influence on the mass of 1000 grains.

The introduction of laumontite-bearing rocks into the soil exerts a significant influence on the growth and productivity of wheat grains, on the straw mass output, and to a lesser extent, on the mass of 1000 grains. These effects increase with increasing the doses of the introduced zeolite-bearing rocks. Thus, while the enrichment of soil in zeolites was 600 g/m^2, the productivity of wheat grain increased against the absolute background (the first variant) at 100% and compared with the control with fertilizers (the second variant) at 40%. The similar situation was observed in case of straw mass productivity, which increased by 60% and 24%, and the mass of 1000 wheat seeds by 6.8% and 14.6%, correspondingly.

Not entirely the usual picture is obtained when laumontite-bearing rocks are used together with mineral fertilizers,[23] but it coincides with the data presented in Ref. [29] devoted to the study of the influence of laumontite-bearing rocks on the germination of wheat seeds.

Application of natural zeolites on the background of mineral fertilizers results in the increase of productivity of agricultural crops,[23,30] while with the

TABLE 13.5 Influence of Laumontite-bearing Rocks and its Mix with Mineral Fertilizers ($N_{90}P_{100}K_{80}$) kg/ha Applied into the Soil on the Yield of Wheat Grain (q/ha); Green Mass of Straw (q/ha) and Mass of 100 Grains (g).

Variants	Yield average of four replications q/ha	RCY, %		Output of green mass of straw of four replications, q/ha	ROMS, %		Mass of 1000 grains average of four replications, g/ha	RMG, %	
		Relative to I variant	Relative to I variant		Relative to I variant	Relative to I variant		Relative to I variant	Relative to I variant
The first series of experiment									
First	30.2	-	-30.3	50.13	-	-23.00	38.3	-	7.6
Second	43.3	43.3	-	65.12	29.9	-	35.6	-7.00	-
Third	40.65	34.6	-6.1	60.00	19.7	-7.9	36.8	-3.9	3.4
Fourth	49.32	63.2	13.9	75.00	49.6	15.2	39.1	2.1	9.8
Fifth	60.65	100.8	40.1	80.50	60.6	23.6	40.8	6.5	14.6
	$LSD_{0.95}$ -3.3 q/ha Sx-2.4%			$LSD_{0.95}$ -0.7 q/ha Sx-3%			$LSD_{0.95}$ -2.4 q/ha Sx-2.1%		
The second series of experiment									
First	30.2	-	-30.3	50.13	-	-23.0	38.3	-	7.6
Second	43.3	43.4	-	65.12	29.9	-	35.6	-7.0	-
Third	40.7	34.8	-6.0	59.9	19.5	-8.0	37.7	-1.6	5.9
Seventh	42.15	39.6	-2.7	60.6	20.9	-7.0	39.0	1.8	9.5
Eighth	47.85	38.4	10.5	74.6	48.9	15.0	39.5	3.1	10.9
	$LSD_{0.95}$ -3.5 q/ha Sx-2.7%			$LSD_{0.95}$ -0.7 q/ha Sx-0.3%			$LSD_{0.95}$ -2.4 q/ha Sx-2.1%		
The second series of experiment									
First	30.2	-	-30.0	50.13	-	-23.0	38.0	-	7.6
Second	43.3	43.4	-	65.12	29.9	-	35.6	-7.0	-
Ninth	41.9	38.7	-3.2	64.0	27.2	-1.7	38.8	0.5	8.1
Tenth	47.5	57.3	9.7	76.5	52.6	17.5	38.8	1.3	9.0
Eleventh	50.0	65.6	15.5	77.5	54.6	19.038	39.1	2.1	9.8
	$LSD_{0.95}$ -3.1 q/ha Sx-2.4%			$LSD_{0.95}$ -0.7 q/ha Sx-1.05%			$LSD_{0.95}$ -2.4 q/ha Sx-2.1%		

increase in the content of mineral fertilizers in a mix, the positive effect of laumontite-bearing rocks is weakened (the third and especially the second series of the experiments). Such important positive effects of laumontite-bearing rocks on the productivity and some other indices of wheat are difficult to explain. First, most probably laumontite favors preservation of moisture in soil for some period of time, which is most important for sowing and germination of winter wheat. Second, laumontite-bearing rocks are characterized by significant cation-exchange capacity compared to clinoptilolite, relatively often used in practice of plant growing. Third, we consider that in the winter-spring period, under the influence of laumontite soil conditioning, the creation of a favorable microbial landscape is apparently taking place.[19,26,31]

The main unfavorable effects exerted on the growth of plantings, which cannot be eliminated by traditional methods, are determined by the stresses of the root system in the soil. The development of the root system of plants contributes to the cultivation of soil, provides conservation of soil aggregates of the size exceeding 250 μm, and prevents erosion processes.

A series of works showed that the application of natural zeolites contributes to the development of plant root systems even on contaminated, infertile soil and results in soil conditioning[30,32] and the creation of optimal structures. The mass of the grass root system, designated for the creation of golf fields at the application of substrates based on clinoptilolite-bearing tuffs and apatite, may be increased by two to five times more than the control.[33]

In our experiments, the increase in plant root mass also took place, calculated for 20 plants taken from every variant and four replications, that is, an average value out of 80 plants of one plant-g/plant. Alongside it, a regularity similar to that observed with the yield of wheat was detected. Maximum growth in the weight of wheat roots took place with the application of only laumontite-bearing rocks to the soil in an amount of 600 g/m^2. Compared to the absolute background (first variant), the increase amounted to 125%, and compared to the control with mineral fertilizers (second variant) to 39%. Apparently, the application of laumontite to the soil contributes to the creation of favorable water regimes and structuring in the zone of root system, that is rhizosphere.

Maximum germination of seeds and further development of plants with a decrease of the possibility of their death are not less important factors in plant growing.

The present work, in connection with the above stated, considers the issue of the influence of laumontite-bearing rocks and their mixture with mineral fertilizers on the example of the germination of wheat seeds and the process of ear formation (Table 13.6). As seen from Table 13.6, the process of ear formation in the first variant (absolute background) is greatly lowered, output of stem forms 53.2%. The application of mineral fertilizers to the soil markedly increases the stem output up to 86.3%. Approximately, such was the case when only laumontite-bearing rock was applied into the soil, but the effect was strengthened with increase in the zeolite dose. Thus, at the amount of 600 g/m^2, it was equal to 93%.

The application of a mixture of laumontite-bearing rock and mineral fertilizer somewhat weakens this effect, but it is higher than in case of only mineral fertilizers. The process of ear formation in all variants, apart from the first one, is high, and the number of stems without ears is insignificant from 1.9–3.5%.

Under the influence of laumontite-bearing rocks and their mixes with mineral fertilizers, an alteration in the biometric indices of wheat (Table 13.6) also takes place.

TABLE 13.6 Influence of Laumontite-containing Rocks and its Mix with Mineral Fertilizers on the Germination, Development, and Some Biometric Indices of Wheat.

Variants	Total quality of stems, pcs/m^2	Quality of stems with ears, pcs/m^2	Quality of stems without ears, pcs/m2	Output of stems, %	Length of stems, cm	Length of ear, cm	Amount of grains per ear	Weight of ear, g	Weight of grain per ear, g
First	319.0	298.0	21.0	53.2	92.5	8.5	32.0	1.75	1.07
Second	518.0	502.5	15.5	86.3	100.0	11.8	40.0	1.59	1.31
Third	514	497.0	17.0	85.7	98.7	10.6	37.0	1.47	1.27
Fourth	532	518.5	14.0	88.7	101.9	11.7	37.5	1.93	1.48
Fifth	557.0	539.0	18.0	92.9	105.2	12.6	43.0	2.23	1.70
Sixth	473.0	462.0	12.0	78.8	102.5	10.7	34.0	1.34	1.16
Seventh	515.0	500.0	15.0	85.8	99.2	11.0	38.0	1.57	1.28
Eighth	528.5	518.5	10.0	88.1	103.0	11.4	38.0	2.06	1.40
Ninth	511.1	498.0	13.0	85.2	97.4	11.5	38.0	1.49	1.31
Tenth	536.0	520.0	16.0	89.3	103.2	11.2	39.0	1.75	1.39
Eleventh	544.5	531.5	13.0	90.7	103.4	11.6	40.0	1.85	1.47

The obtained data given in the table prove that under the influence of mineral fertilizers as well as laumontite-bearing rocks, the stems are lengthened. This somehow may provoke a danger for their lodging, which is not desirable. On the other hand, increase of length of an ear conditions the growth of the amount of grains in the ear. Thus, the average quantity of grains in the ear (in 20 plants of four replications) equals 32 on the pure soil; on the plot where the mineral fertilizers were used—40, and in the fifth variant where only the laumontite-bearing rock was applied in the amount of 600 g/m^2—43. At this time, the mass of grain per ear increased to 1.07, 1.31, and 1.7 g, correspondingly. The obtained data coincide to those in the work.[34]

13.4 CONCLUSION

Thus, summing up the obtained results, we can make the following conclusion: The application of laumontite-bearing rock in plant growing, with the wheat taken as an example, makes positive effect on the quality and quantity indices of this crop. Moreover, their influence is expressed in greater extent than in case of application of mineral fertilizers.

KEYWORDS

- **natural zeolites**
- **laumontite**
- **soil**
- **mineral fertilizer**
- **plant growing**
- **winter wheat**

REFERENCES

1. Alexandratos, N.; Bruinsma, J. World Agriculture Towards 2030/2050. Food and Agriculture Organization of the United Nations, 2012. Available online:www.fao.org/economic/esa
2. Bhatt, R.; Singh, P.; Hossain, A.; et al. Rice–Wheat System in the Northwest Indo-Gangetic Plains of South Asia: Issues and Technological Interventions for Increasing Productivity and Sustainability. *Paddy Water Environ.* **2021,** 1–21.
3. Garai, S.; Mondal, M.; Mukherjee, S. Smart Practices and Adaptive Technologies for Climate Resilient Agriculture. In *Advanced Agriculture*, 1st ed.; Maitra, S., Pramanick, B., Eds.; New Delhi Publishers: Kolkata, India, 2020; pp 327–358.

4. Perez-Caballero, R.; Gil, J.; Benitez, C.; et al. The Effect of Adding Zeolite to Soils In order to Improve the N-K Nutrition of Olive Trees, Preliminary Results. *Am. J. Agric. Biol. Sci.* **2008,** *2*, 321–324.
5. Shahsavari, N.; Jais, H. M.; Shirani Rad, A. H. Responses of Canola Oil Quality Characteristics and Fatty Acid Composition to Zeolite and Zinc Fertilization Under Drought Stress. *Int. J. Agric. Sci.* **2014,** *4*, 49–59.
6. Chen, T.; Guimin, X.; Qi, W.; et al. The Influence of Zeolite Amendment on Yield Performance, Quality Characteristics, and Nitrogen Use Efficiency of Paddy Rice. *Crop. Sci.* **2017,** *57*, 1–11.
7. Ghanbari, M.; Ariafar, S. The Effect of Water Deficit and Zeolite Application on Growth Traits and Oil Yield of Medicinal Peppermint (*Mentha piperita L.). Int. J. Med. Arom. Plants* **2013,** *3*, 33–39.
8. Aainaa, H. N.; Ahmed, O. H.; Majid, N. M. A. Effects of Clinoptilolite Zeolite on Phosphorus Dynamics and Yield of Zea Mays L. Cultivated on an Acid Soil. *PLoS ONE* **2018,** *13*, e0204401.
9. Zheng, J.; Chen, T.; Wu, Q.; et al. Effect of Zeolite Application on Phenology, Grain Yield and Grain Quality in Rice Under Water Stress. *Agric. Water Manag.* **2018,** *206*, 241–251.
10. Hazrati, S.; Tahmasebi-Sarvestani, Z.; Mokhtassi-Bidgoli, A.; et al. Effects of Zeolite and Water Stress on Growth, Yield and Chemical Compositions of Aloe Vera L. *Agric. Water. Manag.* **2017,** *181*, 66–72.
11. Mondal, M., Biswas, B., Garai, S.; et al. Zeolites Enhance Soil Health, Crop Productivity and Environmental Safety. *Agronomy* **2021,** *11*, 448. https://doi.org/10.3390/agronomy11030448
12. Bakhshayesh, B. E.; Delkash, M.; Scholz, M. Response of Vegetables to Cadmium-enriched Soil. *Water* **2014,** *6*, 1246–1256.
13. Nakhli, S. A.; Ahmadizadeh, K.; Fereshtehnejad, M.; et al. Biological Removal of Phenol from Saline Wastewater using a Moving Bed Biofilm Reactor Containing Acclimated Mixed Consortia. *Springer Plus* **2014,** *3*, 112.
14. Leggo, P. J. The Efficacy of the Organo-zeolitic Bio-fertilizer. *Agrotechnology* **2015,** *4*, 1000130.
15. Kosinsky, V. S.; Rubanov, A. M.; Tkachev, V. V., et al. Bases of Agriculture and Plant Growing. Mir, Moskow, 1980.
16. Nikitenko, G. F.; Ed. *Experiments in Field-crop Cultivation.* "Rosselkhozizdat", M., 1982.
17. Ming, D. W.; Allen, E. R. Recent Advances in the United States in the Use of Natural Zeolites Plant Growing. La: Natural Zeolites fot the Third Millenium. Colella, C.; Mumpton, F. A. Eds.; Napoli, Italy, 2000.
18. Andronikashvili, T. G. Some Achievements in Application of Natural Zeolites in Plant Growing in Georgia. *Ann. Agra. Sci.* **2003,** #2.
19. Leggo, P. J. An Investigation of Plant Growing in an Organo-zeolitic Substrate and Ecological Significance. *Plant Soil* **2001,** 219.
20. Kikteva, N. T., Emelina, G. V. Use of Zeolite Tuff in Forestry. In *Natural zeolites in National Economy*; Novosibirsk. Oart 1, 1991.
21. Stoilov, G. I. Results from the Experimental Work on the Application of Natural Zeolites in Plant Growing. In *National Zeolite*; BAS Sofia, 1986.

22. Oglishvili, T.; Shatirishvili, I.; Andronikashvili, T. Effect of Clinoptilolite Containing Tuffs on the Germination of Wheat and Weed. *Bull. Georgian. Acad. Sci.* **1998,** 158 (2).
23. Tsitsishvili, G. V.; Andronikashvili, T. G.; Kardava, M. A. *Natural Zeolite in Plant-Growing*. Metsniereba: Tbilisi, 1992.
24. Skhirtldze, N. M. Ed. *Sedimentary Zeolite of Georgia*. Of Tbilisi State University: Tbilisi, 1991.
25. Mumpton, F. A. La Roca Mogica: Uses of Natural Zeolite in Agriculture and Industry. *Natl. Acad. Sci. USA.* **1999,** 99.
26. Gennaro, M. De'.; Langella A.; Collela C. et al. In *Italian Zeolitized Tuff as Soil Conditioners. Recent Research Advances*. Proceedings of Sofia Zeolite Meeting; 1997; p 95.
27. Ed Kovda, V. A.; Rosanov, B. G. *Soil Science. Part I, "Vysshaya Shkola". M.*, 1988.
28. Pushkina, G. P.; Lian, P. M.; Bushkovskaya, L. M.; et al. Use of Zeolite in Herbs Growing. Chemical-Pharmaceutical Magazine. Folum, M.; 1996, #9.
29. Andronikashvili, T.; Urotadze, S.; Kvernadze, T.; et al. On Evaluation of Comparative Seedling Emergence and Growth of wheat Seed of Application of Laumontite into Soil. *Bull. Georgian. Acad. Sci.* **2001,** *163*, #3.
30. Iskenderov Sh., Mamedova S. N. The Utilization of Natural Zeolites in Azerbaijan for Increasing Yield of Wheat. In *Occurance, Properties and Utilization of Natural Zeolites*; Callo, D.; Sherry, H. S., Eds.; Acad Kiado: Budapest, 1988.
31. Andronikashvili, T. G.; Tsitsishvili, G. V.; Gamisonia, M. K.; et al. In *Effect of Clinoptilolite on Biological Activity of Acidic Soil*. Slovzeo 84, Conference on the Study and Use of Natural Zeolites; Czechoslovakia, Part 2, 1984.
32. Leggo, P. J.; Redesert, B. Use of Organo-zeolite Fertilizer to Sustain Plant Growth and Stabilize Metallurgical and Mine Waste Sites. *Mineral Mag*. **2001,** *65* (5).
33. Andrews, R. D.; Kimi, S. B. *Improvements in Yield and Quality of Crops with Zeoponic Fertilizer Delivert Systems: Turf, Flowers, Vegetables and Grain*; Montpelier: France, 2001.
34. Naskidashvili, P. P.; Naskidashvili, M. P.; Naskidashvili, I. P. The Results of Crossing the Variety of *Arabicum Jakubs* wild Growing Wheat *T. Dicoccoides Schuebl* with Cultural Species of Wheat. *Ann. Agrar. Sci.* **2004,** #4.

CHAPTER 14

Physical and Chemical Characteristics of Pumice from Some Regions of Georgia and the Prospects for its Use in Lightweight Concrete with Environmental Advantages

GIORGI TSINTSKALADZE[1], TEIMURAZ KORDZAKHIA[1], RAJDEN SKHVITARIDZE[2], TINATIN SHARASHENIDZE[1], MARINE ZAUTASHVILI[1], and GIORGI BERIDZE[3]

[1]*Petre Melikishvili Institute of Physical and Organic Chemistry, Ivane Javakhishvili Tbilisi State University, Tbilisi, Georgia*

[2]*Scientific Center "NanoDugabi," Georgian Technical University, Tbilisi, Georgia*

[3]*Al. Janelidze Geological Institute, Department of Petrology, Mineralogy and Lithology, Ivane Javakhishvili Tbilisi State University, Tbilisi, Georgia*

ABSTRACT

Lightweight concrete is a building material with a density not exceeding 2000 kg/m^3. Its structure is determined by the structure of the used components and manufacturing methods. The advantages of these building materials are their low density, good thermal insulation, frost resistance, high thermal resistance, and ease of use of blocks, which is due to their large size and low weight; also important is their wide range, which allows you to choose the optimal material composition suitable for operating conditions.

Environmental Technology and Sustainability: Physical, Chemical, and Biological Technologies for Environmental Protection. Tamara Tatrishvili, Ann Rose Abraham, & A. K. Haghi (Eds.)

In lightweight concrete production, pumice, along with binders, is one of the main components of their composition. Pumice, or porous volcanic glass, is formed during a volcanic eruption in conditions of strong lava boiling. These minerals are frothy, have a low volumetric weight, high porosity, and high toughness. All these properties make them valuable mineral raw materials that are mainly used as additives for lightweight concretes and hydraulic cements. Pumice is abundant in Georgia (more than 35 million m^3). The prospects for the use of these minerals are great, so the study of their physical and chemical properties is relevant and obligatory.

We studied six pumice samples from four different locations in the Javakheti region of Georgia (two Khulgumo samples, two Okami samples, one Modegami sample, and one Paravani sample). Chemical, X-ray diffractometric, infrared spectroscopic, petrographic, and granulometric methods were used for investigations. The chemical and mineralogical compositions of the analyzed samples, peculiarities of their structures, bulk and compacted samples' volume masses, and their granulometric compositions were determined. The slags with the best properties will be recommended as additives for use in lightweight concrete production.

14.1 INTRODUCTION

For the sustainable development of the country, it is necessary to clearly define and work out such priority areas of the economy as the production of highly efficient products using local natural resources.

Pumice reserves in Georgia exceed 35 million m^3,[1] so the prospects for their use are very high. Pumice is a variety of acidic volcanic and magmatic rocks formed by volcanic eruptions in the upper layers of acidic lavas during boiling[2,3] and differing from each other in the content of silicon in minerals as well as in structural and textural features. There are fibrous, vesicular, frothy, and cellular pumice textures. Their structure may be finely porous or coarsely porous. The average porosity is 80%, the density is 2–2.5 g/cm^3, and the hardness according to the Mohs scale is 5–5.6. The volume weight is 0.3–0.9 g/cm^3. Pumice is chemically inert and begins to soften at 1300–1400°C.[4]

Due to the high porosity, pumice is a good thermal insulator, and a large number of closed pores ensure its good frost resistance. Pumice from different deposits has different textures. For their practical use, the size of the pores and the nature of the glassy substances included in their composition are of great importance. According to the SiO_2 content, pumice igneous rocks are divided into acidic igneous (content of SiO_2 is more than 65%), medium igneous (content of SiO_2 is 54–65%), basic igneous (content of SiO_2 is 45–52%), and ultra-basic igneous rocks (content of SiO_2 is below 45%).[5] The minerals that form such igneous rocks in most cases are quartz, feldspar, and mica. Amphibole, plagioclase, monoclinic, and rhombic pyroxenes can also be found in pumice as crystalline inclusions. Pumice is mainly used as **lightweight aggregates** for concretes, as well as a hydraulic additive to cements and lime.[6–10]

Lightweight concretes are made with porous aggregates. Their densities do not exceed 2000 kg/m^3. The structure of such concretes is determined by the structure of their constituent components and manufacturing methods. The composition of this type of building materials is not regulated at the level of state standards; there are only recommendations. Manufacturers independently determine the percentage content of components. The advantages of these building materials are their not very high density, good thermal insulation, frost resistance, and ease of use of blocks, which is due to their large size and low weight. Their wide range, allowing you to select the optimal composition of the material suitable for the operating conditions, is also important. An attention is drawn to their high thermal resistance. Various types of porous lightweight concretes that are used in bearing structures, as well as bearing structures that also perform the function of a thermal insulator, are known today. Of the many types of concrete, some are widely used and some are used in limited quantities. Currently, new varictics of concrete have been obtained that have not yet been used in practice, although there is a prospect of their use in the future.[9,10]

14.2 EXPERIMENTAL METHODS AND MATERIALS

Six samples of four different locations of Javakheti region of Georgia were studied:

- two samples of Khulgumo location of Ninotsminda Municipality—gray and reddish;
- one sample of Paravani location of Ninotsminda Municipality—reddish;
- two samples of Okami location of Akhalkalaki Municipality—gray and reddish;
- one sample of Modegham location of Borjomi Municipality—gray.

Chemical, X-ray diffraction, IR spectroscopic, petrographic, and granulometric analysis methods were used for the study. X-ray diffractometric analysis was carried out on a Dron-4 device (Russia). The X-ray diffractometer was connected to a personal computer via the USB port multimeter AX-18B and the corresponding PC software-PC-Link, which allows the processing of experimental data in Excel format; infrared spectrometric analysis was carried out on an Agilent Cary 630 FTIR Spectrometer in the range of 350–5000 cm^{-1} (USA); chemical analysis was carried out on a Spectroscout XEP-04 (Germany).

14.3 RESULTS AND DISCUSSION

14.3.1 RESULTS OF CHEMICAL ANALYSIS OF PUMICE SAMPLES

The results of the chemical analysis of the studied pumice samples are presented in Table 14.1. In terms of SiO_2 content, the samples have a medium acidic environment. It should be noted that they contain large amounts of iron, magnesium, and calcium.

14.3.2 RESULTS OF PETROGRAPHIC ANALYSIS

Petrographic analysis of the samples showed that the gray sample from the Khulgumo location is a porous rock in the main amorphous mass, in which the smallest plagioclase grains[9] (albite—Na $[AlSi_3O_8]$ and anorthite—Ca $[Al_2Si_2O_8]$) with pore sizes greater than 2 mm are observed. The red Khulgumo sample is a hard porous rock with a pore size of 1–2 mm. Plagioclase grains (albite—Na$[AlSi_3O_8]$ and anorthite—Ca $[Al_2Si_2O_8]$) are mainly observed in the amorphous mass. The rock pores are filled with zeolite $Mx/n[AlxSiyO_2(x+y)]\cdot pH_2O$ by rather large porphyry grains over 1 mm in size. In some places, reddish-brown iron rust is observed.[9]

The red sample of Paravani locality is also a hard porous rock, the main mass of which is amorphous and has a reddish hue under the influence of iron rust. The smallest grains of plagioclase, $Na[AlSi_3O_8]$ and $Ca[Al_2Si_2O_8]$), are occasionally observed in the mass. Relatively large porphyry grains of pyroxene $CaMg(S_{i2}O_6)$ or $CaFe(S_2O_6)$, plagioclase $[Na(AlSi_3O_8]$, and amphibole $Ca_2(Mg,Fe)_5\{S_{i4}O_{11}\}_2\{OH\}_2$ over 1.5 mm in size are often found in the rock.[2,9]

TABLE 14.1 Chemical Analysis of Pumice Samples, %.

Material name, Quarry	Loss when heated	SiO_2	Al_2O_3	Fe_2O_3	CaO	MgO	MnO	Na_2O	K_2O	SO_3	Total
Gray pumice, Khulgumo	1.36	59.75	16.88	6.5	6.27	3.48	0.02	3.14	1.83	0.00	99.20
Red pumice, Khulgumo	0.82	58.38	16.88	7.1	6.46	3.98	0.00	3.51	1.90	0.00	99.06
Red pumice, Paravani	1.83	58.86	17.00	6.3	6.07	3.66	0.01	3.33	2.22	0.01	99.33
Grey pumice, Okami	3.85	55.25	15.58	9.4	6.71	4.20	0.01	2.55	1.35	0.01	98.94
Red pumice, Okami	0.84	53.86	17.07	9.3	7.48	3.94	0.05	3.87	1.56	0.10	98.04
Gray pumice, Modegham	1.81	55.78	15.94	9.4	7.04	3.72	0.05	2.91	1.50	0.01	98.16

The gray-colored sample of the Okami locality is a highly porous rock, the size of which is so large that most of the hewn sample (more than 80%) falls on a porous area. The main amorphous mass is black, grayish areas are found in it (reddish are less common). No small plagioclase grains were observed here, which distinguishes this sample from other studied samples. Large porphyry segments of plagioclase larger than 2 mm are often found in the rock. The red porous rock of the same locality also has large pores. The main amorphous mass is black, reddish areas are found in it. Here, as in the black samples of the same locality, there are no small plagioclase grains. Instead, the rock contains large plagioclase porphyry grains over 3 mm in size (Fig. 14.1).

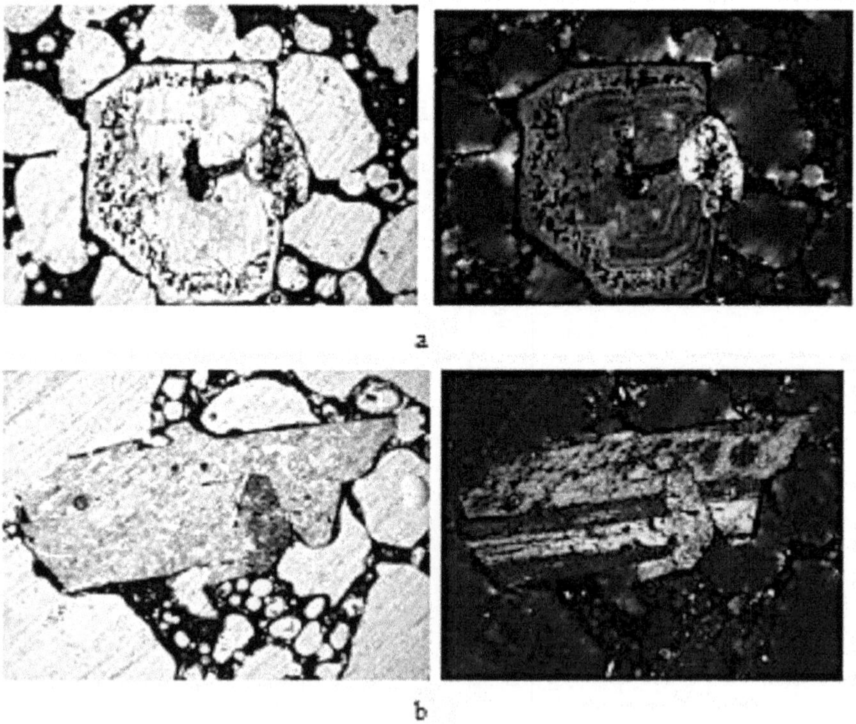

FIGURE 14.1 Petrographic section of pumice of the Okami location: (a) red pumice and (b) gray pumice.

The gray porous sample of Modegham location is constructed from highly chlorinated $(Mg,Fe)Al(AlS_{13}0_{10})(OH)_8$ volcanic glass. Small grains of plagioclase and porphyry segments of the same mineral are developed in the main amorphous mass. Plagioclase is rarely observed in the amorphous mass. Thus, the mineralogical composition of the examined samples is similar, although there are differences as well. Interestingly, the porosity of the samples is significantly different from each other, so the pumice studied has a wide range of applications depending on what properties of the concrete we want to obtain.

14.3.3 RESULTS OF BOTH X-RAY DIFFRACTOMETRIC ANALYSIS AND IR SPECTROSCOPY

X-ray diffraction analysis (XRD analysis) as well as X-ray fluorescence spectral analysis of pumice samples were performed for structural as well as mineralogical studies (Figs. 14.2 and 14.3). XRD analysis showed that all pumice samples contained a large amount of the amorphous phase quartz (SiO_2); both types of feldspar, potassium-sodium-containing (alkaline) and sodium-calcium-containing (so-called plagioclase subgroup), were observed in all samples. In small amounts, samples also contained mica.

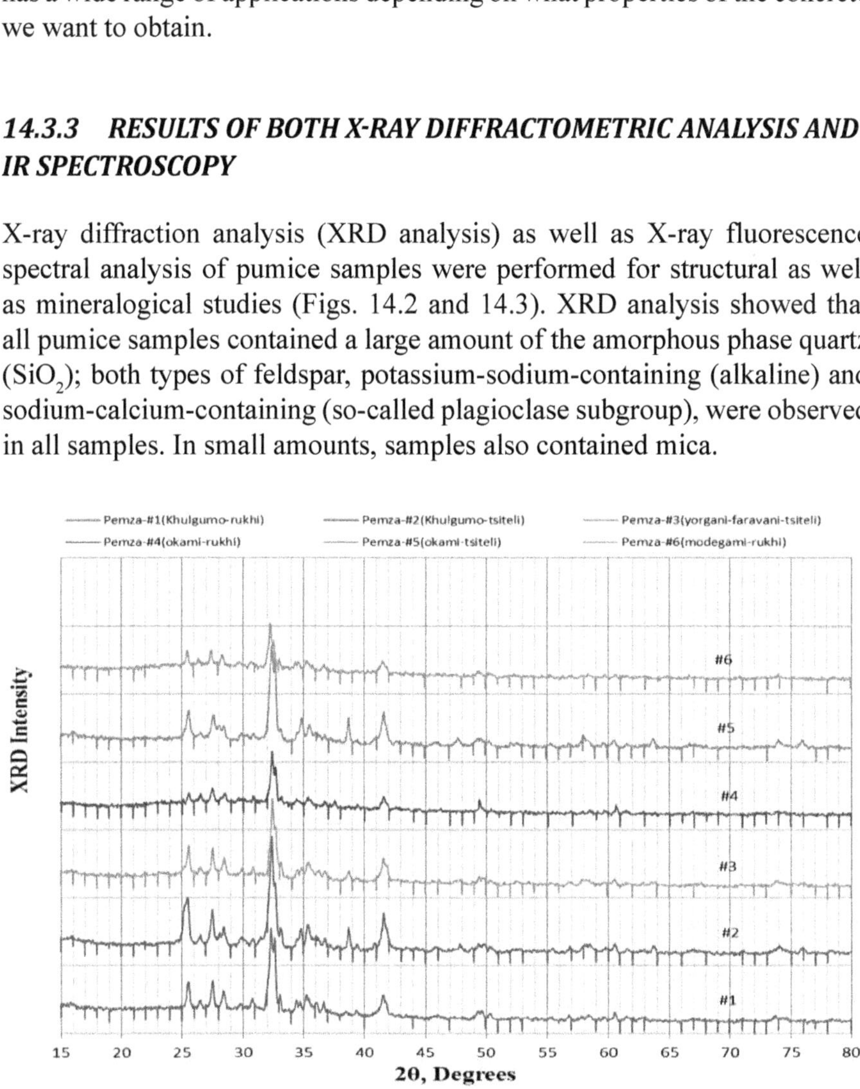

FIGURE 14.2 X-ray diffractograms of pumice samples (with cobalt tube).

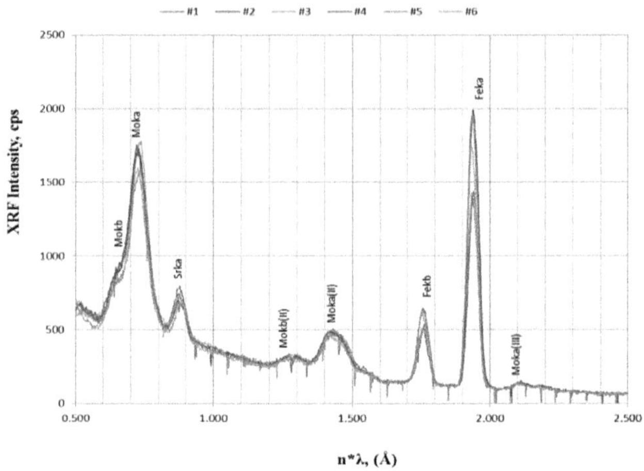

FIGURE 14.3 X-ray diffractograms of pumice samples.

X-ray fluorescence analysis was performed to detect iron atoms in the crystallographic plane. During the experiment, the solid content of iron in pumice samples was established, as was the presence of strontium in all samples, which was not detected by chemical analysis (Figs. 14.2 and 14.3).

IR spectral analysis was performed in the middle and far regions of IR spectrum. Table 14.2 shows the IR spectra of the test samples. Studies have shown that pumice is a mixture of silicates and aluminosilicates of various structures.[11,12] Both amorphous and layered silicates and aluminosilicates are present here.[12] In addition, there are quite large amounts of iron, magnesium, calcium, sodium, and potassium cations in the samples,[10] so the spectra are very diverse.

TABLE 14.2 Vibrational Assignment of the IR Spectrum of Studied Samples.

Vibrational assignment	Pumice samples					
	Grey pumice, Khulgumo	Red pumice, Khulgumo	Red pumice, Paravani	Gray pumice, Okami	Red pumice, Okami	Gray pumice, Modegham
σ Si–O, δ Si–O–Si	465, –	431, 470	441, 466	402, 465	433, 468	410, 466
Mg–O–Si, Fe–O–Si, δ Si–O, δ O–Si–O	543, 580, 629	541, 582, 632	543, 570, 627	537, 577, –	541, 582, 630	541, 574, –
ν Si–O–Si	779	789	778	745	770	–
ν Si–O, ν Si–O–Si(Al)	1028, 1089	1037, 1094	1045, 1077	1035	1018, 1093	1043, 1095
δ H–O–H	1647	1651	1637	1654	1654	1651
ν OH^-	3503	3501	3526	3503	3502	3538

It should be noted that the vibrational frequency of the Me^{2-}(O–Si) bond of Mg- and Fe-substituted pumice coincides in the range of 530–630 cm^{-1},[10,12] so it is difficult to assign particular bands in this region. Here, the band frequencies and their intensities vary depending on the content of bi- and trivalent iron cations.[11] It should be noted that the overall picture of the spectrum of all investigated samples is similar, although the band intensities and frequencies are different due to different chemical and mineralogical compositions of the minerals.

The granulometric composition of the studied porous materials as well as the moisture, density, and bulk mass were determined. The results are presented in Table 14.3.

TABLE 14.3 Physical Characteristics of Porous Materials.

Material name, Quarry	Moisture, %	Bulk mass, kg/m3	Tightened bulk mass, kg/m3
Gray pumice, Khulgumo	1.4	1077	1232
Red pumice, Khulgumo	2.2	993	1182
Red pumice, Paravani	2.7	1078	1233
Gray pumice, Okami	9.0	590	723
Red pumice, Okami	2.1	775	937
Gray pumice, Modegham	3.7	765	902

14.4 CONCLUSION

During the study, it was found that the tested pumice samples have a medium acidic environment (content of SiO_2 is 54–65%), although they differ both in chemical and mineralogical compositions and in pore sizes. Both their bulk and compacted volume weights are different. The experiment showed that according to its results, it is possible to select and recommend porous materials of different compositions and volumes to obtain structural lightweight concretes of appropriate strength and weight.

KEYWORDS

- pumice
- volcanic glass
- lightweight concrete
- porous materials
- additives

REFERENCES

1. Chapter: Pumice and Volcanic Ash. In *Mineral Resources of the USSR of Georgia*, Tbilisi (in Russian), 1993; pp 837–842.
2. Tutberidze, B. D. Chapter: The Geological Structure of the Region, In *Молодой вулканизм восточной части Джавахетского нагорья*, TSU, Tbilisi (in Russian), 1990; pp 8–122.
3. Tutberidze, B. D. Chapter: Volcanism of the Structure and Formation Zone of the Georgian Rock. In *Geology and Petrology of Alpine Late Orogenic Magmatism in the Central Part of the Caucasian Segment*, TSU, Tbilisi (in Russian), 2004; pp 156–168.
4. Shoroog, A.; Hasan, A. Physical Properties of Mesoporous Scoria and Pumice Volcanic Rocks. *J. Phys. Commun.* **2021,** *5* (11), 1–15. https://iopscience.iop.org/article/10.1088/2399-6528/ac3a95
5. Bahri, E.; Ali, S.; Sedef, D.; et al. Characterization of Acidic Pumice and Determination of Its Electrokinetic Properties in Water. *J. Powder Technol.* **2010,** *197* (1–2), 129–135. DOI:10.1016/j.powtec.2009.09.005
6. Parmo, T.; Hafiz R.; Efa S.; et al. In *The Mechanical Properties of Lightweight Concrete Made with Lightweight Aggregate Volcanic Pumice*, Proceedings of the Built Environment, Science and Technology International Conference, BEST ICON, 2018; pp. 167–171. https://www.scitepress.org/Papers/2018/89065/89065.pdf
7. Karthika, V. R. B.; Vidyapriya, K. V.; Nandhini Sri, K.; et al. Experimental Study on Lightweight Concrete Using Pumice Aggregate. *Mater. Today: Proc.* **2021,** *43*, 1606–1613. https://doi.org/10.1016/j.matpr.2020.09.762
8. Parhizkar, T.; Najimi, M.; Pourkhorshidi, A. R. Application of Pumice Aggregate in Structural Lightweight Concrete. *Asian J. Civ. Eng.* **2012,** *13* (1), 43–54. file:///C:/Users/User/Downloads/103820120103.pdf
9. Mücip, T.; Zeynel, Y.; Orhan, İ.; et al. Tolga Depci Effect of Physical, Chemical and Electro-kinetic Properties of Pumice Samples on Radiation Shielding Properties of Pumice Material. *Ann. Nucl. Energy*. **2014,** *65*, 290–298. http://dx.doi.org/10.1016/j.anucene.2013.11.021
10. Keshelava, B.; Skhvitaridze, R.; Tsintskaladze, G.; et al. Eremadze Physico-chemical Research of Some Natural Porous Materials of Georgia. *Sci. Tech. J. Energy.* **2017,** *2* (82), 110–112 (in Russian).
11. Nakamoto, K. Chapter 2: Applications in Inorganic Chemistry. In *Infrared and Raman Spectra of Inorganic and Coordination Compounds*; A Wiley-Interscience Publication John Wiley and Sons: New-York, 1991; pp 149–354. DOI:10.1002/9780470405840
12. Plyusnina I. I. Chapter 4: Basic Regularities of Infrared Spectra of Mineral Classes. In *Infra-red Spectra of Minerals*, Moscow: Publishing House of the Moscow University, 1977; pp 99–163.

PART III

Advanced Functional Materials

CHAPTER 15

Smart Nanomaterials for Sustainable Green Energy Applications: Environmental Technology and Sustainability

KODUVARATHODI VADAKKETHIL ARUNDHATHI[1], NEENAMOL JOHN[1], THOMAS ABRAHAM[2], K. G. AMBADY[3], and BEENA MATHEW[1]

[1]*School of Chemical Sciences, Mahatma Gandhi University, Kottayam, Kerala, India*

[2]*Catholicate College, Pathanamthitta, Kerala, India*

[3]*Department of Special Education, National Institute for the Empowerment of Persons with Intellectual Disabilities, Telangana, India*

ABSTRACT

The design of chemical processes and chemical products that limit or eliminate the use or synthesis of hazardous compounds is known as green chemistry. It applies throughout a chemical product's life cycle, including its design, manufacture, usage, and final disposal. Introducing green chemistry to the nanoworld creates a new era in the scientific field of green nanotechnology. Green nanotechnology contributes nanosized materials that can be synthesized safely and are less hazardous to the environment. Nanosized materials have advantages over conventional materials because of their shape, size, surface morphology, and composition. The synthesis and applications of biobased nanomaterials are a suitable way to reduce their harmful effects on health and the environment. The implementation of 12 principles of green chemistry throughout the research makes the world

Environmental Technology and Sustainability: Physical, Chemical, and Biological Technologies for Environmental Protection. Tamara Tatrishvili, Ann Rose Abraham, & A. K. Haghi (Eds.)

more sustainable and less toxic. This chapter includes fabrication methods of different green nanomaterials and their applications to produce various energy feedstocks, such as solar cells, batteries, and supercapacitors. The enhanced rate of production of sustainable green energy through the green synthesized nanomaterials makes the future world more eco-friendly and energy efficient.

GRAPHICAL ABSTRACT

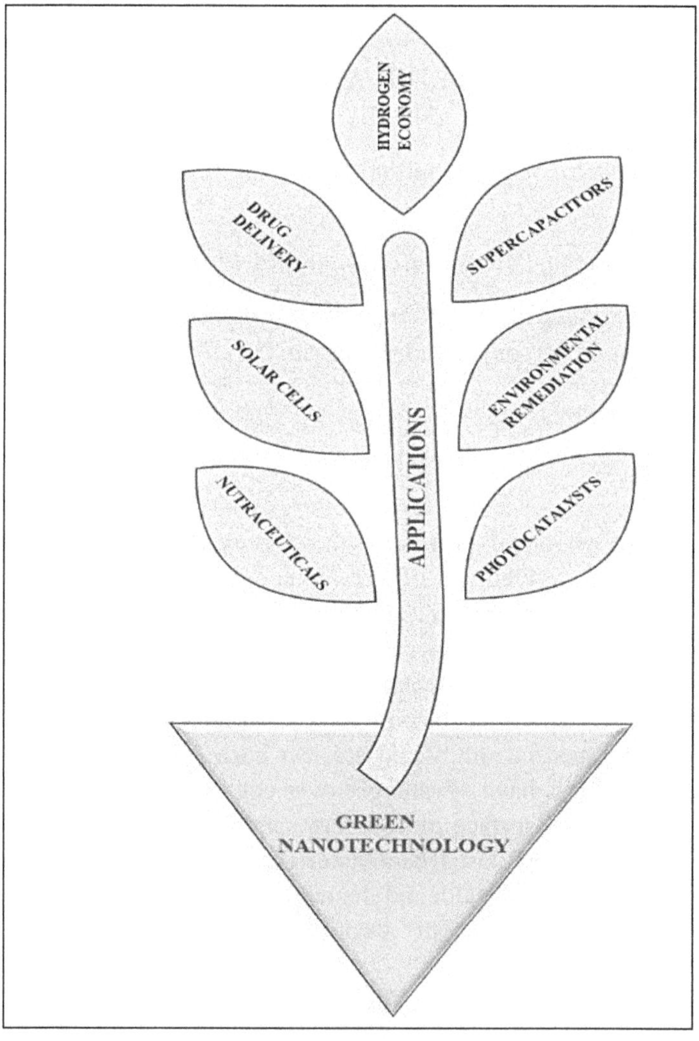

15.1 INTRODUCTION

Development is a socioeconomic and political concept that has to be done with keen observation of the surroundings. We can categorize different countries in the modern world in the aspect ratio of development into developed, developing, and underdeveloped. Day-to-day improvement in science and technology is the main cause of industrial and personal stability of human life in developed countries. The Commission for Science and Technology for Development (CSTD) under the United Nations is the organization to evaluate the development index of different countries based on science and technology. CSTD encourages development with suitable sustainable goals. So, sustainability becomes an unavoidable word in the new scientific era. Scientific innovations should lead to the commercial and sustainable development of different countries. The experiments on different scientific sectors should be successful if they can enrich the economic condition and also reduce the impact of toxic effects on the environment.

Sustainable development is a term that arrived years ago, but the implementation of sustainability is still under question. Preserving resources for future generations is one of the key differences between sustainable development policy and traditional environmental policy, which also attempts to internalize the externalities of environmental degradation.[1] The key point that has to be remembered to achieve sustainability in society is to preserve the environment as much as possible. Irradiation of toxic materials from nature is the primary principle. Industrialization and urbanization lead to the dumping of both industrial and household wastes into natural resources. To avoid these phenomena, we have to obey the 5Rs which include: Reuse, Recycle, Refuse, Repurpose, and Reduce. Beyond the 5R principle, the implementation of principles of green chemistry can introduce more pathways to sustainable development. Green chemistry is a new methodological approach used in chemistry to attain an eco-friendly culture in chemistry. Recently, one-pot synthesis of multicomponent reactions in accordance with the green chemistry principle is very relevant.[2] Nanotechnology has an essential role to play in international sustainability initiatives.[3] The unique features that nanoparticles display due to their tiny size are the primary reason for their diverse applications; however, this may pose a significant threat to the environment.[4]

The cumulative integration of green chemistry principles with nanotechnology introduces a new term known as "Green Nanotechnology."[5] Green nanotechnology is described as the development of clean technologies to minimize human health and potential environmental concerns. Green nanotechnology is, in many respects, the natural progression of nanotechnology, driven by the need to reconcile the vast range of possible and practical nanotechnology applications with the requirement for so-called sustainable development.[6]

This chapter deals with novel innovations in the field of green nanotechnology. It describes the synthesis of green nanoparticles, the application of nanotechnology in the green energy strategy, and the environmental remediation practices that can be achieved using nanomaterials. And at last, we address how existing and future green nanotechnology capabilities match with and support the United Nations' Sustainable Development Goals.

15.2 GREEN NANOTECHNOLOGY: A NEW INITIATIVE

Green nanotechnology introduces a new perspective in material science, that is, the production of safer nanomaterials and the application of nanotechnology for eco-friendly purposes. There are some specific reasons for the rise of such a novel initiative in the material world. The main reasons include[7]:

- The scarcity of numerous natural resources, such as clean water, fossil fuels, and certain minerals
- The environment's limited carrying capacity emphasizes the importance of regulating or reducing emissions and rebuilding damaged areas
- As a basic component of the concept of sustainable development, intra- and intergenerational equity necessitates consideration of the distribution of risks and benefits of new technologies among the population and in the global dimension
- The question of participation's sustainability has implications for the processes of opinion formation and decision-making in designing technology and its interactions with the public.

Nanomaterials having a size in the nanorange (10^{-9} m) have some peculiar properties. The sizes, shapes, and various morphologies associated with nanomaterials, such as nanoparticles, tubes, wires, fibers, etc., make them efficient to produce new materials targeted to several applications. Nanomaterials (NMs) are gaining popularity in technological applications due to their tunable chemical, physical, and mechanical properties, as well as their superior performance when compared to bulkier alternatives. Due to the nanorange size, nanomaterials have a high surface area (surface-to-volume ratio) and associated high reactivity.[8] According to the morphological differences, the nanomaterials are classified into nanowires, nanotubes, nanopillars, nanofibers, nanospheres, nanocubes, and nanopyramids. Variance in the chemical composition again categorizes nanomaterials into single constituent nanoparticles, composites, metallic nanoparticles, ceramic nanomaterials, carbon-based nanomaterials, branched dendrimers, quantum dots, nanogels, polymeric nanomaterials, and core–shell nanomaterials.[9] Due to this wide variety in nanomaterials, different nanomaterials are produced through green synthesis and applied for targeted green applications.

Green chemistry is a sustainable approach to the processing, synthesis, and application of chemical substances to eliminate health and environmental risks.[10] Green chemistry is often taught as a set of 12 Anastas and Warner principles. The principles include guidance for expert chemists to use in the implementation of new synthesis and technological processes.

a. *The 12 main concepts of green chemistry*:11

1. Avoid using wastes as much as you can. Act in advance.
   It is better to avoid waste. "Act in advance" instead of cleaning up waste after it has been created.
2. It is much better to think of using the synthetic methods.
3. Consider using less dangerous chemical synthesis to save human health and our living environment.
4. Try to design safer chemicals with minimum toxicity.
5. Try to consider using much safer solvent as auxiliaries.
6. Design for a clean technology to reach the energy efficiency.
7. Consider using renewable raw materials.
8. Reduce derivatives by minimizing temporary modification processes.
9. Consider selective catalysis reagents.
10. Design the chemical products for degradation for sustainability.
11. Before the formation of hazardous substances, consider using real-time control for pollution prevention.
12. Select safer substances to minimize chemical accidents.

15.3 GREEN SYNTHESIS

Every pathway for the production of different nanoparticles has both advantages and disadvantages. The normal pathways that are used for the production of different NPs include the hydrothermal method, microwave method, chemical vapor deposition, chemical coreduction, magnetic and mechanical stirring, coprecipitation method, and magnetron sputtering. Both bottom-up and top-down synthesis methods are used to produce different nanoparticles. In this, the green synthesis is a bottom-down method of fabrication of various NPs, such as carbon nanotubes and quantum dots. The green synthesis is the pure biological method adopted to prepare the nanoparticles; hence, the source of production is different organisms such as bacteria, fungi, yeast, and plants. The advantages of biological synthesis over conventional techniques include energy efficiency and low-cost production, fewer accidents, a safe product, economics, lesser waste, competitive advantages, and protection of human health and communities.[12]

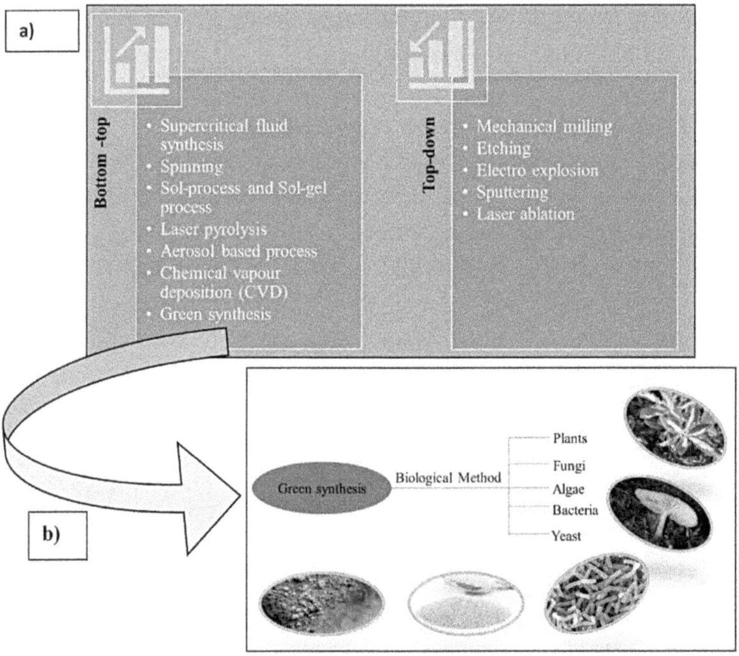

FIGURE 15.1 a) Different methods of preparation of nanomaterials and b) feedstocks for green synthesis.

Source: Adapted from Ref. [12]

15.3.1 BIOSYNTHESIS OF METAL NANOPARTICLES

The production of metal nanoparticles from living organisms increases the possibilities of natural nanofactories. There is a list of things that we have to consider in the green synthesis of metal nanoparticles.[13]

1. Selection of the best organisms
2. Optimal conditions for cell growth and enzyme activity
3. Optimal reaction conditions.

Not only the source of the production of nanoparticles must be eco-friendly, but the substances that are used to facilitate the reactions should also be nontoxic and renewable. Water, the universal solvent, is used as the best solvent in the greener approach to nanoparticle synthesis. Most of the synthetic methods of nanoparticle synthesis widely use organic solvents, which are highly toxic and expensive. Instead of organic solvent, the use of biophilic water as a solvent makes the synthesis more sustainable for the environment. S. Machado et al. produced zerovalent iron nanoparticles from leaf extract using water as the solvent.[14] Other solvents, such as supercritical CO_2, are also used sometimes, but there is some complexity in the isolation of nanoparticles by the usage of CO_2-philic surfactants. The choice of various reducing agents is also a matter of concern in the process of nanoparticle synthesis. Normally, reagents that can be seen as reducing agents are $LiAlH_4$, $NaBH_4$, dimethylformamide, etc. These are strong reducing agents and high-level chemicals. Instead of these, the green alternatives have reducing agents such as β-D glucose, ascorbic acid, light-roasted coffee, and prune juice.[15] Poly(methylhydrosiloxane), waste from the silicon industry, is used to reduce phosphine oxide to phosphine.[16]

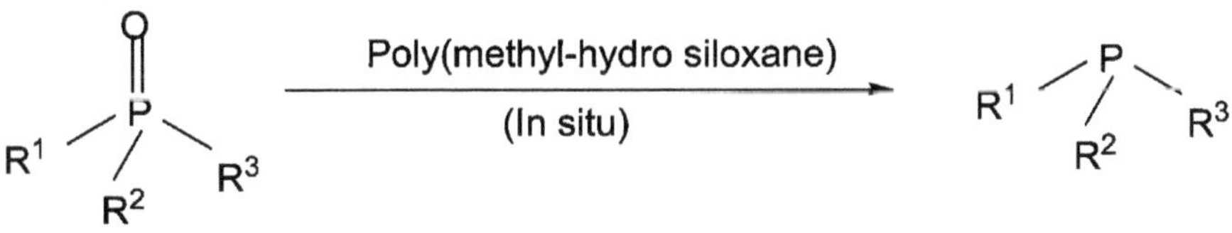

The capping agents are protective stabilizers of nanoparticles in colloidal synthesis because they suppress nanoparticles overgrowth and prevent aggregation and coagulation. Some of the common capping agents are polymer-based capping agents, such as polyethylene glycol (PEG), poly-vinyl pyrrolidone (PVP), and polyvinyl alcohol (PVA). We can also use biocompatible amino acids, carbohydrates, and starch as capping agents.[17,18]

In this chapter, we have to look at the green routes of preparation of some common metal nanoparticles. The inertness of gold metal makes the gold nanoparticle more biocompatible and less reactive.[19] The mixing of cumin with an aqueous solution of sodium tetrachloroaurate produces unprecedented gold nanoparticles. The significance of cumin as a precursor is that it will not allow the produced nanoparticles to aggregate because the cumin phytochemicals get coated on the nanoparticles very strongly.[20] Some of the common plants that we can use for the preparation of AuNPs are *Hibiscus rosa sinensis, Salvia officinalis, Lippia citriodora, Pelargonium graveolens, Punica granatum, and Coriandrum sativum.*[21,22] *Several microorganisms are also used as initial substrates. Both extracellular and intracellular methods can be used for the biogeneration of AuNPs. The extracellular method is common because it reduces the number of processing steps.*[23]

Synthesis of silver nanoparticles (AgNPs) employs *Polyalthia longifolia* leaf extract as a capping and reducing agent, as well as D-sorbitol is utilized to improve the nanoparticles' stability. AgNPs were produced using a *Foeniculum vulgare* extract, and their antibacterial efficacy against *Escherichia coli* and *Staphylococcus aureus* was reported. Both gold and silver nanoparticles can be prepared by eco-friendly route with the usage of *Emblica officinalis* (Indian gooseberry). The primary metabolites of plants can act as reducing and stabilizing agents in the conversion of metallic silver into silver nanoparticles.[24]

Viability, commercial feasibility, eco-friendliness, reliability, no waste generation, and simplicity are the major attractive features of green synthesis. By using *Raphanus sativus* var. longipinnatus leaf extract, the fabrication of spherical cobalt nanoparticles is reported. Face-cubic structured CoNPs are synthesized from a gram-positive bacterium named *Bacillus thuringiensis*. The prepared nanoparticles are efficient against malaria and dengue carriers (*Aedes aegypti* and *Anopheles subpictus*). Similarly, CO_3O_4 nanoparticles can be synthesized from a fungus called *Aspergillus nidulans* with an average size of 20.29 nm through the extracellular method.[25]

El-Belely et al. conducted the cyanobacterial-mediated green synthesis of zinc oxide nanoparticles. The first step is the biomass preparation from the microalga *Arthrospira platensis*, and after that, the biomass is subjected to centrifugation, and the filtrate is used for the preparation of ZnONPs.[26]

For the treatment of eutrophic wastewater, iron nanoparticles are developed from the eucalyptus leaf extracts. The synthesis, as well as application of the work, is in accordance with the 12 principles of green chemistry. Wang et al. collected the pure eucalyptus leaves and washed it thoroughly with deionized water. After that, the leaves were dried under sunlight. The leaf extract is prepared through constant boiling of the leaves in deionized water at 80°C. With the addition of $FeSO_{4,}$ the appearance of a black color was an indication of the reduction of Fe^{+2}.[27] Phosphorous and nitrogen are the leading elements that can cause eutrophication at a higher rate. The usage of the developed FeNPs in the wastewater reduced the pollution within 21 days. For 21 days, the removal efficiency of developed NPs was 71.7% and 30.4% for nitrogen and phosphorous, respectively. The chemical oxygen demand (COD) is an indication of pollution in water. The amount of dissolved oxygen required in water to oxidize chemical organic compounds is known as the chemical oxygen demand (COD). The synthesized FeNPs have 84.5% efficiency in the removal of pollutants. The fungal-based preparation of copper nanoparticles is also reported in the latest era. Copper nanoparticles are synthesized from *Aspergillus niger* strain STA9. The fungal isolates were collected from different biosources, such as rotten fruits, vegetables, and soil.[28] Two main approaches in preparing bionanoparticles are in vitro biosynthesis and in vivo biosynthesis. In vitro biosynthesis uses biological extracts or particular compounds, and in vivo biosynthesis uses living cells. However, in other circumstances, it is difficult to tell whether the synthesis mechanism is particular, that is, biosynthetic, or unspecific, that is, biomimetic.

15.3.2 BIOSYNTHESIS OF ORGANIC AND POLYMER-BASED NANOPARTICLES

As we know, metal/metal oxide nanoparticles are synthesized in a greener way that reduces the toxicity of the preparatory path of inorganic metal/metal oxide nanoparticles. The invention of a novel synthetic route for the preparation of urea from ammonium nitrate breaks down the conventional idea of chemistry. There is a belief that an organic compound can be prepared only from another organic compound. The synthesis of urea from an inorganic compound demolished this concept. Likewise, we can synthesize inorganic metal/metal nanoparticles from organic sources. The organic and natural polymer-based nanoparticles are conventionally prepared from bioresources. The biological polymers present in nature include protein, carbohydrates, and lipids that can be used as precursors for the production of polymer-based nanoparticles. Silk, keratin, collagen, elastin, and corn zein are some of the protein polymer nanoparticles.[29]

15.4 APPLICATION

15.4.1 ENERGY-RELATED APPLICATIONS

15.4.1.1 SOLAR CELL

A solar cell or photovoltaic cell converts solar energy directly into electrical energy. Nowadays, the other methods of energy production are not as sustainable as solar energy. Solar energy is clean, sustainable, and not a cause of atmospheric pollution. The problem faced for enhancing the demand for solar cells is the expense. We have to concentrate our research on the production of less expensive materials that can be used as solar cells or solar panels. In the area of materials chemistry, the innovation of eco-friendly and less expensive materials, which can convert solar energy into electrical energy through a single step, should be a landmark. There exist various semiconductors and nanocomposites which can act as solar energy converters. Silicon due to its high corrosion resistance, long-term durability, and optimal thermal expansion properties served as a peculiar material for solar cells. The first-generation solar cells are mostly made up of silicon; they have high demand due to their efficiency, even though they are expensive. The second-generation solar cells are made of hybrid materials such as copper indium gallium selenide and cadmium telluride (CdTe). The flexibility of these hybrids is attractive. But the third-generation solar cells are now devoid of semiconducting techniques. Quantum dot technology, tandem/multijunction cells, hot carrier cells, up- and down-conversion technologies, and solar thermal technologies are included in it. Organic photovoltaics (OPVs), copper zinc tin sulfide (CZTS), perovskite solar cells, dye-sensitized solar cells (DSSCs), and quantum dot solar cells are some of the third-generation members.

A team from the University of California, Berkeley, invented a pathway to produce solar cells from cheap plastic wastes. It is a very economical,

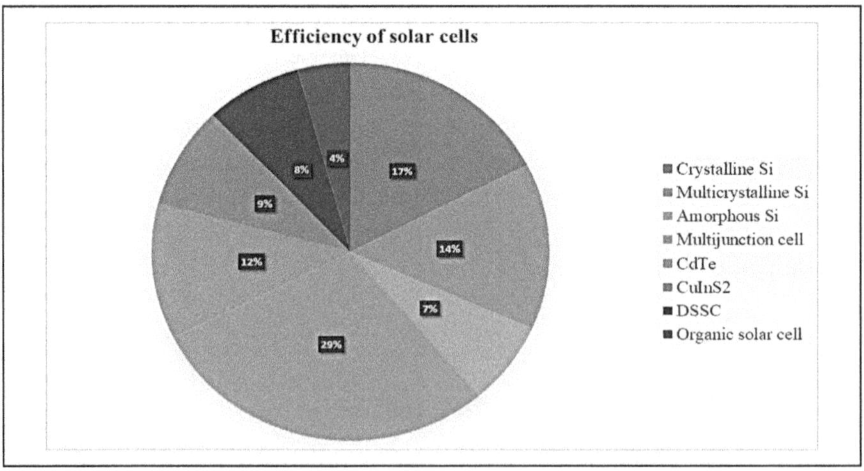

FIGURE 15.2 The graph showing the percentages of efficiency of different types of solar cells.

Source: Adapted from Ref. [30]

sustainable, and renewable energy production method. Plastic is treated as one of the major pollutants in the soil. Recyclability of plastic wastes and conversion of plastic into high-demandable solar cells is a 100% green approach.[30] The efficiency of different types of solar cells is picturized below.

Is it possible to modify the efficiency of a solar cell using green nanomaterials? The question is something interesting and if the answer is yes, then it should be an asset to renewable energy management. There are ways to increase the efficiency of silicon-based solar cells using nanomaterials. Green tea-modified multiwalled carbon nanotubes have a considerable amount of photovoltaic performance. The incorporation of green tea-modified MWCNTs to poly(3,4-ethylene dioxythiophene): Poly(styrene sulfonate) (PEDOT: PSS)/n-Si facilitates the hole transport in PEDOT: PSS. The green tea-modified MWCNTs help to improve the photovoltaic capacity of the device by splitting of excitation and through the suppression of recombination of charge carriers.[31]

The copper oxide (CuO) is a P-type semiconductor. The development of dye-sensitized solar cells using green synthesized copper nanoparticles has been reported. Due to the narrow bandgap (1.2–1.9 eV), CuO nanoparticles have high photoconductivity; hence, they can be used in photothermal applications.[32]

15.4.1.2 HYDROGEN ECONOMY

The future world has to face big issues with their energy requirements due to plenty of reasons. The conservation of energy states that "energy can neither be created nor be destroyed but it can convert from one form to another." Today's energy resources are mainly based on fossil fuels. The burning of fossil fuels induces the emission of greenhouse gases, and as a result, the average temperature of the earth is increasing. Articles show that there is an increase in global surface temperature of about 0.7°C in the past century, and it will increase above 2.4°C by 2050.[33] Various international nongovernmental agencies and the United Nations itself demand suitable methods to reduce the emission of greenhouse gases like CO_2 and CH_4. One of the methods is the reduction of the usage of fossil fuels. Our transport system is completely employed by the usage of fossil fuels. So, as an alternative to fossil fuels, scientists propose hydrogen as a renewable fuel for a sustainable future. Fuel cells are represented as energy-efficient and clean technology that will almost certainly be used in future vehicles.[34]

The production of hydrogen is possible in different ways:

1. Hydrogen from direct chemical reactions

2. Hydrogen from hydrocarbons
3. Hydrogen from water (water electrolysis).

The production of hydrogen from hydrocarbons, that is, from biomass and fossil fuels and how nanotechnology can help to improve the efficiency of hydrogen production are our questions of concern. About 95% of hydrogen is produced from natural gases using steam methane reforming. Partial oxidation, plasma reforming, autothermal reforming, and coal gasification are also used to produce hydrogen from fossil fuels. The steam reformer is used to react fossil fuels with steam at high temperatures. At high temperatures, components of natural gases like methane, propane, gasoline, petrol, and diesel are converted into hydrogen.[35]

$$C_mH_n + mH_2O = mCO + (m+n/2) + H_2$$

Several existing reports show that smart nanomaterials can act as a catalyst in the steam reformer to improve the production of hydrogen. Cargnello et al. reported Pd@CeO_2 core–shell nanostructured dispersible material that can be used as a catalyst precursor in methanol steam reforming operations.[36] Similarly, nanopowdered zirconia is used to catalyze the steam reform reaction of bioethanol.[37] Recently in 2016, Ali and team proposed coke-resistant nickel nanoparticles for steam reformation of methane.[38]

Generally, biomass is a term that represents organic compounds containing carbon, hydrogen, and other elements. Biomass resources are used in the paper, textile, and food industries. The production of hydrogen as a form of energy from biomass is under discussion. The two main ways to produce hydrogen from biomass are[35]:

1. Thermochemical hydrogen production
2. Biological hydrogen production
   2.1 Biophotolysis of water using green algae and blue-green algae (cyanobacteria)
   2.2 Photofermentation
   2.3 Dark fermentation
   2.4 Hybrid reactor.

The gasification of biomass produces hydrogen according to the following equation:

$$\text{Biomass} + O_2 \text{ or } H_2O \rightarrow CO, CO_2, H_2O, H_2, CH_4 + \text{other}$$

Nanomaterials can catalyze the abovementioned reaction and increase the rate of hydrogen production. Yang et al. proposed that zerovalent iron nanoparticles (Fe(0) NPs) can increase the rate of hydrogen fermentation of grass.[39] Mishra et al. produced nickel oxide and cobalt oxide nanoparticles hydrothermally and studied their effect on the enhancement of production of hydrogen from palm oil mill effluent.[40] The nano-Nickel-Lanthanum-Ferrous-γ Al_2O_3(nano-Ni-La-Fe/γ Al_2O_3) can remove the tar, a waste material produced after the biomass steam gasification.[41] The electrochemical production of hydrogen from water is a simple process. By applying external electrical energy, water is split into H_2 and O_2. The produced hydrogen is stored and applied in the fuel cell converting the chemical energy into electrical energy, and it makes the process reversible.

$$H_2 + O_2 \underset{\text{Electrolysis}}{\overset{\text{Fuel cell}}{\rightleftarrows}} H_2O$$

The photoelectrochemical water splitting is possible with the help of Fe_3O_4 nanoparticles. Fe_3O_4 nanoparticle coated on chiral molecule used for water splitting is an approach for generating hydrogen from water.[42] In a photoelectrochemical reaction, the photons enable the semiconductor to promote a redox reaction.

15.4.1.3 BATTERIES

Batteries are the primary energy storage systems. Batteries are a collection of electrochemical cells, and they can be charged by applying an electric current and discharged whenever required. Batteries are used in clocks, watches, remotes for household requirements, and industrially they are used in automobile, medical, aerospace, and telecommunication industries. Due to the large requirement for energy storage in all industries, the novel innovations in battery technology increase the profit of companies. The breakdown of batteries creates solid waste and eliminates a specific amount of chemicals into the environment. As we all know, primary cells are nonrenewable and nonchargeable single usage cells like Leclanché cells, dry cells, and primary lithium cells. But the innovation of secondary cells is more attractive because of the rechargeability and reusability. Nickel–cadmium batteries, lead-acid batteries, and lithium-ion batteries are some examples. A worldwide research is going on in the area of Li-ion batteries due to their high performances, specific energy and volumetric energy density, cyclability, charging rate, stability, and safety.[43]

The necessity of the fabrication of different cells in a greener method is increasing day by day. More and more greener approaches in this field are mandatory because of the requirement of energy storage systems in various industries. And after the usage, these substances should not be considered as a primary waste. The scientific reports suggest that it is possible to develop secondary Li-ion batteries from biomass and biomass-derived materials. A typical Li-ion battery contains different components such as electrode scaffolds, active cathode/anode, binders/additives, and separators/solid-state electrolytes. The preparation of each of these components from natural biomass is possible. This novel idea introduced a new term in green energy which is "green batteries." The 2D nanosheet silk or nanoporous structured rice husk can develop an anode, while glucose is used for the cathode. Nanoplates of D-gluconic acid lactone are used as a coating to cathode and nanorods of bacteria absorbing Sn can act as anode scaffold. Nanofiber agarose is the separator material and starch and silk fibroin make a polymer electrolyte and interlayer, respectively. The binder material used in Li-ion cells is generally derived from any seaweed.[44]

The preparation strategies of different battery components from natural biomass or biomass-derived products are tedious. The problem with this technique is the low yield and some unavoidable greenhouse gas emissions in the processing techniques. So, greener methods adoption in the processing of green batteries is a future idea, and extension of the green method in the synthesis of alkali-metal secondary batteries gives more advantages.

15.4.1.4 SUPERCAPACITORS

Supercapacitor systems based on two-dimensional underpotential deposition reactions are highly reversible, with their behavior resulting from the pseudocapacitance associated with the potential dependence of two-dimensional coverage of electroactive adatoms on an electrode substrate surface. The capacitance of this type can be 10–100 times that of the double-layer capacitance of the same electrode surface. The chemical and associated electrode potentials in such systems are a constant function of the degree of charge, unlike the thermodynamic behavior of single-phase battery reactants, which is a fundamental difference from battery behavior.[45]

The advantages of a supercapacitor over a double-layer electrostatic capacitance are:

1. Low-voltage operation
2. Phase angle as a function of frequency

3. Highly reversible
4. Capacitance not constant with voltage.

The fabrication possibility of supercapacitors using green nanomaterials is a fact to check. Dhanemozhi et al. produced ZnO nanoparticles from a leaf extract and suggested that the produced ZnO nanoparticles are suitable for supercapacitors with the help of characterization techniques. The team collected green tea leaves (*Camellia sinensis*) for the green synthesis of zinc oxide nanoparticles (ZnO Nps). From the Tauc plot, the bandgap of ZnO nanoparticles is concluded as 3.40 eV. Cyclic voltammogram results indicate a good capacitance behavior and low equivalent series resistance, hence a fast diffusion of electrolyte ions into the composite.[46]

Bandgap energy calculation using Tauc plot:

$$\alpha h\nu = A\,(h\nu - E_g)^n$$

where α—absorption coefficient; h—Planck's constant; ν—frequency of the photon; A—constant; E_g—energy band gap; $n = 1/2$—direct transition from the valence band to the conduction band; and $n = 2$—indirect transition from the valence band to the conduction band.

For ZnO nanoparticles, direct transition, so the equation becomes:

$$\alpha h\nu = A\,(h\nu - E_g)^{1/2} \leftrightarrow (\alpha h\nu)^2 = A\,(h\nu - E_g)$$

The $(\alpha h\nu)^2$ versus hν graph gives the energy band gap.

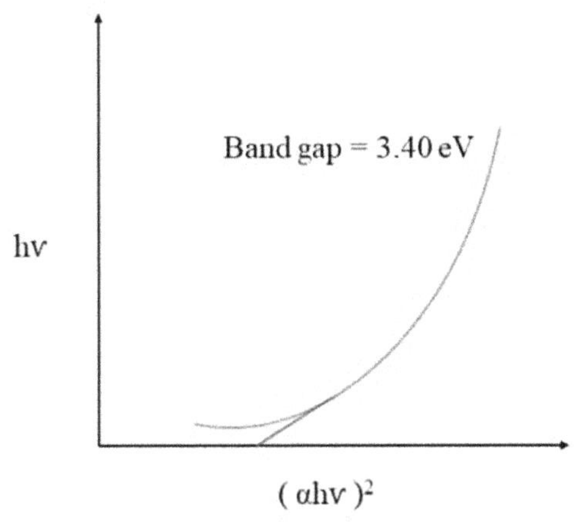

FIGURE 15.3 Tauc plot showing bandgap of ZnO nanoparticles.

Source: Reproduced with permission from Ref. [46]. Copyright Elsevier.

15.4.1.5 TRIBOELECTRIC NANOGENERATORS

Triboelectric nanogenerators (TENG) transform mechanical energy harvested from the environment into electricity, which can then be used to power small devices like sensors or to recharge consumer gadgets. The basic principles of TENGs are triboelectrification and electrostatic induction. TENGs can be considered as energy harvesters because they convert all mechanical energies such as vibration, mechanical triggering, rotation energy, and wind into electrical energy. Since TENGs are devices that convert mechanical energy into electrical energy, they can act as sensors of any physical work. The factor that determines the overall performance of TENGs is the charge density on the two surfaces. Even though the material is the same, the efficiency of different TENGs may be different due to the differences in charge densities. The relative charge density or triboelectric charge density is defined with reference to a standard material.[47]

Reports had shown that TENGs can be prepared using biomaterials such as cellulose and chitosan.[48] Kim et al. proposed a cellulose-based triboelectric nanogenerator, which can act as a folding sensor, humidity sensor, and electronic paper. The dielectric layer of this TENG is made up of nanofiber cellulose and for electrode material and counter triboelectric layer, Ag nanowires are used. The manufactured gadget is disseminated and entirely dissolved in deionized water using a sonicator for 30 min and comes back to the original ingredient without causing any pollution.[49]

15.4.1.6 BIOFUELS

Any fuel derived from biomass can be considered biofuel. Conventional fossil fuels including petrol, diesel, and natural gas create a higher amount of atmospheric pollution due to incomplete combustion (partial oxidation), and it leads to the emission of poisonous gases like carbon monoxide. The new term biofuel has more advantages over conventional fossil fuels because of the higher oxygen content, generally 10–12% greater than fossil fuels. The green approach to biofuels arises due to the renewable and nontoxic nature of biofuels, whereas fossil fuels are nonrenewable and toxic. Both fossil fuels and biofuels have the same source, even though the difference still exists. Biofuels are refined immediately after the organic matter died but in the case of fossil fuels, it takes millions of years after the death of organisms. From this comparison itself, the necessity of a rise in biofuel industries is understood. The advanced biofuels are cellulosic and algae-based fuels that consist of biodiesel, ethanol, and biogas. Nanotechnology has been included in the biofuel production process as a viable solution to the economic infeasibility of advanced biofuels, either by changing the feedstock or boosting the biomass content. The industry is still growing and there exists more to explore.

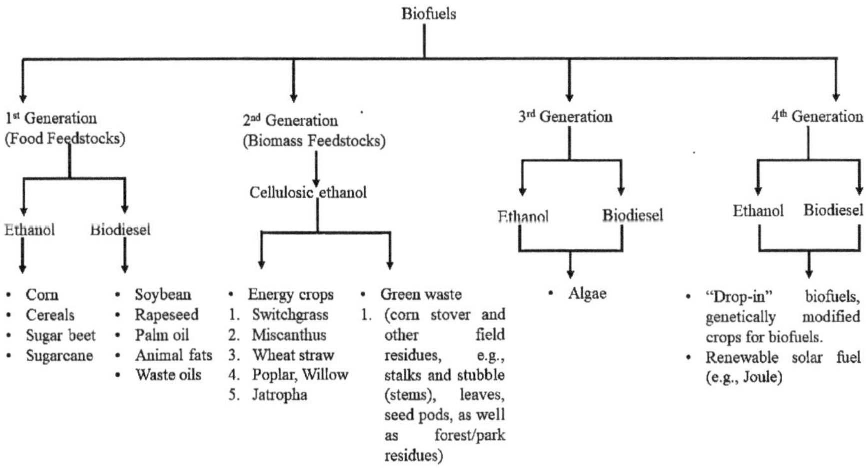

FIGURE 15.4 Classification of biofuels, products, and feedstocks.

Source: Adapted from Ref. [50]

The fundamental premise for utilizing nanotechnology in the biofuel business is to combine science (green and catalytic chemistry) and technical solutions in the pursuit of environmentally friendly energy sources. The specific characteristics of nanomaterials that help to be an efficient candidate in biofuel energy production include high surface areas, high degree of crystallinity, catalytic activity, stability, adsorption capacity, durability, and efficient storage capacity.[51]

Liu et al. confirmed that the use of calcium oxide nanocatalyst has a high catalytic performance for the transesterification of soybean oil to biodiesel than the conventional aluminum-based catalysts such as calcined K_2CO_3/γ-Al_2O_3 and KF/γ-Al_2O_3.[52] Magnetic nanoparticle-based catalysts have been proven to be advantageous due to their ease of separation from reaction media and reuse, resulting in more cost-effective, industrial-scale biodiesel production.[53] Due to the high surface area of nanoparticles, they can load a high amount of enzyme during biocatalysis. Higher enzymatic stability and the possibility of enzyme reusability of nanoparticles make them more efficient in large-scale biofuel production.

15.4.2 OTHER APPLICATIONS

15.4.2.1 WATER REMEDIATION

The 12 principles of green chemistry include the conversion of polluted materials in the surroundings to a clean and clear form. At present, nature is heavily contaminated through water, air, and soil pollution. Water pollution becomes a great threat in front of human life because it will lead to water scarcity subsequently. Even though ¾ of the earth's surface is concentrated by water, the amount of clean water is very little. The underground water and running water is getting polluted day by day through heavy chemical usage and waste disposal of water resources. The recyclability of wastewater through the purification process and desalination of saltwater are two possible mechanistic pathways to reduce water scarcity in the future.

The process of water remediation can be achieved with the help of nanoparticles. Zerovalent iron nanoparticles (n-Fe(0)) are used to clean high-contaminated dark-brown color distillery wastewater having a foul smell and higher values of biological oxygen demand (BOD), chemical oxygen demand (COD), and total suspended solids (TSS).[55] The nanosized iron particles are prepared in accordance with green chemistry principles. The contaminated distillery water was treated with both n-Fe(0) and ferrous sulfate heptahydrate, and the result shows that nanosized zerovalent iron particles are more efficient in purification.

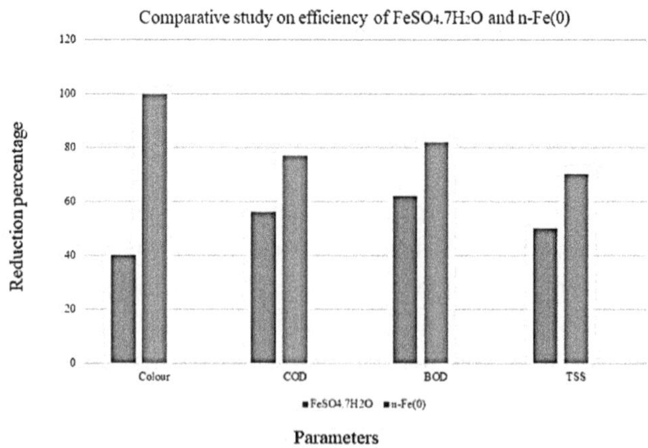

FIGURE 15.5 Comparative study on the efficiency of FeSO4, 7H2O, and nanosized zerovalent iron particles.

Source: Reproduced with permission from Ref. [55]. Copyright Elsevier.

The nanoparticles have the ability to adsorb heavy metal ions.[56] Exploration of this property of different nanoparticles helps to remove the presence of heavy metal ions from water. Carbon-based nanomaterials, zerovalent metals, metal oxide-based nanomaterials, and nanocomposites are used for this purpose. Najafabadi et al. suggested that nanofibrous chitosan/graphene oxide composite is an efficient adsorbent of Cu^{+2}, Pb^{+2}, and Cr^{+6}.[57]

15.4.2.2 DESALINATION

Desalination is the process of removal of minerals from saltwater and extracting clean drinkable water through reverse osmosis. In reverse osmosis, by applying external force, the natural flow of solvent is changed against the concentration gradient. The solvent molecules start to flow from low concentration to high concentration. Cao et al. developed silver nanoparticles using watermelon peel extract as a reducing agent and concluded that the developed AgNps are the perfect material in solar-driven interfacial evaporation for seawater desalination.[58]

A solar still is a device that uses solar energy to desalinate salt water or dirty water. Designing a solar still using suitable nanomaterials is a brilliant idea for the desalination process. The abovementioned system is a perfect green approach because desalination using solar energy is a completely green procedure. Cuprous and aluminum oxide nanoparticles, Fe_3O_4 multiwalled carbon nanotubes, and copper oxide with nanostructured coated absorber plates (NCAP) are the different nanomaterials that are used in solar still for increasing productivity and performance.[59]

15.4.2.3 NANOMEDICINE

Nanomedicine is a new generic term that encapsulates the concept of environmentally safe chemical aspects in the field of medicine using nanotechnology. Nowadays, nanomedicine gets hype in medicinal chemistry with the usage of different nanometal particles, polymers, and biological materials. Safe chemical reactions and different biomaterials, such as plant extract and microorganisms, are synthesized in green pathways for developing novel green synthesized drug delivery systems.[60] The discussion is going on the basis of the synthesis and functioning of green synthesized nanomedicinal systems and their future applications in nanomedicine. Zhang et al. invented nanometal iron oxide modified with folic acid and coated with polyethylene glycol, which acts as a targeting agent for intracellular uptake. The synthesized drug delivery system incorporates several materials and hence has a high atom economy.[61]

Quantum dots and carbon dots have been utilized successfully for active and passive drug delivery. Zhu et al. prepared a functional shell composed of 2-deoxyglucose (DG)-polyethylene glycol (PEG) connected with the compound of lipoic acid, lysine, and 9-poly-D-arginine (LA–Lys–9R) by a hydrazone bond and a core of CdTe quantum dots (QDs) to deliver the hypoxia-inducible factor-1 RNA into hypoxic cancer cells.[62] For imaging-guided delivery, Feng et al. developed a tumor extracellular microenvironment-responsive drug nanocarrier based on cisplatin (IV) prodrug-loaded charge-convertible CDs. The charge-convertible CDs' strong tumor-inhibition efficacy and negligible side effects demonstrate their capabilities as a smart drug nanocarrier with enhanced therapeutic effects.[63]

15.5 FUTURE PERSPECTIVES AND CONCLUSIONS

Green nanotechnology is a future vision in the area of nanoscience. The chapter detailed the basic agendas of green chemistry in the field of nanotechnology. Instead of toxic materials, green synthesis and application of biobased green nanomaterials are discussed in accordance with the 12 principles of green chemistry. The various advantages of using nanosized materials have already been discussed. They have several advantages over the conventional alternatives due to their small size and large surface area. But this peculiar property of nanomaterials becomes a threat. NP emissions to the environment can happen at any time. Due to its small size, it is easy to enter the life system which causes various threats to organisms. The NP can undergo oxidation and have the ability to form highly reactive and toxic reactive oxygen species (ROS). There is still a need to design NPs that are nontoxic and have more ecologically favorable qualities. The future of green nanotechnology relies on this. Synthesis of less reactive and nontoxic nanomaterials from a biobased feedstock for energy-related applications helps to achieve a sustainable socioeconomic energy goal. In the medicinal and environmental industries, improvements in green nanomaterials will reach a new high point, and green manufacturing will make significant progress.

KEYWORDS

- **green nanotechnology**
- **biobased nanomaterials**
- **sustainable energy**
- **green chemistry**
- **eco-friendly**
- **nontoxic materials**

REFERENCES

1. Emas, R. The Concept of Sustainable Development: Definition and Defining Principles. *Brief GSDR*, **2015,** *2015*, 10–13140.
2. Cao, X.; Zhu, L.; Bai, Y.; et al. Green One-step Synthesis of Silver Nanoparticles and Their Biosafety and Antibacterial Properties. *Green Chem. Lett. Rev.* **2022,** *15* (1), 26–32.
3. Pokrajac, L.; Abbas, A.; Chrzanowski, W.; et al. Nanotechnology for a Sustainable Future: Addressing Global Challenges with the International Network4Sustainable Nanotechnology. *UCLA SchLaw, Public Law Res. Pap.* **2022,** (22–03), 15.
4. Martínez, G.; Merinero, M.; Pérez-Aranda, M.; et al. Environmental Impact of Nanoparticles' Application as An Emerging Technology: A Review. *Materials* **2021,** *14*(1), 166.
5. Martins, R.; Kaczerewska, O. B. Green Nanotechnology: The Latest Innovations, Knowledge Gaps, and Future Perspectives. *Appl. Sci.* **2021,** *11* (10), 4513.
6. Ciambelli, P.; La Guardia, G.; Vitale, L. Nanotechnology for Green Materials and Processes. In *Studies in Surface Science and Catalysis;* Vol. 179, Elsevier; pp 97–116.
7. Fleischer, T.; Grunwald, A. Making Nanotechnology Developments Sustainable. A Role for Technology Assessment? *J. Cleaner Prod.* **2008,** *16* (8–9), 889–898.
8. Khin, M. M.; Nair, A. S.; Babu, V. J.; et al. A Review on Nanomaterials for Environmental Remediation. *Energy & Environ. Sci.* **2012,** *5* (8), 8075–8109.
9. Saleh, T. A. Nanomaterials: Classification, Properties, and Environmental Toxicities. *Environ. Tech. & Innov.* **2020,** *20*, 101067.
10. Ghernaout, D.; Ghernaout, B.; Naceur, M. W. Embodying the Chemical Water Treatment in the Green Chemistry—A Review. *Desalination* **2011,** *271* (1–3), 1–10.
11. Abdussalam-Mohammed, W.; Ali, A. Q.; et al. Green Chemistry: Principles, Applications, and Disadvantages. *Chem. Methodol.* **2020,** *4*, 408–423.
12. Ijaz, I.; Gilani, E.; Nazir, A.; et al. Detail Review on Chemical, Physical and Green Synthesis, Classification, Characterizations and Applications of Nanoparticles. *Green Chem. Let. Rev.* **2020,** *13* (3), 223–245.
13. Iravani, S. Green Synthesis of Metal Nanoparticles Using Plants. *Green. Chem.* **2011,** *13* (10), 2638–2650.

14. Machado, S.; Pinto, S. L.; Grosso, J. P.; et al. Green Production of Zero-valent Iron Nanoparticles Using Tree Leaf Extracts. *Sci. Total Environ.* **2013,** *445*, 1–8.
15. Dinn, J.; Liu, J. L.; Bashir, S. Use of Natural Products as Green Reducing Agents to Fabricate Highly Effective Nanodisinfectants. *J. Agric. Food Chem.* **2013,** *61* (9), 2019–2027.
16. Hamstra, D. F.; Lenstra, D. C.; Koenders, T. J.; et al. Poly (Methylhydrosiloxane) As a Green Reducing Agent in Organophosphorus-catalysed Amide Bond Formation. *Org. Biomol. Chem.* **2017,** *15* (30), 6426–6432.
17. Ghiyasiyan-Arani, M.; Salavati-Niasari, M.; Masjedi-Arani, M.; et al. An Easy Sonochemical Route for Synthesis, Characterization and Photocatalytic Performance of Nanosized FeVO4 in the Presence of Aminoacids as Green Capping Agents. *J. Mater. Sci.: Mater. Electron.* **2018,** *29* (1), 474–485.
18. Eghbali-Arani, M.; Sobhani-Nasab, A.; Rahimi-Nasrabadi, M.; et al. Green Synthesis and Characterization of SmVO4 Nanoparticles in the Presence of Carbohydrates as Capping Agents with Investigation of Visible-light Photocatalytic Properties. *J. Electron. Mater.* **2018,** *47* (7), 3757–3769.
19. Sperling, R. A.; Gil, P. R.; Zhang, F.; et al. Biological Applications of Gold Nanoparticles. *Chem. Soc. Rev.* **2008,** *37* (9), 1896–1908.
20. Katti, K.; Chanda, N.; Shukla, R.; et al. Green Nanotechnology from Cumin Phytochemicals: Generation of Biocompatible Gold Nanoparticles. *Int. J. Green. Nanotechnol: Biomed.* **2009,** *1* (1), B39–B52.
21. Philip, D. Green Synthesis of Gold and Silver Nanoparticles using Hibiscus Rosa Sinensis. *Phys. E: Low-Dimensional Sys. Nanotubes* **2010,** *42* (5), 1417–1424.
22. Elia, P.; Zach, R.; Hazan, S.; et al. Green Synthesis of Gold Nanoparticles Using Plant Extracts as Reducing Agents. *Int. J. Nanomed.* **2014,** *9*, 4007.
23. Ahmed, S.; Ikram, S. Biosynthesis of Gold Nanoparticles: A Green Approach. *J. Photochem. Photobiol. B: Biol.* **2016,** *161*, 141–153.
24. Gauttam, P.; Sharma, P. Green Nanotechnology: A Review on Green Synthesis of Bio-inspired Silver Nanoparticles—An Ecofriendly Approach.
25. Waris, A.; Din, M.; Ali, A.; et al. Green Fabrication of Co and Co_3O_4 Nanoparticles and Their Biomedical Applications: A Review. *Open Life Sci.* **2021,** *16* (1), 14–30.
26. El-Belely, E. F.; Farag, M.; Said, H. A.; et al. Green Synthesis of Zinc Oxide Nanoparticles (ZnO-NPs) Using Arthrospira Platensis (Class: Cyanophyceae) and Evaluation of their Biomedical Activities. *Nanomaterials* **2021,** *11* (1), 95.
27. Wang, T.; Jin, X.; Chen, Z.; et al. Green Synthesis of Fe Nanoparticles Using Eucalyptus Leaf Extracts for Treatment of Eutrophic Wastewater. *Sci. Total Environ.* **2014,** *466*, 210–213.
28. Noor, S.; Shah, Z.; Javed, A.; et al. A Fungal Based Synthesis Method for Copper Nanoparticles with the Determination of Anticancer, Antidiabetic and Antibacterial Activities. *J. Microbiol. Methods* **2020,** *174*, 105966.
29. DeFrates, K.; Markiewicz, T.; Gallo, P.; et al. Protein Polymer-based Nanoparticles: Fabrication and Medical Applications. *Int. J. Mol. Sci.* **2018,** *19* (6), 1717.
30. Chaware, P. J.; Bhat, A. P.; Rewatkar, K. G. Solar Cell Review: The Green Environmental Energy Resource with Nanomaterial. **2015**.
31. Khatri, I.; Tang, Z.; Liu, Q.; et al. Green-tea Modified Multiwalled Carbon Nanotubes for Efficient Poly (3, 4-ethylenedioxythiophene): Poly (stylenesulfonate)/n-silicon Hybrid Solar Cell. *Appl Phys Lett.* **2013,** *102* (6), 063508.

32. Andualem, W. W. Green Synthesis of Cuo Nanoparticles for the Application of Dye Sensitized Solar Cell. **2020**.
33. Watts, R. G. Global Warming and the Future of the Earth. Synthesis Lectures on Energy and the Environment: Technology, Science, and Society # 1. **2007**.
34. Hultman, M.; Nordlund, C. Energizing Technology: Expectations of Fuel Cells and the Hydrogen Economy, 1990–2005. *Hist. Technol.* **2013,** *29* (1), 33–53.
35. Lang, Y.; Arnepalli, R. R.; Tiwari, A. A Review on Hydrogen Production: Methods, Materials and Nanotechnology. *J. Nanosci. Nanotechnol.* **2011,** *11* (5), 3719–3739.
36. Cargnello, M.; Wieder, N. L.; Montini, T.; et al. Synthesis of Dispersible Pd@ CeO2 core− shell Nanostructures by Self-assembly. *J. Am. Chem. Soc.* **2010,** *132* (4), 1402–1409.
37. Machocki, A.; Denis, A.; Grzegorczyk, W.; et al. Nano-and Micro-powder of Zirconia and Ceria-supported Cobalt Catalysts for the Steam Reforming of Bio-ethanol. *Appl. Surf. Sci.* **2010,** *256* (17), 5551–5558.
38. Ali, S.; Al-Marri, M. J.; Abdelmoneim, A. G.; et al. Catalytic Evaluation of Nickel Nanoparticles in Methane Steam Reforming. *Int. J. Hydrogen Energy* **2016,** *41* (48), 22876–22885.
39. Yang, G.; Wang, J. Improving Mechanisms of Biohydrogen Production from Grass Using Zero-valent Iron Nanoparticles. *Bioresour. Technol.* **2018,** *266*, 413–420.
40. Mishra, P.; Thakur, S.; Mahapatra, D. M.; et al. Impacts of Nano-metal Oxides on Hydrogen Production in Anaerobic Digestion of Palm Oil Mill Effluent–A Novel Approach. *Int. J. Hydrogen Energy* **2018,** *43* (5), 2666–2676.
41. Li, J.; Xiao, B.; Yan, R.; et al. Development of a Nano-Ni-La-Fe/Al2O3 Catalyst to be Used for Syn-gas Production and Tar Removal After Biomass Gasification. *Bio. Resour.* **2009,** *4* (4), 1520–1535.
42. Zhang, W.; Banerjee-Ghosh, K.; Tassinari, F.; et al. Enhanced Electrochemical Water Splitting with Chiral Molecule-coated Fe3O4 Nanoparticles. *ACS Energy Let.* **2018,** *3* (10), 2308–2313.
43. Deng, D. Li-ion Batteries: Basics, Progress, and Challenges. *Energy Sci. Eng.* **2015,** *3* (5), 385–418.
44. Jin, C.; Nai, J.; Sheng, O.; et al. Biomass-based Materials for Green Lithium Secondary Batteries. *Energy Environ. Sci.* **2021,** *14* (3), 1326–1379.
45. Conway, B. E. Transition from "Supercapacitor" to "Battery" Behavior in Electrochemical Energy Storage. *J. Electrochem. Soc.* **1991,** *138* (6), 1539.
46. Dhanemozhi, A. C.; Rajeswari, V.; Sathyajothi, S. Green Synthesis of Zinc Oxide Nanoparticle Using Green Tea Leaf Extract for Supercapacitor Application. *Mater Today: Proc.* **2017,** *4* (2), 660–667.
47. LináWang, Z. Triboelectric Nanogenerators as New Energy Technology and Self-powered Sensors–Principles, Problems and Perspectives. *Faraday Discuss.* **2014,** *176*, 447–458.
48. Cheedarala, R. K. Highly Transparent Chitosan-PVA Blended Membrane for Strong Mechanical Stiffness and as Robust Bio-material Energy Harvester through Contact-Separation Mode TEG. *Frontiers Nanotechnol.* **2021,** *3*, 30.
49. Kim, I.; Jeon, H.; Kim, D.; et al. All-in-one Cellulose Based Triboelectric Nanogenerator for Electronic Paper Using Simple Filtration Process. *Nano. Energy* **2018,** *53*, 975–981.
50. Ziolkowska, J. R. Introduction to Biofuels and Potentials of Nanotechnology. In *Green Nanotechnology for Biofuel Production;* Springer: Cham, 2018; pp. 1–15.

51. Nizami, A. S.; Rehan, M. Towards Nanotechnology-based Biofuel Industry. *Biofuel Res. J.* **2018,** *5* (2), 798–799.
52. Liu, X.; He, H.; Wang, Y.; et al. Transesterification of Soybean Oil to Biodiesel Using CaO as a Solid Base Catalyst. *Fuel* **2008,** *87* (2), 216–221.
53. Mapossa, A. B.; Dantas, J.; Silva, M. R.; et al. Catalytic Performance of NiFe2O4 and Ni0. 3Zn0. 7Fe2O4 magnetic Nanoparticles During Biodiesel Production. *Arab J. Chem.* **2020,** *13* (2), 4462–4476.
54. Kim, K. H.; Lee, O. K.; Lee, E. Y. Nano-immobilized Biocatalysts for Biodiesel Production from Renewable and Sustainable Resources. *Catalysts* **2018,** *8* (2), 68.
55. Srivastava, K.; Srivastava, A.; Singh, P.; et al. Remediation of Distillery Waste Water Using Zero Valent Iron Nanoparticles. *Curr. Res. Green Sus. Chem.* **2021,** *4*, 100072.
56. Yang, J.; Hou, B.; Wang, J.; et al. Nanomaterials for the Removal of Heavy Metals from Wastewater. *Nanomater.* **2019,** *9* (3), 424.
57. Hadi Najafabadi, H.; Irani, M.; Roshanfekr Rad, L.; et al. Removal of Cu2+, Pb2+ and Cr6+ from Aqueous Solutions Using a Chitosan/graphene Oxide Composite Nanofibrous Adsorbent. *RSC Adv.* **2015,** *5* (21), 16532–16539.
58. Cao, H.; Cui, T.; Wang, W.; et al. Green-synthesizing Ag Nanoparticles by Watermelon Peel Extract and Their Application in Solar-driven Interfacial Evaporation for Seawater Desalination. *Mat. Res. Express* **2020,** *7* (4), 045005.
59. Elgendi, M.; Al Dhaheri, M.; Al Baeek, A.; et al. *Nanotechnology in Solar Still: Conceptual Design.* In 2021 6th International Conference on Renewable Energy: Generation and Applications (ICREGA); IEEE, 2021, February; pp. 189–193.
60. Jahangirian, H.; Lemraski, E. G.; Webster, T. J.; et al. A Review of Drug Delivery Systems Based on Nanotechnology and Green Chemistry: Green Nanomedicine. *Int. J. Nanomed.* **2017,** *12*, 2957.
61. Zhang, Y.; Sun, C.; Kohler, N.; et al. Self-assembled Coatings on Individual Monodisperse Magnetite Nanoparticles for Efficient Intracellular Uptake. *Biomed. Microdevices* **2004,** *6* (1), 33–40.
62. Zhu, H.; Zhang, S.; Ling, Y.; et al. pH-responsive Hybrid Quantum Dots for Targeting Hypoxic Tumor siRNA Delivery. *J. Control Release* **2015,** *220*, 529–544.
63. Feng, T.; Ai, X.; An, G.; et al. Charge-convertible Carbon Dots for Imaging-guided Drug Delivery with Enhanced In Vivo Cancer Therapeutic Efficiency. *ACS Nano.* **2016,** *10* (4), 4410–4420.

CHAPTER 16

Green Technologies and Sustainable Energy Materials

AISWARYA T. SOMAN, MUHAMMED SHAH S., ARYA UTHAMAN, HIRAN MAYOOKH LAL, and SABU THOMAS

School of Energy Materials, Mahatma Gandhi University, Kerala, India

ABSTRACT

In the present scenario, the topic has gained relevance due to the climate crisis and increasing population. Our planet, people, and economy are craving sustainable solution. COP26 has addressed the situation as code red and calls for net zero emission. The climatic crisis has created devastating destruction around the world, depletion of fossil fuels urges humans to look for alternative sources, and the entire geopolitics plays around it. Biodegradable batteries will provide great relief to the environment as the ordinary batteries create e-waste as they are less recyclable. For the existence of every creature on this planet, clean technology and the use of sustainable materials is the armour for the environment. This chapter discusses present clean technologies like biodegradable batteries and other sustainable battery material innovations. Further, it discusses the sustainable energy materials for hydrogen storage.

16.1 INTRODUCTION

The global problems of environmental degradation made society think about green tech and explore sustainable energy materials; hence, green initiatives are made to tackle the problem. The main goal is to replace the harmful materials and use other alternatives which will not deplete the natural resources. Sustainable materials reduce the carbon footprint. The level of

Environmental Technology and Sustainability: Physical, Chemical, and Biological Technologies for Environmental Protection. Tamara Tatrishvili, Ann Rose Abraham, & A. K. Haghi (Eds.)

carbon dioxide rises up day by day in the earth's atmosphere because of the overwhelming use of fossil fuels to generate electricity. This caused greenhouse gas emissions. Green technology is environmentally friendly, which will solve all the alarming situations that the planet undergoes. Now the world is running to EVs, hydrogen, and solar. Replacing the damage caused by fossil fuels, energy is the right field to start the cleansing. Sustainable energy materials will provide hope for humanity for a sustainable future. Research fields are heavily occupied with these material works.

Battery technologies play a vital role in the present world. They have many applications from cell phones to medical devices. EVs are the new trend in automobile sector because they can reduce the carbon consumption, as a result the demand for energy materials rises; hence, battery technology will revolutionize the way of living. For an eco-friendly approach biodegradable battery is an upcoming promising solution. These batteries use materials that degrade over time; hence, it has got applications in medical field. To decarbonize the environment hydrogen can rise to green hydrogen. The merits belonging to hydrogen are: it is light weight, has high energy density, and has no emission of harmful chemical by-products. This chapter will speak about the materials used in Li-ion, Li-Sulfur, Li-oxygen batteries, biodegradable batteries, and hydrogen.

16.2 BATTERIES

A battery is a device that can convert chemical energy into electrical energy through an electrochemical redox reaction. The basic unit inside a battery is a cell, and it contains three main parts. They are two electrodes and an electrolyte (between these electrodes). When two electrodes are connected in a circuit, it starts to create ions, and these ions take part in a reaction with electrolytes. Hence, the electron jumps to move from one terminal to another through the external circuit. This is how the battery works. Batteries require two different materials for their electrode, then only they can conduct the electricity. Hence, the material chosen for the electrodes should be dissimilar.

Batteries are of different capacities and voltages and even different shapes and sizes. Hence, they are classified into primary batteries (nonrechargeable) and secondary batteries (rechargeable). There are three main kinds of primary batteries; they are zinc-carbon, alkaline, and lithium. These batteries do not contain liquids. Hence, they are known as dry cells. Zinc-carbon batteries are the standard batteries invented by Georges Leclanche.

In this battery, the positive electrode is made out of a carbon rod. They are surrounded by powdered carbon and manganese (IV) oxide. The negative electrode is zinc alloy, and the electrolyte is ammonium chloride compared to alkaline batteries, alkaline batteries can store more energy and last longer. The positive electrode encompasses manganese (IV) oxide, and the negative electrode is zinc, but the electrolyte is a concentrated alkaline solution with potassium hydroxide. The other category is the button, which is more often used in quartz watches and hearing aids. The negative electrode is made of zinc or lithium, and the positive electron is manganese oxide, silver oxide or copper oxide, and alkaline electrolytes.

Secondary batteries (rechargeable) are the following categories, which include lead-acid, nickel-cadmium, and nickel-metal-hydride (NiMH). Lead-acid batteries are the most commonly recycled consumer product. Because of their closed-loop life cycle, lead-acid batteries are a sustainable solution. All components of a discharged automotive battery are collected, recycled, and sold to battery manufacturers by a certified recycler. The cradle-to-cradle method ensures that your car, truck, boat, or motorbike battery has been and will be recycled indefinitely. This makes the lead-acid battery an economic and environmental success. Lead-acid batteries help to promote sustainability in a variety of ways, including lowering greenhouse gas emissions in automobiles with start-stop engines and reducing fuel usage by up to 10%. Lead-acid batteries are leading the way in fulfilling the world's expanding energy demands as the most environmentally sustainable battery technology. A new lead-acid battery contains more than 80% recycled lead battery material, making it the greenest energy storage technology available.

Nickel-cadmium is one of the best options for sustainability since it can be charged and discharged hundreds of times. NiMH has become more popular because it can replace NiCd batteries (because Cd causes severe pollution), and it is an excellent alternative for NiCd, which is often used topped up with quick recharge. New set of batteries like lithium-ion, lithium-sulfur, lithium-oxygen, and sodium ion emerge as a result of sustainability and their capacity to ensure power from mobile battery to medical devices. Researches try different type of materials to increase the capacities of these batteries.

16.2.1 Li-ION BATTERIES

Li-ion batteries are made up of materials such as cobalt, graphite, and lithium. These materials are considered critical, and also, they withstand temperature up to 80°C which allow them to have intense activity with less loss of battery life; hence, it has become a ubiquitous factor in today's energy storage industry. Battery prototypes are evaluated using electrochemical performance, cell balance, efficiency, and cycle life, and the final evaluation is conducted through its favorable characteristics.[1] Cathode materials can be changed to make a sustainable way of recycling like hydrophobic eutectic solvents (HESs), which is a green solvent.[2] Hanada et al.[2] demonstrated a novel cathode recycling process which is an environmental harmonic recycling process. In a Li-ion battery, Li+ is the agent ion, and the host network compounds are metal chalcogenide transition metal oxide and polyanion compounds. Their intercalate compounds can be divided into several crystal structures, such as layered, spinal, olivine, and tavorite.[3] Different types of chalcogenides, $LiTiS_2$ (LTS), were widely studied due to their high gravimetric energy density combined with long cycle life, and it was made commercialized by EXXON.[4] $LiCO_2$ introduced Goodenough is the first and the most commercially successful form of layered transition metal oxide cathode.[5] $LiNiO_2$ has the same crystal structures as $LiCoO_2$ and a similar theoretical specific capacity of 275 mAh/g. The high energy density and lower cost compared to Co-based materials are the main research driving forces.[3]

***Cathode materials*:** The commonly used cathode material is $LiCoO_2$, Li-Mn-O, $LiFePO_4$, and lithium layered metal oxide. Li-Mn-O is low cost and has desirable electrochemical properties. Its different forms make it suitable for interacting with small helium and lithium ions. The lambda form of spinal (Mn_2O_4) makes it for the intercalation of Li-ion. This spinal high thermal threshold, excellent rate capability, and environmental impacts are advantages. Lithium metal oxide also is used as cathode materials. $LiNi_{0.5}Mn_{0.5}O_2$ and $Li_{1.2}Cr_{0.4}Mn_{0.4}O_2$ are joint solid solution compounds.[6] The best option is an application that requires fast charging and discharging layered metal oxides. Olivines ($LiFePO_4$) have a flat discharge plateau and capacity in the range of 150–160 mAh/g.[7] Using nanocomposite of $LiFePO_4$ with conductive carbon matrix[8,9] advances the synthesizing olivines with other transition metals.

Searching for new cathode materials, researchers developed a new compound called polyanionic. These compounds occupy lattice positions and increase cathode redox potential while also stabilizing its structure[6,10] $LiFeSO_4F$(LFSF) is an exciting cathode material because of its high cell voltage and specific capacity (151 mAh/g).[11]

Anode materials: Most commonly used anode materials are carbon(graphite) and lithium alloyed metals. Commercialized lithium alloyed metal is oxide spinel Li_4Ti_5-O_{12}.[12] Use the minimum potential intercalation electrode to avoid the production of dendrites. The carbon material used as the anode is graphite, and they have sheets peeled in hexagonal or rhombohedral arrangement.[13] Graphitic carbon is used as anode material in commercial Li-ion cells. Intermetallic elements like Al_3Ni, Fe-Sn, Sn-Sb, and Sn-Cu show results as anode materials. Carbonaceous anodes show improvement in charge–discharge efficiency and discharge capacity.[14] Natural Graphite has been modified in different ways to improve its properties, like "kish" graphite.[1,4] To develop an artificial graphitic anode, it requires heat treatment. Non-graphitic carbon has a graphene domain but does not have the structural order shown by graphene; hence, they are disordered carbon.[14] Li-Al is the first alloy type anode developed for Li-ion batteries. Lithium–titanium oxide is a material used as an alternative for carbon anodes.

Commercial Li-ion batteries use graphite-based anode, and the other materials are graphene, Li-carboxylates, CNT, Li-M (M-Si, Ge, Sn, etc.) alloys, and lithium transition metal oxides. Carbon has an attractive balance of relatively low cost, abundance, moderate energy density, power density, and cycle life compared to any other intercalation-type anode materials, which enabled carbon anode enabled Li-ion battery to become commercially viable more than 20 years ago.[3] Graphitic carbon has large graphite grains and achieves close to theoretical charge capacity. However, this carbon does not combine with PC. Hard carbons smell graphitic grains with disordered orientation and are much less susceptible to exfoliation. These grains have nanovoids between them which result in isotropic volume expansion; impurities like hydrogen atoms provide extra capacity to carbon-based anodes. Lithium titanium oxide (LTO) has been successfully commercialized because it allows the combination of superior thermal stability, and also it is highly safe as it prevents the Li dendrite formation. Figure 16.1 shows the range of average discharge potentials and specific capacities for all types of electrodes.

Electrolyte materials: A sustainable battery always relies on a suitable electrolyte consisting of salt and solvent combinations, for example, polyethylene oxide (PEO). In liquid electrolyte, the ion conduction of the electrolyte is field-trial which is guided by the concepts of dielectric and viscosity constants.[15] Aqueous electrolytes (organic and ionic liquids), gelled electrolytes, and solid organic and inorganic electrolyte materials

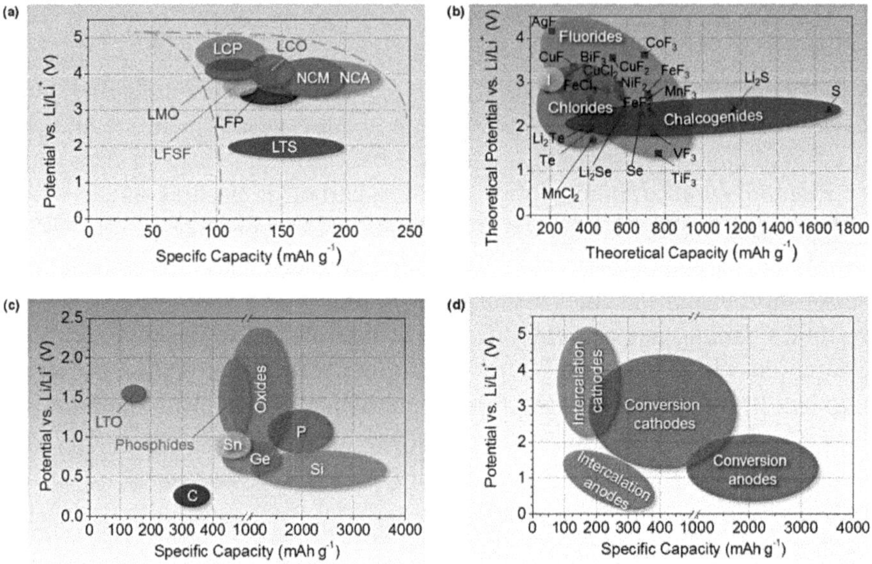

FIGURE 16.1 Approximate range of average discharge potentials and specific capacity of some of the most common (a) intercalation-type cathodes (experimental), (b) conversion-type cathodes (theoretical), (c) conversion type anodes (experimental), and (d) an overview of the average discharge potentials and specific capacities for all types of electrodes.

Source: Reproduced with permission from Ref. [3]. Copyright 2014 Elsevier.

are commonly used electrolytes. The significant electrolytes currently used comprise a mixture of cyclic and linear carbonates and electrolyte additives such as vinylene and ethylene carbonates or sulfides are used for electrolyte solvents to improve cyclic performance.[16]

16.2.2 SODIUM-ION BATTERIES

Sodium-ion battery is one of the sustainable solutions available since it can be found worldwide, unlike other batteries. This can unhook main challenges faced with Lithium-ion battery (LIB) technology, as it is a scarce resource and having geopolitical constraints. On one hand, Sodium-ion battery (SIB) provides good environmental and electrochemical performance, which also has a lower demand for scarce resources.[17] The abundance of resources and the lower cost give the pull to switch from lithium-ion batteries to sodium-ion batteries. The components and electrical storage mechanisms of SIB and LIB are the same.[18] Aluminum is one of the cost-effective alternatives to copper as a current collector for SIBs. The reported cathode material for SIBs is tunnel-type transition metal oxides, transition metal sulfides, fluorides, oxyanionic compounds, Prussia blue analogs, and polymers. Non-graphitic carbonaceous materials are taken as the anode in SIB. The commonly used electrolyte is $NaClO_4$ or $NaPF_6$ salts in carbonate ester solvents, mainly in propylene carbonate (PC). Aqueous electrolytes are more favorable for SIB than organic electrolytes.[18]

***Cathode materials*:** High specific energy, high-rate capability, and excellent cycling life are the chemical features of cathode material. The mainly used materials are transition-metal oxide, polyanionic compounds, metal hexacyanometalates, and organic compounds[19] transition-metal oxides like layered Na_xMO_2(M=Co, Mn, Fe, Cr, Ni, etc.). Delmas et al.[19] classified this into O^{3-} and P^{2-} phases depending upon the location of Na^+ between M layers.[20] Na^+ and M are packed in ABCABC arrangement. Na_xCoO_2 is the potential cathode to SIBs. Due to its analog to $LiCoO_2$ in LIBs[21] high practical capacities can be given by layered $Na_xMnO_{2.}$[22] Other interesting host material is $NaFeO_2$, as Fe is an abundant resource with highly active $F^{3+}/_{4+}$ couple[23] layered oxides like $NaCrO_2$, $NaNiO_2$, NaV_xO_y demonstrated as electrochemically active in SIBs. Tunnel Na_xMO_2 is one of the extensively investigated Na storage cathodes with a theoretical specific capacity of 121 mAh/g.[24] Liu's group synthesized single-crystalline $Na_{0.44}MnO_2$ nanowires through a polymer-pyrolysis process, exhibiting a high reversibility capacity for SIB application.[25] Polyanionic compounds are very promising SIB cathodes because of their structural diversity and stability and also the substantial inductive effect of the anions.

Polyanionic compounds mainly contain olivine $NaFePO_4$, NASICON $Na_3V_2(PO_4)_3$, fluorophosphates (Na_2MPO_4F, $Na_3(VO_x)_2(PO_4)_2F_{3-2x}$), pyrophosphates ($Na_2MP_2O_7$, $Na_4M_3(PO_4)_2P_2O_7$) (M=Fe, Co, Mn), and sulfates ($Na_2Fe_2(SO_4)_3$, Na, Fe $(SO_4)_2 \cdot 2H_2O$.NASICON $Na_3V_2(PO_4)_3$ provides a reversible and stable $V^{3+}/_{4+}$redox reaction, which occurs at 3.4 V. Carbon coating and creating a porous structure can improve the electrochemical activity of $Na_3V_2(PO_4)_3$. To enhance the electrochemical performance of $Na_3V_2(PO_4)_3$ cationic substitutions can also be used. For the high-performance cathode material, fluorophosphate is an excellent option. Metal hexacyanometalates, metal fluoride, organic cathode materials. All organic Na ion full batteries based on $Na_4C_8H_2O_6$ displayed an average operation voltage of 1.8 V and a practical energy density of about 65 Wh/Kg.[26] Figure 16. 2 shows the electrode materials used in SIBs.

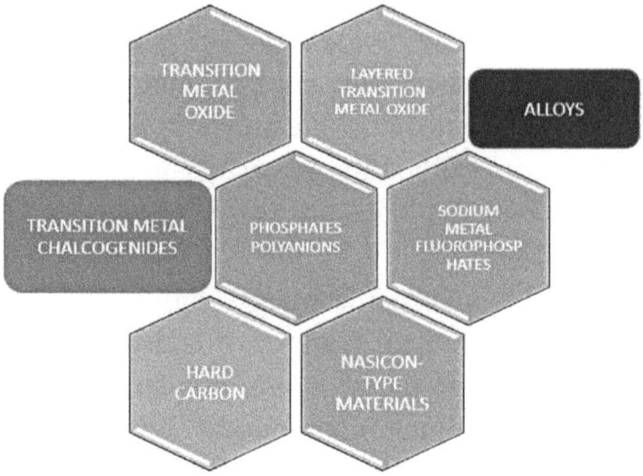

FIGURE 16.2 Electrodes used in SIBs.

Anode materials: Available anode materials consist of carbonaceous materials, alloy-type materials (Sn, Sb, P, Ge, SnSb, Sn_4P_3, etc.), metal oxides/sulfides based on conversion reaction (Fe_2O_3, CuO, Co_3O_4, MoS_2, SnS_2, etc.), titanium-based compounds ($Na_2Ti_3O_7$, TiO_2, etc), and organic compounds. Well-known anodes for SIBs are carbonaceous materials. Adelhelms group first reported about the graphite as SIB anode, since it showed electrochemical activity in the diglyme-based electrolyte.[27] Sodium storage behavior in natural in got explored by Chen's group they ether-based electrolyte found out its exceptional rate capability, unprecedented cyclability. Due to significant lattice distance and highly disordered structure, hard carbon is favorable for sodium storage performance of hard carbon can enhance significantly by rational morphological and structural engineering.[28] Tailored nanostructures and doping heteroatoms can improve carbon's electrochemical performance, reactivity, and electric conductivity. Zhou et al. prepared s doped N rich nanosheets with H_2S/Ar mixed gas. Hence, it showed high capacity and excellent rate performance as a SIB anode.[29] Graphene is a popular 2-D material with a large surface area, good flexibility, high chemical stability, and electronic conductivity.[30] Dai et al. used 3D N-doped graphene foams as the SIB anode. They achieved a high initial reversible capacity.[31] Soft carbon has higher electronic conductivity, representing[28] a graphitize non-graphitic carbon.

Another type of material that is authentic to use as an anode for SIB is alloy-type material. This family mainly contains Sn ($Na_{15}Sn_4$, 847 mAh/g), Sb (Na_3Sb, 660 mAh/g), P (Na_3P, 2596 mAh /g), Ge (NaGe, 369 mAh/g), Bi (Na_3Bi, 385 mAh/g). Using the density functional theory (DFT), sodium insertion into the crystalline tin is explained in four steps by caviar and cedar that successively[32] fabricates a composite of nano-Sn and carbon can change (buffer) the volume fluctuation and suppress the accumulation. Darice et al. investigated the sodium storage mechanism of antimony (Sb), using in situ XRD.[28] Other metals like Ge, Bi, In, and Pb can form an alloy with sodium for SIB anode by designing carbon-containing composites, and porous micro/nanostructure can improve the electrochemical performance of a metal oxide. Metal sulfide is the other potential candidate metal sulfide, including layered metal disulfides and nonlayered metal sulfides FeS/FeS_2, Sb_2S_3, NiS/Ni_3S_2, CoS/Co_9S_8. Various titanium-based compounds TiO_2, $Li_4Ti_5O_{12}$, $Na_2Ti_3O_7$, $Na_2Ti_6O_{13}$, $Na_{0.66}Li_{0.22}Ti_{0.78}O_2$, $NaTi_2(PO_4)_3$ are used at low cost, with no toxicity, low strain, and excellent cyclability. Sustainable anode materials are organic compounds that are abundant in biomass.

Electrolyte Materials: The common solutes like ($NaClO_4$, $NaPF_6$, and $NaSO_3CF_3$) use the solvents like easter-based propylene carbonate (PC), ethylene carbonate (EC), diethyl carbonate (DEC), either based ethylene glycol dimethyl ether (DEGDME), and tetra ethylene glycol dimethyl ether (TEGDME) in an organic electrolyte system. For Na_2So_4 solute H_2O is taken as the solvent (aqueous electrolytes). Polymer electrolytes (PEO-$NaClO_4$, PEO-$NaCF_3SO_3$, PMMA-SiO_2) show Na-ion transmissibility and inorganic NASICON-type to have great potential due to its 3D channel. Due to super-sonic conductivity, Na_3PS_4-series solid-state electrolytes have great attention. On the other hand, $Na_{10}Sn\ P_2$, Na_3PSe_4, and $Na_3P_{.62}A_{.38}S_4$ electrolytes show excellent ionic conductivity. Low cost, long lifespan, environmental friendliness, and easy manipulator are the merits of Sodium-ion batteries' aqueous electrolytes. NASICON-type $Na_3Zr_2Si_2PO_{12}$ solid-state electrolyte with $Na_3V_2(PO)_3$ electrodes are best for all-solid-state batteries and are considered the batteries with more minor hazards without liquid leakage flammability and volitation.

16.2.3 LITHIUM-SULFUR BATTERIES

Lithium-sulfur is considered a high-energy rechargeable system. It has a sulfur cathode and lithium anode with multi-electron conversion electro chemistry.[33] Li-S battery is closed; hence, it prevents exposure to atmospheric impurities and potential explosive danger. Nazar et al. made a highly ordered, nanostructured, and mesoporous carbon host encapsulating sulfur.[34] This made the Li-S battery to be high capacity and stable cycling. New electrolytes like chains, glymes, solids, glass type media improve lithium-sulfur batteries. The overall reaction is

$$16Li + S_8 \rightarrow 8Li_2S \qquad (16.1)^{33}$$

Using the in situ method makes it possible to find out the reaction species that are formed during discharge and charge of a Lithium-sulfur battery.[35] A bifunctional microporous carbon paper is inserted between the cathode and separator[36] of a traditional lithium-sulfur battery to achieve high capacity, long cycle life, and rapid charge rate.[37] Lithium-sulfur may become an alternative to Lithium-ion batteries because of their high energy density and lower production cost.[38]

***Cathode materials*:** Cathode materials for Lithium-Sulfur batteries are elemental Sulfur as the direct conversion of Sulfur and lithium is impeded by the conductivity issue. To tackle this issue, polysulfides are added. Polysulfides are in the form of sulfur chain and with terminal lithium ions; hence, the mobility and solubility of inorganic electrolytes make it easy for active material to access both electrons. Polysulfide plays the role of a redox mediator.[33] Due to superior electrical conductivity, high surface area, tunable porous structure, and excellent electrochemical stability, carbon nanotubes (CNTs), carbon nanofibers,[39,40] 2D graphene/carbon nano sheets,[41,42] and 3D porous carbon[43,44] are used as the sulfur host electrolyte solvents[45] anion[46] and current rates[47] affect the ionic strength of Li^+ complexes. Adopting polar and conductive host materials in the sulfur composite is a growing research field. Polar conductors possess capabilities to enhance the surface electrochemical kinetics and solution-solid phase transfer.

Other sulfur allotropes are short sulfur chain (S_2-S_4)[38,47] chemical covalent-bounded organosulfur. pPAN@s is one of the earliest cathode materials for rechargeable Li-S batteries that were based in Wang et al.[41] The high quality of pPAN@s and HS average discharge voltage is of 1.8 V, the specific energy of a cathode is said to be 1080 Wh/Kg. The formation of polysulfide is inherently stopped. Hence, it can stabilize both the cathode and the lithium anode.

***Anode materials*:** With the elemental cyclo-S8, metallic lithium is the anode material for Li-S batteries. As Li has a high theoretical specific capacity of 3860 mAh/g and low negative electrochemical potential of –3.040 V, it lays the extraordinary specific energy for Li-S batteries. To overcome the challenges of Li-metal anode, nonmetallic anode materials are used in, that is intercalation type graphite and other carbon,[48–50] alloy-type silicon,[51] tin[52,53] or germanium.[54]

In high-loading Li-S batteries, corrosive polysulfides and high areal current density inevitably exaggerated the formation of thick, mossy, and insulating interphase.[33] Lithium metal anode and magnified by the high loading of active materials as be classified as: (1) intrinsic kinetics issues of Li plating/stripping. (2) Exotic issues because of unique Li-S redox chemistry. Nanostructured Cu-based current collector emerges as one of the 3D-lithium hosts. Yang et al. synthesized a Cu skeleton through in situ dehydration and reduction of Cu $(OH)_2$ fibers on a Cu foil and then allowed a lithium metal to be deposited and accommodated within the 3D skeleton without the dendrite formation.[55] Other 3D porous Cu-based current collectors can be accessed by chemically leading Zn from Cu–Zn alloy[56] or ordinary Cu nanowires into free-standing networks.[57] The purpose of replacing the lithium metal anode with nonmetal is to improve the safety Li-S battery. Hagen et al. employed electrochemically lithiated silicon microwire arrays as the anode that allowed fuel cells to be operational in different electrolytes.[58]

Compound technologies made up of lithium compounded with other materials to form a composite lithium anode, and the primary compounded materials hence can be classified into six categories; they are C-Li, Si-C-Li, Sn-C, Li co compound, metal compounds, and other compounds.[59,60] The electrical and mechanical properties of carbon materials made them the materials for lithium-sulfur batteries. Fan et al.[61] first introduced the composite anode for Li-S battery by using stabilized lithium metal powered (SLMP) to compound hard carbon.[60] Li-S battery with the presence of hard carbon showed long cycle life, better safety performance, and higher specific capacity. Instead of hard carbon graphite can be used. The anode with SEI-coated graphene (SC4) structure is endowed with a dendrite's inhibition behavior and dendrite-free morphology. Silicon materials can be used as anode compound materials. The Li-S battery with a novel prelithiated Si microwire anode was highly stable. Also during charging and discharging it shows remarkable coulometric efficiency and capacity. Kimet et al. prepared an advanced Si anode lithiated through direct contact in the ether-based liquid electrolytes.[62] The theoretical capacity of the lithiated carbon-coated silicon anode was 40. The battery with lithium/Sn-C composite anode nullifies the intrinsic safety issues of an ordinary battery. The lithium anode is alloy-coated and has excellent rate capability and meager charge transfer resistance. The Li-S battery with lithium-Al alloy coating layer exhibits significant cycle performance with a high coulombic efficiency. A 3D nano-structure lithium anode can overcome the dendrite problem present in an anode with a superior specific area that has been constructed.[53] Jin et al.

created a 3D lithium metal compound as an anode on the matrix of bamboo-derived 3D hierarchical porous carbon (HPC), decorated by ZnO quantum dots via a cost-effective and sustainable method.[63] Carbon materials like graphite and hard carbon help to improve lithium-sulfur batteries Li anode. For example, Zhang et al. attempted to use commercial graphite as an anode to form Li sulfidemicroporous carbon cathode to form a Li-S cell.[64] Tin materials are also considered anode materials of Li-S batteries because of their high theoretical specific capacity.

Electrolyte materials: The electrolyte for Li-S batteries is broadly classified into: (1) nonaqueous liquid electrolytes, (2) ionic liquids, (3) solid polymer, and (4) glass-ceramic electrolytes. Rauh et al.[65] reported electrolytes with high basicity dissolve many lithium polysulfides, and sulfur solubility in ions in dimethyl/sulfoxide or ether-like tetrahydrofuran. DME has high lithium polysulfide dissolution ability, and it provides a considerable number of active materials.[66] Nonaqueous liquid electrodes like 1,3 dioxolane and glyme solvents exhibit higher sulfur solubility. When TEGDME and DOL/DME are used as electrolytes, Li-S cells showed 1000 mAh/g discharge capacity, but when PC/EC/DEC and EMS/DEC are employed as electrolytes, the cell delivered had no ability. As per Wang et al.[67] a combination of electrolyte DME: DOL in a 2:1 ratio can provide 1200 mAh/g discharge capacity at its initial stage and retain the 800 mAh/g after 20 cycles. Properties like non-flammability, overall electrochemical stability, nonvolatility, high ionic conductivity, and environmental friendliness made ionic liquids electrolyte.[59] The ternary compound operates as an electrolyte in Li-S cells. Wang et al. designed an ionic liquid-based organic electrolyte which was a combination of 1,2 dimethoxyethane and N-methyl-N-propylpiperidium bis (trifluoromethane Sulfonyl) inside (PP13-TFSI).[67] This electrolyte shows high coulombic efficiency, capacity retention, and suppressed internal shuttling of polysulfides. Polymer electrolytes hide the dissolution of polysulfides in Li/sulfur cells, resulting in cyclability and enhanced capacity retention. Lecuyer et al. analyzed the structural evolution of Li-S cells based on PEO-based dry polymer electrolytes.[68]

16.2.4 LITHIUM–OXYGEN BATTERIES

One of the most studied types of metal-air batteries is the Li-O_2 battery, which consists of a metallic lithium anode as the negative electrode, air-O_2 as the active cathode mass, and a Li^+ containing electrolyte solution. The cathode in these systems is a composite electronically conducting porous matrix that allows oxygen gas and Li ions in the electrolyte solution phase to make electrochemical contact. At the anode, it works by dissolving and depositing lithium metal, while at the cathode, it works by performing an oxygen reduction reaction (ORR)/oxygen evolution reaction (OER).[69]

Aprotic, aqueous, hybrid, and solid-state batteries are the four types of Li-O_2 systems currently being investigated. In terms of the electrolyte species involved, the four categories differ. The latter, in turn, controls the electrochemical events that occur during energy storage and release.

A lithium anode, a porous air cathode, and an electrolyte of a lithium salt mixed in a nonaqueous; organic solvent makes up the aprotic lithium-oxygen battery (LOB).[70] PolyPlus Battery Company's researchers replaced the nonaqueous solvent with an aqueous solution. They used an Ohara glass Li-ion conducting membrane to prevent the Li anode from corrosion, which eliminated the solid, insoluble discharge product of Aprotic LOB.[71] In contrast to solid Li_2O_2, using an aqueous electrolyte created soluble lithium hydroxide (LiOH) as the discharge product. This system has yet to be commercialized.

The hybrid lithium-oxygen battery was created as part of another attempt to develop an aqueous system.[72] The anode side of the storm was filled with an aprotic solvent, while the air electrode was in contact with an aqueous electrolyte in the suggested design. A solid electrolyte membrane separated the two electrolyte solutions. Because the hybrid LOB avoided direct connection between the lithium and the solid electrolyte membrane, the lithium-ion conductivity was increased. The design, however, adds complexities and results in differing diffusion kinetics in the two solutions.

No liquid electrolytes were used in the development of solid-state LOBs. Around 2010, solid-state lithium ion-conducting materials used in LOBs started to gain traction—the amount of Li_2O_2 stored in the cathode limits an all-solid-state LOB.[73] Only the aprotic structure of an LOB has shown significant rechargeability; hence, it has drawn the most attention globally.

Cathode materials: In contrast to lithium-ion batteries, the complete mass of the storage ingredient Li_2O_2 is fully generated and destroyed in every cycle in lithium-air batteries. The cathode's role is to offer a framework in which this can happen. The cathode must have high electronic conductivity and open porosity to allow for Li_2O_2 storage and O2 and Li^+ movement. Carbon cathodes are widely used in Li-O_2 batteries because of their excellent conductivity, lightweight, low cost, and adaptable structure.

Carbons are typically chosen because they have a wide surface area and pore volume in which the Li_2O_2 ORR product can be hosted. Carbon blacks like Super P, Vulcan XC 72R, and Ketjen Black were used in early lithium-air studies.[74–76] These materials have a comparatively large specific pore volume despite having a moderate surface area. Despite their many advantages, carbonaceous materials were shown to be unstable in lithium-air batteries. However, when redox mediators are used to charging Li-O_2 batteries, the degradation of the battery components is significantly reduced.[77] Carbon cathodes are more stable when redox mediators are used. When soluble discharge redox mediators are used, the products form in solution and have no intermediate interaction with the carbon, decreasing the decomposition potential. Redox mediators force the cells to charge at 3.6 V, avoiding the excessively positive potentials that cause carbon oxidation. Surface area loss from Li_2CO_3 passivation has a lower impact on the redox mediators, which would normally raise OER (oxygen evolution reaction) overpotentials.

Carbon has been replaced with alternative stable cathode materials. Peng et al. demonstrated that a cell with an Au cathode can improve cycle efficiency as a proof of concept.[78] The evolution of CO2 upon charge was significantly reduced in the absence of carbon, indicating that decomposition was significantly inhibited. Metal carbides, particularly TiC, were identified as viable cathode materials during a search for more practical alternatives.[79] According to these findings, the surface of TiC creates an oxide layer that is stable during cycling. This was supported by theoretical work that showed a TiC surface in contact with Li_2O_2 would result in a stable oxide layer termination.[80]

Amanchukwu et al. investigated the stability of polymers as binders in cathodes by measuring their reactivity with Li_2O_2.[81] The polymeric binders PAN, PVC, PVdF, PVdF-HFP, and PVP were found to be unstable to Li2O2, confirming prior research that PVdF was unstable.[82] PMMA, PTFE, and Nafion polymers, on the other hand, were shown to be chemically stable in these conditions.

Anode materials: Li-O_2 batteries are of great interest because they have theoretical specific energy that is 4–5 times higher than traditional LIBs. The use of metallic lithium as the anode, which has a very high theoretical specific capacity (~3860 mAh/g) and the highest negative reduction potential (–3.040 V versus the standard hydrogen electrode), is responsible for the high energy density. To date, work has been done to introduce lithium as a practical anode to meet the growing demand for advanced, lightweight, high-energy batteries. Still, unfortunately, all of this metallic anode's impressive characteristics come at the expense of three important factors, namely reactivity, efficiency, and safety, which prevent its successful application in rechargeable energy storage devices.

On the one hand, lithium's high Fermi energy level and negative reduction potential provide high operational voltage benefits. These features, however, render the lithium electrode chemically reactive with practically all liquid electrolyte solutions. Despite the fact that Li anodes are the preferred choice, various studies have focused on alternate anodes such as lithiated Si, Sn, Al, and carbons.

Lithiated Si anodes were mostly used to meet the anode capacity requirements. Before being utilized in Li-O_2 cells, these alternative anodes are lithiated electrochemically or chemically. It was long known that Li-O_2 cells with alternate anodes behaved less reversibly than Li-O_2 cells with Li metal anodes. This is due to the fact that these lithiated alternative anodes contain a substantially lower amount of Li, which soon deteriorates and becomes inactive due to side reactions on both the anode and cathode sides.[83] As a result, using alternate lithiated anodes (such as Li-Si) may require surface protection.

Alternative anodes can be useful in investigations focusing on the evaluation of cathode reversibility from an experimental/didactic viewpoint.[84] Protected Li-metal as the anode is, however, unavoidable for practical purposes. Several requirements must be completed to replace Li-metal anodes with safer, alternative materials such as Li-Si alloys (which have a lower active Li concentration). To keep the limited Li source operating at full capacity throughout lengthy cycling, such electrodes should be protected against side reactions, and fully reversible cathodes should be utilized. It is also critical to develop efficient prelithiation methods.

***Electrolyte materials*:** Organic carbonates were used as solvents in early Li-O_2 battery studies due to their dominance in Li-ion batteries.[85] The extreme instability of organic carbonates in Li-O_2's highly reactive environment prompted research into alternative solvents and accelerated the search for better, more stable solvents. Read et al. studied the use of ethers as electrolyte solvents, building on their previous expertise with carbonate instability.[86] They described a Li-O_2 cell that demonstrated good stability and rate capabilities using DME (dimethoxyethane) and DOL (1,3-dioxolane). Ether-based electrolytes are more stable than carbonates, owing to the ether solvents' stronger resistance to nucleophilic attack and higher cathodic stability.[87]

Without glyme-based electrolyte solutions, advancements in Li-O_2 batteries would not have been conceivable.[88] It is worth noting that the reactivity of different glymes to the oxygen species produced in Li-O_2 cells varies. Liu et al. recently published a Li-O_2 cell based on super-concentrated salt/DMSO (dimethyl sulfoxide) that demonstrated good stability and reversibility even with an unprotected Li anode.[89] Because of their supposed stability toward the superoxide radical, amide-based solvents such as N, N-dimethylacetamide, dimethylformamide (DMF), and N-methyl-2-pyrrolidone (NMP) were studied next. Unfortunately, the amide-based solvent cycle stability was discovered to be ineffective for oxygen reduction applications. Another popular polar solvent, acetonitrile, also had some success due to its stability toward superoxide.[90] However, acetonitrile did not extensively explore for Li-O_2 batteries because of its high volatility and reactivity toward lithium metal. In conclusion, practically all aprotic solvents degrade in the extremely reactive environment of LiO_2 batteries. At the metallic lithium anodes, four types of degradation mechanisms may be identified: nucleophilic attack, H atom abstraction, acid/base reactions, and reductive degradation.

Li salts are an essential component of electrolyte solutions that significantly impact oxygen reduction reaction (ORR)/oxygen evolution reaction (OER). In Li-O_2 batteries, $LiNO_3$ is one of the most studied salts. $LiNO_3$ serves a dual purpose in addition to promoting solution-mediated development. It catalyzes the OER while also protecting the carbon electrode from oxidation.[91] The highest discharge capacity was found in LiTFSI (lithium bis-trifluoromethanesulfonimide), followed by LiTf (Lithium triflate) and $LiPF_6$ (Lithium hexafluorophosphate). Because of its chemical and electrochemical stability and high ionic conductivity, LiTFSI is one of the most commonly utilized salts for Li-O_2 batteries.

16.3 BIODEGRADABLE BATTERIES

Most battery components, when disposed of in nature, have a deterioration time of over 100 years, causing serious environmental problems. The battery electrodes are made up of metal electrodes. Still, in the case of a biodegradable battery, the electrodes contain polypeptides, which alter to shift electrons with redox-active groups. This biodegradable protein can utilize for single-use purposes like wearable sensors. Lutkenhaus et al. made the first protein battery that is fully biodegradable, where both of its electrodes made with the polymer used glutamic acid units.[92]

Ordinary batteries are not all recyclable; hence, biodegradable batteries, made out of sustainable materials, have an image impact on society. Scientists developed a biodegradable battery dissolved in water, and others produced a biodegradable battery from a wood-based foam substance called aerogel. The application of these materials comes under wearable devices and car electronics.

Traditionally, the focus has been placed on durable and long-lasting electronics that have gained considerable attention recently. These electronics have colossal applications in agriculture, remediation, and sustainability.[93] The fundamental operation is similar to nonbiodegradable counterparts, and it also has anode and electrolytes. Magnesium can work as the anode material and iron as the cathode material. These materials are suitable components for biodegradability and low cost.

16.3.1 SYNTHESIS

Mg and Fe are the anode and cathode materials, respectively. Biodegradable substrate: A 2.5-um-thick chitosan film was spin-coated on a silicon wafer to serve as a biodegradable substrate. The mask was placed on the chitosan layer with paper clips as a reinforcement to present his alignment. Fe layer of 150 nm sputtered. The advantage of this fabrication approach is that it is cheap, nontoxic, and highly scalable. The fabrication process is depicted in Figure 16.3.

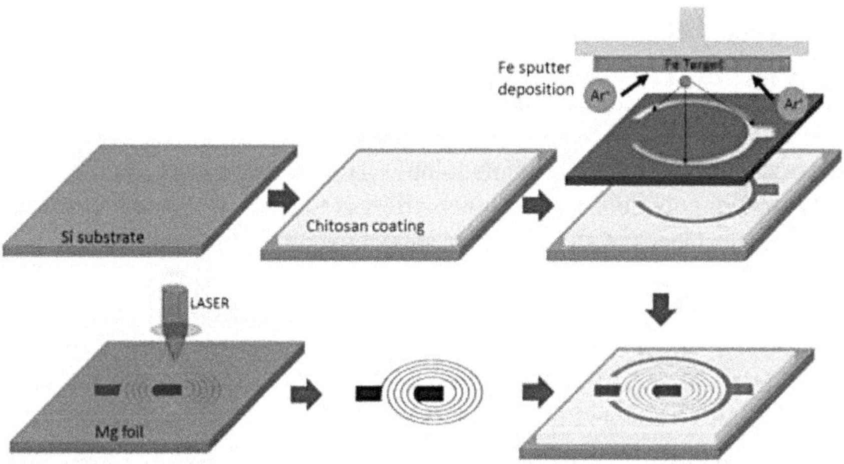

FIGURE 16.3 Fabrication process.

Source: Reproduced with permission from Ref. [93]. Copyright 2016 Elsevier.

An eco-friendly and biodegradable sodium ion secondary development through extensive material screening was followed by the synthesis assembly with an unconventional biodegradable separator, electrolyte, and package.[94] Attempts to make the fully biodegradable battery have been made; however, it has been only realized in the primary battery system, which could not be recharged.[95] Moreover, the chemical and biological evolution of the commercial battery components and their constituent materials after disposal has not been studied in depth, and it remains unclear.[94] Attempts have been made to apply a few biodegradable,[96–99] bioabsorbable,[98] or transient[100] materials as electrodes in the conventional primary battery system[101]; hence, their relatively fast decomposition can partly reduce the accumulation of overall battery waste.[102] Electrode chemistries derived from natural redox-active organics have often been explored, serving as a first step toward realizing eco-friendly batteries.

The biodegradable material chosen can entirely or partially degrade into simple compounds through the action of microorganisms like fungi, bacteria, and algae, with water and oxygen in the natural environment. For the material selection, the following criteria are adopted: (1) material should decompose or degrade naturally within 1 year, (2) the degradation products should be nontoxic, (3) the naturally stable materials and nontoxic (4) the materials are selected in the range of sodium-ion battery[94] depending upon the above criteria $Na4Fe_3(PO_4)(P_2O_7)$ (NFP)[103] and proprotein-based carbon[104,105] with binders of cellulose derivatives selected as a composite electrode, cellulose acetate (CA) as cathode, and carboxymethylcellulose (CMC) for the anode.[94]

Organic batteries which utilize organic and polymeric active material instead of metal or metal oxides represent a great deal to overcome the technical and economic restrictions of the established metal-based system.[106] Materials like quinones are one among the most popular organic active materials for electrochemical storage.[107] Polyquinoyl sulfide represents quinine-based redox-active polymers. Today, much effort is focused on organic electrodes. Organic substances of all sizes, from small molecules, from polymers to polymers, can be used as battery electrodes. Low solubility polymers tend to have a long life. On the other hand, low molecular weight organics can usually provide high energy.[108,109] Bio-derived electrochemically reactive materials are C, H, O, N, P, and S. A wide range of metallic anodes (Li, Na, Ca, Mg, and Zn) can be coupled to organic cathodes.[110] Biodegradable batteries can be designed using biodegradable fiber conductors joined with polydopamine or polypyrrole composites materials on the anode and MnO_2 as the cathode, biodegradable chitosan as the separator, and body fluid as the electrolyte.[111]

16.4 HYDROGEN STORAGE

In the present situation, it is predicted that we will run out of fossil fuels in this century. Oil can last up to 50 years, natural gas up to 53 years, and coal up to 114 years. The top five fossil fuel options are renewable, nuclear, hydrogen, biomass, and geothermal. Renewable energy is electricity from natural resources that can regenerate themselves, such as wind, solar, tidal, and hydroelectric. Renewable energy requires energy storage, transportation, and conversion into the energy valuable form for the demand. This is one of the uses of hydrogen in renewable energy infrastructure as an energy storage medium.

Different techniques are used to store hydrogen. Gas compression and liquefaction are two traditional ways of storing hydrogen.[112] However, because of the safety concerns with these systems, hydrogen storage options are being explored. Solid-state hydrogen storage materials are being developed.[113] On these surfaces, materials are absorbed or adsorbed.

The efficiency of these hydrogen storage materials is dependent on two essential aspects, considerably gravimetric and volumetric density, as the material should be both light and compact.[114] The gravimetric density of a material is measured by dividing the weight of hydrogen stored by the system's weight. In contrast, volumetric density is calculated by dividing the mass of hydrogen held by the unit volume of the system.[115] There are different approaches for hydrogen storage, including using high pressure, compressed gases, cryogenic liquid hydrogen storage, solid-state, and electrochemical.[116] The solid-state method does not satisfy gravimetric capacity, thermodynamics, and kinetics criteria. Electrochemical storage of hydrogen is a good alternative where hydrogen is generated in situ and stored easily at ambient temperature and pressure.[117]

16.4.1 ELECTROCHEMICAL HYDROGEN STORAGE

Hydrogen is the largest energy carrier, but the synthesis is done primarily to use it completely, so you need to carefully combine.[118] Conventional hydrogen storage for gaseous and liquid hydrogen has various drawbacks; therefore, electrochemical hydrogen storage is better.[119] In an electrochemical hydrogen storage process, atomic hydrogen adsorbs onto hydrogen storage material on the electrochemical decomposition of an aqueous medium.[120] In the process, molecular dissociation of hydrogen into atomic hydrogen is absent. So, it overcomes the limitations of hydrogen storage.[121] The applications of the electrochemical method are very high due to its availability and accuracy.[122] One of the essential aspects that further enhances electrochemical hydrogen storage is that hydrogen ions are stored as roaming charges.[123]

Three methods of electrochemical hydrogen storage, namely

1. **Galvanostatic method of charge–discharge:** In this method, the electrode on which the sample is deposited is first charged to the required current density. Then, the current is turned off so that the potential is relaxed to equilibrium. The electrodes are then discharged at the same current density until the potential reaches the break point, at which point the current is turned off again. At the time of charging, the potential increases slowly and continuously, then it stabilizes at specific possible values where the generation of hydrogen gas bubbles is visually observed. The discharge potential shows an accurate likely response at turn-off potential.

   Nevertheless, it is complicated to determine the starting point of hydrogen generation, so it is difficult to measure the charge capacity accurately. Therefore, the discharge capacity gives accurate results regarding the amount of hydrogen energy stored in the material.[124] A charge/discharge process is performed using a three-electrode system in room temperature electrolyte. The working electrode is charged to some specific capacity, held, and discharged to cut off potential.[125]
2. **Cyclic voltammetry, chronoamperometry, and linear voltammetric method:** Voltammetry methods fall into the electroanalytical techniques for which analytical evidence is obtained by varying the potential and determining the resulting current. Chronoamperometry is one of the simplest voltammetry methods but one of the most commonly used. In its most basic form, chronoamperometry consists of applying a single voltage step at time t_0 and then measuring the current resulting from the applied potential. The simplicity of chronoamperometry makes it an ideal analytical method as an example of basic voltammetry operations. Chronoamperometry is performed to saturate hydrogen on the film, and linear voltammetry is performed to desorb hydrogen.

   Electronic Impedance Spectroscopy: EIS measurements help identify electrochemical processes that lead to galvanostatic or cyclic voltammetry studies. The main advantage of EIS is that the data provide capacitive results, which offers excellent information about the static charge accumulation obtained by this method. The double-layer capacitance determined by the impedance measurement is the same as the double-layer charge or discharge current calculated by the cyclic voltammogram in the same potential range. The electric double layer current is equal to the product of the electric double-layer capacitance, and the scan rate is in the identical potential range as cyclic voltammetry. Hence, this electric double-layer process is directly connected to the amount of charge stored.

EIS measurement is a very effective tool for determining the interfacial properties of an electrode and measuring electron transfer on the electrode surface.[126] Hydrogen storage capacity of various metal alloys and graphene was evaluated by EIS measurement. The work done by Li and his group was essential; it assessed the high-rate discharge ability of the hybrid assembly electrode prepared of hydrogen storage alloy and reduced graphite oxide.[125,127] The hydrogen diffusivity was easily evaluated by impedance measurements.

16.4.2 MECHANISM OF ELECTROCHEMICAL HYDROGEN STORAGE

Basically, two mechanisms determine electrochemical hydrogen storage. The first is adsorption to the outside, and the second is an insertion into the bulk of the sample.[128]

The primary mechanism of electrochemical hydrogen sorption is a three-step process that involves the reactions of Volmer, Tafel, and Heyrovsky.[129] The first stage involved in the charge transfer mechanism, as shown in eq 16.2, occurs at the electrode/electrolyte interface.

$$H_2O + M + e^- \leftrightarrow MH_{ads} + OH^- \qquad (16.2)^{127}$$

Equation 16.2 represents the Volmer reaction in which water is reduced. Hydroxyl ions are generated, and even the adsorption of hydrogen atoms on the electrode surface occurs.[130] Atoms are adsorbed on the host material to form subsurface hydrogen (H_{ss}).

$$MH_{ads} \leftrightarrow MH_{ss} \qquad (16.3)^{127}$$

This subsurface hydrogen will eventually diffuse into the bulk as H_{abs}.

$$MH_{ss} \leftrightarrow MH_{abs} \qquad (16.4)^{127}$$

Sidewise, Tafel reaction occurs even where adsorbed hydrogen atoms recombine into gaseous hydrogen.[131]

$$2\ MH_{ads} \leftrightarrow H_{2(g)} + 2\ M \qquad (16.5)^{127}$$

The Heylovsky reaction occurs when a hydrogen atom forms a hydrogen molecule formed by dissociating a hydrogen atom from a water molecule.

$$MH_{ads} + H_2O + e^- \leftrightarrow M + H_{2(g)} + OH^- \qquad (16.6)^{127}$$

16.4.3 ELECTROCHEMICAL HYDROGEN STORAGE MATERIALS

16.4.3.1 HYDROGEN ADSORPTION ON PLATINUM

A system that has been extensively studied on the adsorption of hydrogen on the platinum surface exact mechanism is not yet fully understood. In acidic media, anions are adsorbed in addition to hydrogen.[132] This process significantly alters the electrochemical behavior of the system depending on the anions involved but maintains hydrogen adsorption. The type of anion does not affect hydrogen absorption. But, the role of anions in hydrogen adsorption is ignored. In the basic medium, cations are also antagonists.[133] Cation depends on absorption and desorption rates but is not affected by the hydrogen adsorption.

Generally, adsorption occurs in two steps. (1) Reaction species are electro catalyzed to form adatoms, (2) then adatoms reach active adsorption sites through surface diffusion. The rate-determining step for excellent catalysts such as Pt is the second step directly controlled by the surface structure.

Theoretically, Pt has an excellent ability to store hydrogen or direct electrical energy. Pt is too expensive to use for this purpose, except for the fact that hydrogen adsorption to porous Pt is too complex.[134]

16.4.3.2 DIFFUSION OF HYDROGEN INTO PALLADIUM

The adsorption of hydrogen on Pd is similar to that of Pt but takes an extra step in which the hydrogen diffuses deep into the mass. The hydrogen storage capacity of Pd is not limited to its surface, and Pd can easily reach the composition of $PdH_{0.8}$ by diffusing the adsorbed hydrogen in the mass. The solid-state diffusion of hydrogen in the Pd lattice structure is certainly the rate-determining step, and the capacity strongly depends on the thickness and time of the Pd. Another interesting feature is that solid-state diffusion occurs at a specific potential, similar to intercalation materials.

Hydrogen ions are tiny to diffuse within the palladium lattice, but they cause large changes in the lattice volume.[135] The hydrogenated Pd may be used as an electrochemical reference electrode. In any case, palladium is too expensive for large-scale hydrogen storage. Pd has been claimed to be not a suitable catalyst for hydrogen spillover in hydrogen storage systems.[136] However, Honarpazhouh et al. reported that Pd nanoparticles function as an active site for hydrogen adsorption in porous silicon or porous silicon/graphene oxide nano-composites.[137]

16.4.3.3 ALLOYS

The ability of an alloy to store hydrogen and hydrogen incorporated into the alloy is based on the crystal structure of the alloy. The connections between hydrogen and metal atoms at interstitial sites are primarily involved in the storage of hydrogen. In metal hydride batteries, the hydrogen storage alloy mainly functions as the negative electrode. A variety of metal alloys can be used for this purpose.

Nonmetal (B, P, and Si) alloys containing cobalt are effective catalysts for hydrolysis, improving electrochemical reversible hydrogen storage properties.[138] Synthesis of alloys is done by ball milling process. A Co-S sample with a milling period of 10 h yielded 385 mAh/g of capacity. Anode and cathode peaks are analyzed using cyclic voltammetric methods. The cathode peak is due to hydrogen adsorption on the CoS electrode surface, and the anode peak is due to the electrochemical oxidation of oxygen. Temperature-programmed desorption experiments were performed to investigate hydrogen's absorption/desorption properties.

For a long time, electrochemical hydrogen storage in alloys has been used. Similarly, the A_2B type alloy is gaining popularity because of the presence of substantial interstitial hydrogen storage sites. Ti_2Ni alloy was created using the induction melting process, and various characterizations such as SEM, TEM, XRD, and XPS were used to confirm its formation.[139] Furthermore, the electrode was prepared by pressing alloy and nickel powder under a pressure of 15 MPa to convert it into a pellet. Then, the charge–discharge measurements were done to calculate the amount of hydrogen storage. The crystalline phase of Ti_2Ni alloy had a higher hydrogen storage capacity than the amorphous phase. Mechanical milling reduced the hydrogen storage capacity of both the crystalline and amorphous phases, which they believe is due to destructed storage sites during the milling process.

The nanocrystalline phase was useful for hydrogen storage, while the amorphous phase improved the kinetics of hydrogen absorption–desorption and increased corrosion resistance. As a result, the alloy is suitable for use in Ni-MH batteries since it is corrosion-resistant, and the crystalline phase has a high hydrogen discharge efficiency of 450 mAh/g.

16.4.3.4 METAL CHALCOGENIDES

Metal sulphides from the chalcogenide family are commonly used in electrochemical hydrogen storage. MoS_2 has a well-defined structure and electrochemical behavior, and it has been extensively investigated in a variety of electrochemical systems.[140] Metal sulphides have the benefit of being able to create nanostructures with large surface areas while maintaining a rigid lattice. As a result, they can store a lot of hydrogen and charge and discharge it quickly.[141]

Metal selenides have recently gained a lot of attention because they have a better electrical conductivity than metal sulphides, which is an important need in electrochemical systems.[142] Metal selenides have extremely high electrochemical hydrogen storage capabilities.[143] The fundamental disadvantage remains, however, poor cyclability. In general, increasing the number of bigger atoms in transition metal chalcogenides allows for more flexibility in manipulating material properties by changing the 2D structure.[142]

16.4.3.5 CARBON NANOMATERIALS

Carbon nanotubes' popularity in the 1990s led to the identification of novel applications for carbon nanomaterials. Nutzenadel et al. demonstrated an electrochemical method for storing hydrogen.[144] Carbon has a low atomic mass compared to metal hydrides; hence, its theoretical specific capacity is substantially higher. However, surface adsorption is used to store hydrogen on carbon nanomaterials, and hydrogen does not diffuse deep within the solid except through graphitic layers and the 002-*d* interlayer. In any instance, carbon nanostructures' specific capacity for hydrogen storage is substantially lower than their theoretical capacities and is comparable to metal hydrides.[145] Carbon nanomaterials, on the other hand, have three distinct advantages over metal hydrides: (1) lower material cost, (2) faster hydrogen storage and release, and (3) significantly improved cyclability.

The specific capacity of hydrogen storage is related to the specific surface area because it is based on adsorption. It is important to remember that the real surface area as measured by conventional methods differs greatly from the electrochemically accessible area. For electrochemical hydrogen storage and capacitive behavior in supercapacitors, the pore size of mesoporous carbon materials is far more critical than the overall specific surface area.[146] The reactivity of carbon atoms is another important factor that is often neglected. The majority of hydrogen adsorption occurs at active sites. Charge transfer over active sites differs significantly, substantially altering the electrochemical reaction pathway. Catalysts can be used to induce hydrogen adsorption on carbon nanomaterials as an alternative to carbon reactive sites.

The mechanism of hydrogen storage in carbon nanomaterials is still debated, and most investigations have emphasized practical applications over fundamental research. Although hydrogen chemisorption appears to be the preferable method, physisorption is thought to be the dominating mechanism in some circumstances. Electrochemical hydrogen storage in SWCNTs is via physisorption, whereas chemisorption is the primary process in MWCNTs, according to Raman Spectro electrochemical investigations.[147]

Hydrogen spillover, which is a common mechanism in organic catalytic synthesis, is a possible mechanism in hydrogen storage systems.[148] In electrochemical systems, the mechanism is not exactly the normal spillover effect, but electrochemical hydrogen storage is somehow involved in some spillover effect. Note that the source of the H adatom in the electrochemical system is not H2. However, similar to the spillover mechanism, H adatoms can spill on the surface of the electrode or, in some cases, deep inside.

It is difficult to precisely estimate the capacity of hydrogen storage because metallic impurities in carbon nanostructures are electrochemically active.[149] Changing the acid treatment for carbon purification can also have a significant impact on carbon nanotube electrochemical hydrogen storage. Normal sp^2 carbons are incapable of capturing hydrogen from the electrolyte solution, and H adatoms are the result of catalytic activities of metallic impurities, edge atoms, or altered sp^2 carbon atoms. The total carbon nanostructure then has the ability to accommodate H adatoms. Carbon nanotubes, it should be noted, are not stand-alone materials in the future of electrochemical hydrogen storage.

16.4.3.6 METAL OXIDES

Metal oxides are the most popular electroactive materials for alkali metal insertion and extraction in secondary batteries, although they are not good for hydrogen storage. Like the cathode materials used in lithium batteries, metal oxides have strong cyclability for the insertion and extraction of hydrogen atoms. The rigid metal oxide structure, which remains intact when a cation is added or captured, is the reason for this. This huge lattice structure (when compared to the inserting cation) comes with a significant disadvantage: low specific capacity. In general, metal oxides have a small specific surface area, so hydrogen storage is caused by solid diffusion rather than surface adsorption. The rate capability is noticeably reduced as a result of this. Compared to crystalline materials, a mixture is better for electrochemical hydrogen storage. This also gives you the option of fine-tuning the hydrogen storage capacity. The influence of the Cu/Al ratio on the hydrogen capacity of CuO/Al2O3 nanocomposite was explored by Gholami and Salavati Niasari, as the specific capacity might be in the region of 20–240 mA h/g. Furthermore, by altering the chemical composition, the reaction potential was altered.

Perovskite-type oxides have also been proposed as potential electrochemical hydrogen storage possibilities.[150] Like other metal oxides, they have low specific capacity due to their large lattices and poor rate capability due to their low electrical conductivity and diffusion coefficient. On the other hand, their rigid lattice architectures can ensure that they are cyclable. It has been claimed that the specific capacity can be significantly increased at higher temperatures. Figure 16.4 shows structure models of hydrogen storage materials. These materials can safely store high density hydrogen at room temperature compared to gaseous and liquid hydrogen storage systems.[151]

16.5 CONCLUSION

The race for sustainable materials for energy storage is filled with both concern and inspiration. Li^+ ions components are more abundant materials. The criteria for active battery materials are green features, energy efficiency, material availability, operational lifetime, and cost. The other materials that come into the list of abundant materials are Na, A, Mg, Al, Zn, and organic compounds. These materials are more environmentally compatible, but they offer low energy densities than Li batteries. At present, lithium-sulfur batteries have good energy cycle life and rate capability. Lithium oxygen is an emerging system. These systems possess the highest theoretical energy

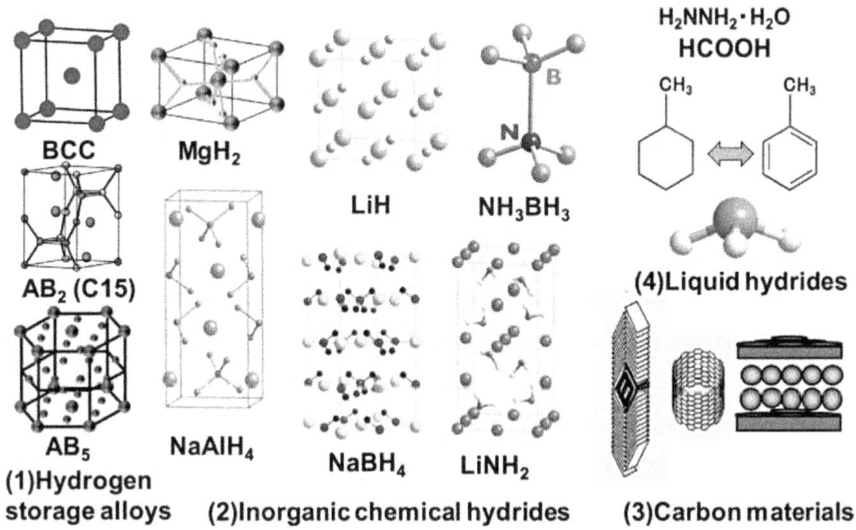

FIGURE 16.4 Structure models of hydrogen storage materials.

Source: Reproduced with permission from Ref. [151]. Copyright 2019 Elsevier.

density and cost-effectiveness; both are new frontiers in sustainable energy storage in EVs and renewable energy plans. Electrochemical hydrogen storage can be the basis for different types of power sources as well as storing hydrogen as a fuel. Hence, it will be a part of the future energy system. This chapter discusses the different sustainable battery systems with their materials for an anode, cathode, electrolyte, and sustainable stability and an eco-friendly approach for biodegradable battery and it is a synthesis method. Developing practically sustainable materials will make a promising sustainable future.

KEYWORDS

- **energy storage devices**
- **batteries**
- **sustainable materials**
- **biodegradable batteries**
- **hydrogen storage**

REFERENCES

1. di Lecce, D.; Verrelli, R.; Hassoun, J. Lithium-ion Batteries for Sustainable Energy Storage: Recent Advances towards New Cell Configurations. *Green Chem.* **2017,** *19* (15), 3442–3467. DOI: 10.1039/c7gc01328k
2. Hanada, T.; Goto, M. Cathode Recycling of Lithium-ion Batteries based on Reusable Hydrophobic Eutectic Solvents. *Green Chem.* [Online] 2022. DOI: 10.1039/D1GC04846E
3. Nitta, N.; Wu, F.; Lee, J. T.; Yushin, G. Li-ion Battery Materials: Present and Future. *Mater. Today* **2015,** *18* (5), 252–264. DOI: 10.1016/J.MATTOD.2014.10.040
4. Whittingham, M. S. Electrical Energy Storage and Intercalation Chemistry. *Science* **1976,** *192* (4244), 1126–1127. DOI: 10.1126/SCIENCE.192.4244.1126
5. Mizushima, K.; Jones, P. C.; Wiseman, P. J.; Goodenough, J. B. LixCoO2 ($0<x\leq1$): A New Cathode Material for Batteries of High Energy Density. *Solid State Ion.* **1981,** *3–4*, 171–174. DOI: 10.1016/0167-2738(81)90077-1
6. Hummel, R. E. Electronic Properties of Materials. In *Electronic Properties of Materials* [Online] 2011. DOI: 10.1007/978-1-4419-8164-6
7. Shukla, A. K.; Prem Kumar, T. Materials for Next-Generation Lithium Batteries. *Curr. Sci.* **2008,** *94*.
8. Rong, P.; Pedram, M. An Analytical Model for Predicting the Remaining Battery Capacity of Lithium-ion Batteries. *IEEE Trans. Very Large Scale Integr. Syst.* **2006,** *14* (5), 441–451. DOI: 10.1109/TVLSI.2006.876094
9. *Safety of Lithium-Ion Batteries. The European Association for Advanced Rechargeable Batteries*, 2013. www.rechargebatteries.org
10. Nanjundaswamy, K. S.; Padhi, A. K.; Goodenough, J. B.; et al. Synthesis, Redox Potential Evaluation and Electrochemical Characteristics of NASICON-Related-3D Framework Compounds. *Solid State Ion.* **1996,** *92* (1-2), 1–10. DOI: 10.1016/S0167-2738(96)00472-9
11. Recham, N.; Chotard, J. N.; Dupont, L.; et al. A 3.6 v Lithium-based Fluorosulphate Insertion Positive Electrode for Lithium-ion Batteries. *Nat. Mater.* **2010,** *9* (1), 68–74. DOI: 10.1038/NMAT2590
12. Kam, K. C.; Doeff, M. M. Aliovalent Titanium Substitution in Layered Mixed Li Ni-Mn-Co Oxides for Lithium Battery Applications. *J. Mater. Chem.* **2011,** *21* (27), 9991–9993. DOI: 10.1039/C0JM04193A
13. On the Environmental Competitiveness of Sodium-ion Batteries under a Full Life Cycle Perspective – A Cell-Chemistry Specific Modelling Approach. https://en.x-mol.com/paper/article/1459639715016638464 (accessed June 8, 2022).
14. Mekonnen, Y.; Sundararajan, A.; Sarwat, A. I. In *A Review of Cathode and Anode Materials for Lithium-ion Batteries*, Conference Proceedings - IEEE SOUTHEASTCON, 2016, July 2016. DOI: 10.1109/SECON.2016.7506639
15. Chen, X.; Shen, W.; Vo, T. T.; Cao, Z.; Kapoor, A. In *An Overview of Lithium-ion Batteries for Electric Vehicles*, 10th International Power and Energy Conference, IPEC 2012, [Online] 2012; pp 230–235. DOI: 10.1109/ASSCC.2012.6523269
16. Bhatt, M. D.; O'Dwyer, C. Recent Progress in Theoretical and Computational Investigations of Li-ion Battery Materials and Electrolytes. *Phys. Chem. Chem. Phys.* **2015,** *17* (7), 4799–4844. DOI: 10.1039/C4CP05552G

17. Liu, Y.; Liu, X.; Wang, T.; Fan, L. Z.; Jiao, L. Research and Application Progress on Key Materials for Sodium-ion Batteries. *Sustain. Energy Fuels* **2017,** *1* (5), 986–1006. DOI: 10.1039/C7SE00120G
18. Zhao, J.; Zhao, L.; Dimov, N.; Okada, S.; Nishida, T. Electrochemical and Thermal Properties of α-NaFeO2 Cathode for Na-ion Batteries. *J. Electrochem. Soc.* **2013,** *160* (5), A3077–A3081. DOI: 10.1149/2.007305JES
19. Delmas, C.; Braconnier, J. J.; Fouassier, C.; Hagenmuller, P. Electrochemical Intercalation of Sodium in NaxCoO2 Bronzes. *Solid State Ion.* **1981,** *3–4*, 165–169. DOI: 10.1016/0167-2738(81)90076-X
20. Lei, Y.; Li, X.; Liu, L.; Ceder, G. Synthesis and Stoichiometry of Different Layered Sodium Cobalt Oxides. *Chem. Mater.* **2014,** *26* (18), 5288–5296. DOI: 10.1021/CM5021788/ASSET/IMAGES/LARGE/CM-2014-021788_0009.JPEG
21. Su, D.; Wang, C.; Ahn, H. J.; Wang, G. Single Crystalline Na0.7MnO2 Nanoplates as Cathode Materials for Sodium-Ion Batteries with Enhanced Performance. *Chem. A Eur. J.* **2013,** *19* (33), 10884–10889. DOI: 10.1002/CHEM.201301563
22. Parant, J. P.; Olazcuaga, R.; Devalette, M.; Fouassier, C.; Hagenmuller, P. Sur quelques nouvelles Phases de formule NaxMnO2 ($x \leq 1$). *J. Solid State Chem.* **1971,** *3* (1), 1–11. DOI: 10.1016/0022-4596(71)90001-6
23. Carbone, L.; Greenbaum, S. G.; Hassoun, J. Lithium Sulfur and Lithium Oxygen Batteries: New Frontiers of Sustainable Energy Storage. *Sustain. Energy Fuels* **2017,** *1* (2), 228–247. DOI: 10.1039/C6SE00124F
24. Cao, Y.; Xiao, L.; Wang, W.; et al. Reversible Sodium Ion Insertion in Single Crystalline Manganese Oxide Nanowires with Long Cycle Life. *Adv Mater.* **2011,** *23* (28), 3155–3160. DOI: 10.1002/ADMA.201100904
25. Wang, S.; Wang, L.; Zhu, Z.; Hu, Z.; Zhao, Q.; Chen, J. All Organic Sodium-ion Batteries with Na4C8H 2O6. *Angew. Chem. Int. Ed.* **2014,** *53* (23), 5892–5896. DOI: 10.1002/ANIE.201400032
26. Jache, B.; Adelhelm, P. Use of Graphite as a Highly Reversible Electrode with Superior Cycle Life for Sodium-Ion Batteries by Making Use of Co-Intercalation Phenomena. *Angew. Chem. Int. Ed.* **2014,** *53* (38), 10169–10173. DOI: 10.1002/ANIE.201403734
27. Yang, J.; Zhou, X.; Wu, D.; Zhao, X.; Zhou, Z. S-Doped N-Rich Carbon Nanosheets with Expanded Interlayer Distance as Anode Materials for Sodium-Ion Batteries. *Adv. Mater.* **2017,** *29* (6). DOI: 10.1002/ADMA.201604108
28. Xiang, X.; Zhang, K.; Chen, J. Recent Advances and Prospects of Cathode Materials for Sodium-Ion Batteries. *Adv. Mater.* **2015,** *27* (36), 5343–5364. DOI: 10.1002/ADMA.201501527
29. Dai, L. Functionalization of Graphene for Efficient Energy Conversion and Storage. *Acc. Chem. Res.* **2013,** *46* (1), 31–42. DOI: 10.1021/AR300122M/ASSET/IMAGES/MEDIUM/AR-2012-00122M_0017.GIF
30. Yan, Y.; Yin, Y. X.; Guo, Y. G.; Wan, L. J. A Sandwich-Like Hierarchically Porous Carbon/Graphene Composite as a High-Performance Anode Material for Sodium-Ion Batteries. *Adv. Energy Mater.* **2014,** *4* (8), 1301584. DOI: 10.1002/AENM.201301584
31. Chevrier, V. L.; Ceder, G. Challenges for Na-ion Negative Electrodes. *J. Electrochem. Soc.* **2011,** *158* (9), A1011. DOI: 10.1149/1.3607983
32. Hwang, J. Y.; Myung, S. T.; Sun, Y. K. Sodium-ion Batteries: Present and Future. *Chem. Soc. Rev.* **2017,** *46* (12), 3529–3614. DOI: 10.1039/C6CS00776G

33. Ji, X.; Lee, K. T.; Nazar, L. F. A Highly Ordered Nanostructured Carbon-Sulphur Cathode for Lithium-Sulphur Batteries. *Nat. Mater.* **2009,** *8* (6), 500–506. DOI: 10.1038/NMAT2460
34. Han, S. C.; Song, M. S.; Lee, H.; Kim, H. S.; Ahn, H. J.; Lee, J. Y. Effect of Multiwalled Carbon Nanotubes on Electrochemical Properties of Lithium/Sulfur Rechargeable Batteries. *J. Electrochem. Soc.* **2003,** *150* (7), A889. DOI: 10.1149/1.1576766/META
35. Liu, J.; Yuan, L.; Yuan, K.; et al. SnO2 as a High-Efficiency Polysulfide Trap in Lithium–sulfur Batteries. *Nanoscale* **2016,** *8* (28), 13638–13645. DOI: 10.1039/C6NR02345B
36. Su, Y. S.; Manthiram A. Lithium-Sulphur Batteries with a Microporous Carbon Paper as a Bifunctional Interlayer. *Nat. Commun.* **2012,** *3*. DOI: 10.1038/NCOMMS2163
37. Ma, L.; Hendrickson, K. E.; Wei, S.; Archer, L. A. Nanomaterials: Science and Applications in the Lithium-sulfur Battery. *Nano Today* **2015,** *10* (3), 315–338. DOI: 10.1016/J.NANTOD.2015.04.011
38. Peng, H. J.; Huang, J. Q.; Cheng, X. B.; Zhang, Q. Review on High-Loading and High-Energy Lithium–Sulfur Batteries. *Adv. Energy Mater.* **2017,** *7* (24), 1700260. DOI: 10.1002/AENM.201700260
39. Zheng, G.; Yang, Y.; Cha, J. J.; Hong, S. S.; Cui, Y. Hollow Carbon Nanofiber-Encapsulated Sulfur Cathodes for High Specific Capacity Rechargeable Lithium Batteries. *Nano Lett.* **2011,** *11* (10), 4462–4467. DOI: 10.1021/NL2027684/SUPPL_FILE/NL2027684_SI_001.PDF
40. Wang, J. Z.; Lu, L.; Choucair, M.; Stride, J. A.; Xu, X.; Liu, H. K. Sulfur-graphene Composite for Rechargeable Lithium Batteries. *J. Power Sources* **2011,** *196* (16), 7030–7034. DOI: 10.1016/j.jpowsour.2010.09.106
41. Wang, J.; He, Y. S.; Yang, J. Sulfur-Based Composite Cathode Materials for High-Energy Rechargeable Lithium Batteries. *Adv. Mater.* **2015,** *27* (3), 569–575. DOI: 10.1002/ADMA.201402569
42. Schuster, J.; He, G.; Mandlmeier, B.; et al. Spherical Ordered Mesoporous Carbon Nanoparticles with High Porosity for Lithium-sulfur Batteries. *Angew. Chem. Int. Ed. Engl.* **2012,** *51* (15), 3591–3595. DOI: 10.1002/ANIE.201107817
43. Lee, J. T.; Zhao, Y.; Thieme, S.; et al. Sulfur-infiltrated Micro- and Mesoporous Silicon Carbide-Derived Carbon Cathode for High-Performance Lithium Sulfur Batteries. *Adv. Mater.* **2013,** *25* (33), 4573–4579. DOI: 10.1002/ADMA.201301579
44. Aetukuri, N. B.; McCloskey, B. D.; Garciá, J. M.; Krupp, L. E.; Viswanathan, V.; Luntz, A. C. Solvating Additives Drive Solution-Mediated Electrochemistry and Enhance Toroid Growth in Non-aqueous Li-O_2 Batteries. *Nat. Chem.* **2015,** *7* (1), 50–56. DOI: 10.1038/NCHEM.2132
45. Burke, C. M.; Pande, V.; Khetan, A.; Viswanathan, V.; McCloskey, B. D. Enhancing Electrochemical Intermediate Solvation Through Electrolyte Anion Selection to Increase Nonaqueous Li-O_2 Battery Capacity. *Proc. Natl. Acad. Sci. U S A* **2015,** *112* (30), 9293–9298. DOI: 10.1073/PNAS.1505728112/SUPPL_FILE/PNAS.1505728112.SAPP.PDF
46. Adams, B. D.; Radtke, C.; Black, R.; Trudeau, M. L.; Zaghib, K.; Nazar, L. F. Current Density Dependence of Peroxide Formation in the Li–O2 Battery and its Effect on Charge. *Energy Environ. Sci.* **2013,** *6* (6), 1772–1778. DOI: 10.1039/C3EE40697K
47. Xin, S.; Gu, L.; Zhao, N. H.; et al. Smaller Sulfur Molecules Promise Better Lithium'sulfur Batteries. *J. Am. Chem. Soc.* **2012,** *134* (45), 18510–18513. DOI: 10.1021/JA308170K/SUPPL_FILE/JA308170K_SI_001.PDF

48. Li, Z.; Zhang, S.; Terada, S.; et al. Promising Cell Configuration for Next-Generation Energy Storage: Li2S/Graphite Battery Enabled by a Solvate Ionic Liquid Electrolyte. *ACS Appl. Mater. Interfaces* **2016,** *8* (25), 16053–16062. DOI: 10.1021/ACSAMI.6B03736
49. Brückner, J.; Thieme, S.; Böttger-Hiller, F.; et al. Carbon-Based Anodes for Lithium Sulfur Full Cells with High Cycle Stability. *Adv. Funct. Mater.* **2014,** *24* (9), 1284–1289. DOI: 10.1002/ADFM.201302169
50. Lv, D.; Yan, P.; Shao, Y.; et al. High Performance Li-ion Sulfur Batteries Enabled by Intercalation Chemistry. *Chem. Commun.* **2015,** *51* (70), 13454–13457. DOI: 10.1039/C5CC05171A
51. Yang, Y.; McDowell, M. T.; Jackson, A.; Cha, J. J.; Hong, S. S.; Cui, Y. New Nanostructured Li2S/Silicon Rechargeable Battery with High Specific Energy. *Nano Lett.* **2010,** *10* (4), 1486–1491. DOi: 10.1021/NL100504Q/SUPPL_FILE/NL100504Q_SI_001.PDF
52. Hassoun, J.; Scrosati, B. A High-Performance Polymer Tin Sulfur Lithium Ion Battery. *Angew. Chem. Int. Ed.* **2010,** *49* (13), 2371–2374. DOI: 10.1002/ANIE.200907324
53. Duan, B.; Wang, W.; Wang, A.; Yu, Z.; Zhao, H.; Yang, Y. A New Lithium Secondary Battery System: The Sulfur/lithium-ion Battery. *J. Mater. Chem. A* **2013,** *2* (2), 308–314. DOI: 10.1039/C3TA13782A
54. Cai, W.; Zhou, J.; Li, G.; et al. B,N-Co-doped Graphene Supported Sulfur for Superior Stable Li-S Half Cell and Ge-S Full Battery. *ACS Appl. Mater. Interfaces* **2016,** *8* (41), 27679–27687. DOI: 10.1021/ACSAMI.6B08852
55. Yang, C. P.; Yin, Y. X.; Zhang, S. F.; Li, N. W.; Guo, Y. G. Accommodating Lithium into 3D Current Collectors with a Submicron Skeleton Towards Long-Life Lithium Metal Anodes. *Nat. Commun.* **2015,** *6*. DOI: 10.1038/NCOMMS9058
56. Yun, Q.; He, Y. B.; Lv, W.; et al. Chemical Dealloying Derived 3D Porous Current Collector for Li Metal Anodes. *Adv. Mater.* **2016,** *28* (32), 6932–6939. DOI: 10.1002/ADMA.201601409
57. Lu, L. L.; Ge, J.; Yang, J. N.; et al. Free-standing Copper Nanowire Network Current Collector for Improving Lithium Anode Performance. *Nano Lett.* **2016,** *16* (7), 4431–4437. DOI: 10.1021/ACS.NANOLETT.6B01581/SUPPL_FILE/NL6B01581_SI_001.PDF
58. Hagen, M.; Quiroga-González, E.; Dörfler, S.; et al. Studies on Preventing Li Dendrite Formation in Li–S Batteries by using Pre-lithiated Si Microwire Anodes. *J. Power Sources* **2014,** *248*, 1058–1066. DOI: 10.1016/J.JPOWSOUR.2013.09.144
59. Galiński, M.; Lewandowski, A.; Stepniak, I. Ionic Liquids as Electrolytes. *Electrochim. Acta.* **2006,** *51* (26), 5567–5580. DOI: 10.1016/J.ELECTACTA.2006.03.016
60. Zhao, H.; Deng, N.; Yan, J.; et al. A Review on Anode for Lithium-sulfur Batteries: Progress and Prospects. *Chem. Eng. J.* **2018,** *347*, 343–365. DOI: 10.1016/J.CEJ.2018.04.112
61. Fan, K.; Tian, Y.; Zhang, X.; Tan, J. Short Communication. *J. Electroanal. Chem.* **2016,** *760*, 80–84. DOI: 10.1016/J.JELECHEM.2015.10.020
62. Kim, H. S.; Jeong, T. G.; Kim, Y. T. Electrochemical Properties of Lithium Sulfur Battery with Silicon Anodes Lithiated by Direct Contact Method. *J. Electrochem. Sci. Technol.* **2016,** *7* (3), 228–233. DOI: 10.5229/JECST.2016.7.3.228

63. 3D Lithium Metal Embedded within Lithiophilic Porous Matrix for Stable Lithium Metal Batteries. https://www.infona.pl/resource/bwmeta1.element.elsevier-06bba9a1-e5fa-3306-bd1d-4dc69677e5e0 (accessed June 8, 2022).
64. Zang, J.; Ryu, S.; Pugno, N.; et al. Multifunctionality and Control of the Crumpling and Unfolding of Large-Area Graphene. *Nat. Mater.* **2013,** *12* (4), 321–325. DOI: 10.1038/NMAT3542
65. Rauh, R. D.; Abraham, K. M.; Pearson, G. F.; Surprenant, J. K.; Brummer, S. B. A Lithium/Dissolved Sulfur Battery with an Organic Electrolyte. *J. Electrochem. Soc.* **1979,** *126* (4), 523–527. DOI: 10.1149/1.2129079/XML
66. Tachikawa, N.; Yamauchi, K., Takashima, E.; Park, J. W.; Dokko, K.; Watanabe, M. Reversibility of Electrochemical Reactions of Sulfur Supported on Inverse Opal Carbon in Glyme–Li Salt Molten Complex Electrolytes. *Chem. Commun.* **2011,** *47* (28), 8157–8159. DOI: 10.1039/C1CC12415C
67. Wang, W.; Wang, Y.; Huang, Y.; et al. The Electrochemical Performance of Lithium-sulfur Batteries with LiClO 4 DOL/DME Electrolyte. *J. Appl. Electrochem.* **2010,** *40* (2), 321–325. DOI: 10.1007/S10800-009-9978-Z
68. Lécuyer, M.; Gaubicher, J.; Deschamps, M.; Lestriez, B.; Brousse, T.; Guyomard, D. Structural Changes of a Li/S Rechargeable Cell in Lithium Metal Polymer Technology. *JPS* **2013,** *241*, 249–254. DOI: 10.1016/J.JPOWSOUR.2013.04.119
69. Jung, K. N.; Kim, J.; Yamauchi, Y.; Park, M. S.; Lee, J. W.; Kim, J. H. Rechargeable Lithium–air Batteries: A Perspective on the Development of Oxygen Electrodes. *J. Mater. Chem. A* **2016,** *4* (37), 14050–14068. DOI: 10.1039/C6TA04510C
70. Tan, P.; Jiang, H. R.; Zhu, X. B.; et al. Advances and Challenges in Lithium-air Batteries. *Appl. Energy* **2017,** *204*, 780–806. DOI: 10.1016/J.APENERGY.2017.07.054
71. US7282295B2 - Protected Active Metal Electrode and Battery Cell Structures with Non-aqueous Interlayer Architecture, Google Patents. https://patents.google.com/patent/US7282295B2/en (accessed June 8, 2022).
72. McCloskey, B. D.; Scheffler, R.; Speidel, A.; Girishkumar, G.; Luntz, A. C. On the Mechanism of Nonaqueous Li-O2 Electrochemistry on C and its Kinetic Overpotentials: Some Implications for Li-air Batteries. *J. Phys. Chem. C* **2012,** *116* (45), 23897–23905. DOI: 10.1021/JP306680F/ASSET/IMAGES/MEDIUM/JP-2012-06680F_0008.GIF
73. Shimonishi, Y.; Zhang, T.; Imanishi, N.; et al. A Study on Lithium/Air Secondary Batteries-Stability of the NASICON-type Lithium Ion Conducting Solid Electrolyte in Alkaline Aqueous Solutions. *JPS* **2011,** *196* (11), 5128–5132. DOI: 10.1016/J.JPOWSOUR.2011.02.023
74. Ogasawara, T.; Débart, A.; Holzapfel, M.; Novák, P.; Bruce, P. G. Rechargeable LI2O2 Electrode for Lithium Batteries. *J. Am. Chem. Soc.* **2006,** *128* (4), 1390–1393. DOI: 10.1021/JA056811Q
75. Black, R.; Lee, J. H.; Adams, B.; Mims, C. A.; Nazar, L. F. The Role of Catalysts and Peroxide Oxidation in Lithium-Oxygen Batteries. *Angew. Chem. Int. Ed. Engl.* **2013,** *52* (1), 392–396. DOI: 10.1002/ANIE.201205354
76. Lu, Y. C.; Gasteiger, H. A.; Parent, M. C.; Chiloyan, V.; Shao-Horn, Y. The Influence of Catalysts on Discharge and Charge Voltages of Rechargeable Li-oxygen Batteries. *Electrochem. Solid State Lett.* **2010,** *13* (6). DOI: 10.1149/1.3363047
77. Liang, Z.; Lu, Y. C. Critical Role of Redox Mediator in Suppressing Charging Instabilities of Lithium-Oxygen Batteries. *J. Am. Chem. Soc.* **2016,** *138* (24), 7574–7583. DOI: 10.1021/JACS.6B01821/SUPPL_FILE/JA6B01821_SI_001.PDF

78. Peng, Z.; Freunberger, S. A.; Chen, Y.; Bruce, P. G. A Reversible and Higher-Rate Li-O2 Battery. *Science* **2012,** *337* (6094), 563–566. DOI: 10.1126/SCIENCE.1223985
79. Ottakam Thotiyl, M. M.; Freunberger, S. A.; Peng, Z.; Chen, Y.; Liu, Z.; Bruce, P. G. A Stable Cathode for the Aprotic Li-O2 Battery. *Nat. Mater.* **2013,** *12* (11), 1050–1056. DOI: 10.1038/NMAT3737
80. Wang, Z.; Sun, J.; Cheng, Y.; Niu, C. Adsorption and Deposition of Li2O2 on TiC{111} Surface. *J. Phys. Chem. Lett.* **2014,** *5* (21), 3919–3923. DOI: 10.1021/JZ501775A/SUPPL_FILE/JZ501775A_SI_001.PDF
81. Amanchukwu, C. V.; Harding, J. R.; Shao-Horn, Y.; Hammond, P. T. Understanding the Chemical Stability of Polymers for Lithium-air Batteries. *Chem. Mater.* **2015,** *27* (2), 550–561. DOI: 10.1021/cm5040003
82. Black, R.; Oh, S. H.; Lee, J. H.; Yim, T.; Adams, B.; Nazar, L F. Screening for Superoxide Reactivity in Li-O2 Batteries: Effect on Li2O2/LiOH Crystallization. *J. Am. Chem. Soc.* **2012,** *134* (6), 2902–2905. DOI: 10.1021/JA2111543
83. Kwak, W. J.; Shin, H. J.; Reiter, J.; et al. Understanding Problems of Lithiated Anodes in Lithium Oxygen Full-Cells. *J. Mater. Chem. A* **2016,** *4* (27), 10467–10471. DOI: 10.1039/C6TA03013K
84. Zhang, W.; Shen, Y.; Sun, D.; et al. Promoting Li2O2 Oxidation via Solvent-Assisted Redox Shuttle Process for Low Overpotential Li-O2 Battery. *Nano Energy* **2016,** *30*, 43–51. DOI: 10.1016/J.NANOEN.2016.09.031
85. Read, J.; Mutolo, K.; Ervin, M.; et al. Oxygen Transport Properties of Organic Electrolytes and Performance of Lithium/Oxygen Battery. *JElS* **2003,** *150* (10), A1351. DOI: 10.1149/1.1606454
86. Read, J. Ether-Based Electrolytes for the Lithium/Oxygen Organic Electrolyte Battery. *J. Electrochem. Soc.* **2006,** *153* (1), A96. DOI: 10.1149/1.2131827
87. McCloskey, B. D.; Bethune, D. S.; Shelby, R. M.; Girishkumar, G.; Luntz, A. C. Solvents Critical Role in Nonaqueous Lithium-Oxygen Battery Electrochemistry. *J. Phys. Chem. Lett.* **2011,** *2* (10), 1161–1166. DOI: 10.1021/JZ200352V/SUPPL_FILE/JZ200352V_SI_001.PDF
88. Sharon, D.; Hirshberg, D.; Afri, M.; Frimer, A. A.; Aurbach, D. The Importance of Solvent Selection in Li-O 2 Cells. *Chem. Commun.* **2017,** *53* (22), 3269–3272. DOI: 10.1039/C6CC09086A
89. Liu, B.; Xu, W.; Yan, P.; et al. Stabilization of Li Metal Anode in DMSO-Based Electrolytes via Optimization of Salt–Solvent Coordination for Li–O2 Batteries. *Adv. Energy Mater.* **2017,** *7* (14), 1602605. DOI: 10.1002/AENM.201602605
90. McCloskey, B. D.; Bethune, D. S.; Shelby, R. M.; et al. Limitations in Rechargeability of li-o 2 Batteries and Possible Origins. *J. Phys. Chem. Lett.* **2012,** *3* (20), 3043–3047. DOI: 10.1021/JZ301359T/SUPPL_FILE/JZ301359T_SI_001.PDF
91. Walker, W.; Giordani, V.; Uddin, J.; Bryantsev, V. S.; Chase, G. V.; Addison, D. A Rechargeable Li-O2 Battery using a Lithium Nitrate/ N,N-dimethylacetamide Electrolyte. *J. Am. Chem. Soc.* **2013,** *135* (6), 2076–2079. DOI: 10.1021/JA311518S/SUPPL_FILE/JA311518S_SI_001.PDF
92. Lutkenhaus, J. L.; Hammond, P. T. Electrochemically Enabled Polyelectrolyte Multilayer Devices: From Fuel Cells to Sensors. *Soft Matter.* **2007,** *3* (7), 804–816. DOI: 10.1039/B701203A

93. Edupuganti, V.; Solanki, R. Fabrication, Characterization, and Modeling of a Biodegradable Battery for Transient Electronics. *J. Power Sources* **2016,** *336*, 447–454. DOI: 10.1016/J.JPOWSOUR.2016.11.004
94. Lee, M. H.; Lee, J.; Jung, S. K.; et al. A Biodegradable Secondary Battery and its Biodegradation Mechanism for Eco-Friendly Energy-Storage Systems. *Adv. Mater.* **2021,** *33* (10). DOI: 10.1002/ADMA.202004902
95. Pablo Esquivel, J.; Alday, P.; Ibrahim, O. A.; et al. A Metal-Free and Biotically Degradable Battery for Portable Single-Use Applications. *Adv. Energy Mater.* **2017,** *7* (18), 1700275. DOI: 10.1002/AENM.201700275
96. Tao, H.; Hwang, S. W.; Marelli, B.; et al. Silk-based Resorbable Electronic Devices for Remotely Controlled Therapy and In Vivo Infection Abatement. *Proc. Natl. Acad. Sci. U S A* **2014,** *111* (49), 17385–17389. DOI: 10.1073/PNAS.1407743111
97. Son, D.; Lee, J.; Lee, D. J.; et al. Bioresorbable Electronic Stent Integrated with Therapeutic Nanoparticles for Endovascular Diseases. *ACS Nano* **2015,** *9* (6), 5937–5946. DOI: 10.1021/ACSNANO.5B00651
98. Boutry, C. M.; Nguyen, A.; Lawal, Q. O.; Chortos, A.; Rondeau-Gagné, S.; Bao, Z. A Sensitive and Biodegradable Pressure Sensor Array for Cardiovascular Monitoring. *Adv. Mater.* **2015,** *27* (43), 6954–6961. DOI: 10.1002/ADMA.201502535
99. Kang, S. K.; Murphy, R. K. J.; Hwang, S. W.; et al. Bioresorbable Silicon Electronic Sensors for the Brain. *Nature* **2016,** *530* (7588), 71–76. DOI: 10.1038/NATURE16492
100. Hwang, S. W.; Tao, H.; Kim, D. H.; et al. A Physically Transient Form of Silicon Electronics. *Science* **2012,** *337* (6102), 1640–1644. DOI: 10.1126/SCIENCE.1226325
101. Lei, T.; Guan, M.; Liu, J.; et al. Biocompatible and Totally Disintegrable Semiconducting Polymer for Ultrathin and Ultralightweight Transient Electronics. *Proc. Natl. Acad. Sci. U S A* **2017,** *114* (20), 5107–5112. DOI: 10.1073/PNAS.1701478114
102. Ojha, N.; Pradhan, N.; Singh, S.; et al. Evaluation of HDPE and LDPE Degradation by Fungus, Implemented by Statistical Optimization. *Sci. Rep.* **2017,** *7*. DOI: 10.1038/SREP39515
103. Fernández-Ropero, A. J.; Zarrabeitia, M.; Reynaud, M.; Rojo, T.; Casas-Cabanas, M. Toward Safe and Sustainable Batteries: Na4Fe3(PO4)2P2O7 as a Low-Cost Cathode for Rechargeable Aqueous Na-Ion Batteries. *J. Phys. Chem. C* 2018, *122* (1), 133–142. DOI: 10.1021/ACS.JPCC.7B09803/SUPPL_FILE/JP7B09803_SI_001.PDF
104. Cho, S. Y.; Yun, Y. S.; Lee, S.; et al. Carbonization of a Stable β-Sheet-Rich Silk Protein into a Pseudographitic Pyroprotein. *Nat. Commun.* **2015,** *6*. DOI: 10.1038/NCOMMS8145
105. Liu, P.; Li, Y.; Hu, Y. S.; Li, H.; Chen, L.; Huang, X. A Waste Biomass Derived Hard Carbon as a High-Performance Anode Material for Sodium-ion Batteries. *J. Mater. Chem. A* **2016,** *4* (34), 13046–13052. DOI: 10.1039/C6TA04877C
106. Friebe, C.; Lex-Balducci, A.; Schubert, U. S. Sustainable Energy Storage: Recent Trends and Developments toward Fully Organic Batteries. *ChemSusChem* **2019,** *12* (18), 4093–4115. DOI: 10.1002/CSSC.201901545
107. Miroshnikov, M.; Divya, K. P.; Babu, G.; et al. Power from Nature: Designing Green Battery Materials from Electroactive Quinone Derivatives and Organic Polymers. *J. Mater. Chem. A* **2016,** *4* (32), 12370–12386. DOI: 10.1039/C6TA03166H
108. Yu, X.; Manthiram, A. Sustainable Battery Materials for Next-Generation Electrical Energy Storage. *Adv. Energy Sustain. Res.* **2021,** *2* (5), 2000102. DOI: 10.1002/AESR.202000102

109. Liang, Y.; Yao, Y. Positioning Organic Electrode Materials in the Battery Landscape. *Joule* **2018,** *2* (9), 1690–1706. DOI: 10.1016/J.JOULE.2018.07.008
110. Chen, H.; Cong, G.; Lu, Y. C. Recent Progress in Organic Redox Flow Batteries: Active Materials, Electrolytes and Membranes. *J. Energy Chem.* **2018,** *27* (5), 1304–1325. DOI: 10.1016/J.JECHEM.2018.02.009
111. Mei, T.; Wang, C.; Liao, M.; et al. A Biodegradable and Rechargeable Fiber Battery. *J. Mater. Chem. A* **2021,** *9* (16), 10104–10109. DOI: 10.1039/D1TA01507A
112. Zhu, J.; Dai, L.; Yu, Y.; Cao, J.; Wang, L. A Direct Electrochemical Route from Oxides to TiMn2 Hydrogen Storage Alloy. *Chinese J. Chem. Eng.* **2015,** *23* (11), 1865–1870. DOI: 10.1016/j.cjche.2015.08.033
113. Chen, W.; Zhu, Y.; Yang, C.; et al. Significantly Improved Electrochemical Hydrogen Storage Properties of Magnesium Nickel Hydride Modified with Nano-nickel. *JPS* **2015,** *280*, 132–140. DOI: 10.1016/J.JPOWSOUR.2015.01.089
114. Singer, J. P.; Mayergoyz, A.; Portet, C.; Schneider, E.; Gogotsi, Y.; Fischer, J. E. Enhanced Volumetric Hydrogen Storage Capacity of Porous Carbon Powders by Forming Peels or Pellets. *Micropor. Mesopor. Mater.* **2008,** *116* (1-3), 469–472. DOI: 10.1016/J.MICROMESO.2008.05.005
115. Ramimoghadam, D.; Gray, E. M. A.; Webb, C. J. Review of Polymers of Intrinsic Microporosity for Hydrogen Storage Applications. *Int. J. Hydrog. Energy* **2016,** *41* (38), 16944–16965. DOI: 10.1016/J.IJHYDENE.2016.07.134
116. Niaz, S.; Manzoor, T.; Pandith, A. H. Hydrogen Storage: Materials, Methods and Perspectives. *Renew. Sustain. Energy Rev.* **2015,** *50*, 457–469. DOI: 10.1016/J.RSER.2015.05.011
117. Honarpazhouh, Y.; Astaraei, F. R.; Naderi, H. R.; Tavakoli, O. Electrochemical Hydrogen Storage in Pd-coated Porous Silicon/Graphene Oxide. *Int. J. Hydrog. Energy* **2016,** *28* (41), 12175–12182. DOI: 10.1016/J.IJHYDENE.2016.05.241
118. Jin, R.; Li, G.; Zhang, Z.; Yang, L. X.; Chen, G. Carbon Coated Flower like Bi_2S_3 Grown on Nickel Foam as Binder-Free Electrodes for Electrochemical Hydrogen and Li-ion Storage Capacities. *Electrochimica Acta* **2015,** *173*, 458–464. DOI: 10.1016/J.ELECTACTA.2015.05.021
119. Konda, S. K.; Chen, A. One-step Synthesis of Pd and Reduced Graphene Oxide Nanocomposites for Enhanced Hydrogen Sorption and Storage. *Electrochem. Commun.* **2015,** *60*, 148–152. DOI: 10.1016/J.ELECOM.2015.08.023
120. Guo, G. F.; Huang, H.; Xue, F. H.; et al. Electrochemical Hydrogen Storage of the Graphene Sheets Prepared by DC Arc-Discharge Method. *Surf. Coat. Technol.* **2013,** *228* (SUPPL.1), S120–S125. DOI: 10.1016/J.SURFCOAT.2012.07.016
121. Li, M. M.; Yang, C. C.; Jing, W. T.; Jin, B.; Lang, X. Y.; Jiang, Q. In Situ Grown Co3O4 on Hydrogen Storage Alloys for Enhanced Electrochemical Performance. *Int. J. Hydrog. Energy* **2016,** *21* (41), 8946–8953. DOI: 10.1016/J.IJHYDENE.2016.01.129
122. Konda, S. K.; Ostrom, C. K.; Chen, A. Synthesis and Electrochemical Study of Cd@Pd Core/Shell Nanomaterials for Hydrogen Sorption and Storage. *Int. J. Hydrog. Energy* **2015,** *46* (40), 16365–16374. DOI: 10.1016/J.IJHYDENE.2015.09.051
123. Eftekhari, A.; Fang, B. Electrochemical Hydrogen Storage: Opportunities for Fuel Storage, Batteries, Fuel Cells, and Supercapacitors. *Int. J. Hydrog. Energy* **2017,** *42* (40), 25143–25165. DOI: 10.1016/J.IJHYDENE.2017.08.103

124. Adams, B. D.; Wu, G.; Nlgro, S.; Chen, A. Facile Synthesis of Pd-Cd Nanostructures with High Capacity for Hydrogen Storage. *J. Am. Chem. Soc.* **2009,** *131* (20), 6930–6931. DOI: 10.1021/JA901798U
125. Eftekhari, A.; Fang, B. Electrochemical Hydrogen Storage: Opportunities for Fuel Storage, Batteries, Fuel Cells, and Supercapacitors. *Int. J. Hydrog. Energy* **2017,** *42* (40), 25143–25165. DOI: 10.1016/J.IJHYDENE.2017.08.103
126. Xiang, D.; Tang, R.; Su, Q.; Yin, L. Crystalline NiFex Nanoparticles Homogeneously Embedded in Ordered Mesoporous Carbon for Improved Electrochemical Hydrogen Storage Performance. *CrystEngComm* **2013,** *15* (27), 5442–5451. DOI: 10.1039/C3CE40369F
127. Kaur, M.; Pal, K. Review on Hydrogen Storage Materials and Methods from an Electrochemical Viewpoint. *J. Energy Storage* **2019,** *23*, 234–249. DOI: 10.1016/J.EST.2019.03.020
128. Chang, C.; Gao, P.; Bao, D.; et al. Ball-milling Preparation of One-dimensional Co–carbon Nanotube and Co–carbon Nanofiber Core/Shell Nanocomposites with High Electrochemical Hydrogen Storage Ability. *J. Power Sources* **2014,** *255*, 318–324. DOI: 10.1016/J.JPOWSOUR.2014.01.034
129. Bordbar, M.; Alimohammadi, T.; Khoshnevisan, B.; Khodadadi, B.; Yeganeh-Faal, A. Preparation of MWCNT/TiO2–Co Nanocomposite Electrode by Electrophoretic Deposition and Electrochemical Study of Hydrogen Storage. *Int. J. Hydrog. Energy* **2015,** *40* (31), 9613–9620. DOI: 10.1016/J.IJHYDENE.2015.05.138
130. Gholami, T.; Salavati-Niasari, M. Effects of Copper:Aluminum Ratio in CuO/Al2O3 Nanocomposite: Electrochemical Hydrogen Storage Capacity, Band Gap and Morphology. *Int. J. Hydrog. Energy* **2016,** *41* (34), 15141–15148. DOI: 10.1016/J.IJHYDENE.2016.06.191
131. Ostrom, C. K.; Chen, A. Synthesis and Electrochemical Study of Pd-based Trimetallic Nanoparticles for Enhanced Hydrogen Storage. *J. Phys. Chem. C* **2013,** *117* (40), 20456–20464. DOI: 10.1021/JP405923N/SUPPL_FILE/JP405923N_SI_001.PDF
132. Jaaf-Golze, K. Al.; Kolb, D. M.; Scherson, D. On the Voltammetry of Curves of Pt (111) in Aqueous Solutions. *J. Electroanal. Chem. Interfacial Electrochem.* **1986,** *200* (1-2), 353–362. DOI: 10.1016/0022-0728(86)90067-7
133. McCrum, I. T.; Janik, M. J. [논문]pH and Alkali Cation Effects on the Pt Cyclic Voltammogram Explained Using Density Functional Theory. *J. Phys. Chem. C Nanomater. Interfaces* **2016,** *120* (1), 457–471. DOI: 10.1021/ACS.JPCC.5B10979
134. García, G.; Koper, M. T. M. Stripping Voltammetry of Carbon Monoxide Oxidation on Stepped Platinum Single-Crystal Electrodes in Alkaline Solution. *Phys. Chem. Chem. Phys.* **2008,** *10* (25), 3802–3811. DOI: 10.1039/B803503M
135. [PDF] Palladium-Hydrogen System STRUCWRES NEAR PHASE TRANSITION AND CRITICAL POEW!3 | Semantic Scholar. https://www.semanticscholar.org/paper/Palladium-Hydrogen-System-STRUCWRES-NEAR-PHASE-AND-Lewis/5766b367d413648d9590d929cb644ce65b05b52a (accessed June 9, 2022).
136. Konda, S. K.; Ostrom, C. K.; Chen, A. Synthesis and Electrochemical Study of Cd@Pd Core/shell Nanomaterials for Hydrogen Sorption and Storage. *Int. J. Hydrog. Energy* **2015,** *46* (40), 16365–16374. DOI: 10.1016/J.IJHYDENE.2015.09.051
137. Gao, P.; Yang, S.; Xue, Z.; et al. High Energy Ball-milling Preparation of Co-B Amorphous Alloy with High Electrochemical Hydrogen Storage Ability. *J. Alloys Compd.* **2012,** *539*, 90–96. DOI: 10.1016/J.JALLCOM.2012.06.008

138. Wang, Q.; Jiao, L.; Du, H.; et al. Electrochemical Hydrogen Storage Property of Co–S Alloy Prepared by Ball-Milling Method. *Int. J. Hydrog. Energy* **2010,** *35* (15), 8357–8362. DOI: 10.1016/J.IJHYDENE.2009.12.002
139. Zhao, X.;, Zhou, J.; Shen, X.; Yang, M.; Ma, L. Structure and Electrochemical Hydrogen Storage Properties of A2B-type Ti–Zr–Ni Alloys. *Int. J. Hydrog. Energy* **2012,** *37* (6), 5050–5055. DOI: 10.1016/J.IJHYDENE.2011.12.010
140. Lukowski, M. A.; Daniel, A. S.; Meng, F.; Forticaux, A.; Li, L.; Jin, S. Enhanced Hydrogen Evolution Catalysis from Chemically Exfoliated Metallic MoS2 Nanosheets. *J. Am. Chem. Soc.* **2013,** *135* (28), 10274–10277. DOI: 10.1021/JA404523S/SUPPL_FILE/JA404523S_SI_001.PDF
141. Ye, L.; Wu, C.; Guo, W.; Xie, Y. MoS2 Hierarchical Hollow Cubic Cages Assembled by Bilayers: One-Step Synthesis and their Electrochemical Hydrogen Storage Properties. *Chem. Commun.* **2006,** (45), 4738–4740. DOI: 10.1039/B610601C
142. Eftekhari, A. Molybdenum Diselenide (MoSe2) for Energy Storage, Catalysis, and Optoelectronics. *Appl. Mater. Today* **2017,** *8*, 1–17. DOI: 10.1016/J.APMT.2017.01.006
143. Sun, Z.; Liufu, S.; Chen, X.; Chen, L. Controllable Synthesis and Electrochemical Hydrogen Storage Properties of Bi2Se3 Architectural Structures. *Chem. Commun.* **2010,** *46* (18), 3101–3103. DOI: 10.1039/B924655J
144. Nützenadel, C.; Züttel, A.; Chartouni, D.; Schlapbach, L. Electrochemical Storage of Hydrogen in Nanotube Materials. *Electrochem. Solid State Lett.* **1999,** *2* (1), 30–32. DOI: 10.1149/1.1390724/XML
145. Eftekhari, A.; Jafarkhani, P. Curly Graphene with Specious Interlayers Displaying Superior Capacity for Hydrogen Storage. *J. Phys. Chem. C* **2013,** *117* (48), 25845–25851. DOI: 10.1021/JP410044V
146. Eftekhari, A.; Fan, Z. Ordered Mesoporous Carbon and its Applications for Electrochemical Energy Storage and Conversion. *Mater. Chem. Front.* **2017,** *1* (6), 1001–1027. DOI: 10.1039/C6QM00298F
147. Martin, J. B.; Kinloch, I. A.; Dryfe, R. A. W. Are Carbon Nanotubes Viable Materials for the Electrochemical Storage of Hydrogen? *J. Phys. Chem. C* **2010,** *114* (10), 4693–4703. DOI: 10.1021/JP905203C/SUPPL_FILE/JP905203C_SI_001.PDF
148. Li, Y.; Yang, R. T. Significantly Enhanced Hydrogen Storage in Metal-organic Frameworks via Spillover. *J. Am. Chem. Soc.* **2006,** *128* (3), 726–727. DOI: 10.1021/JA056831S/SUPPL_FILE/JA056831SSI20051125_100241.PDF
149. Jones, C. P.; Jurkschat, K.; Crossley, A.; Compton, R. G.; Riehl, B. L.; Banks, C. E. Use of High-Purity Metal-Catalyst-Free Multiwalled Carbon Nanotubes to Avoid Potential Experimental Misinterpretations. *Langmuir* **2007,** *23* (18), 9501–9504. DOI: 10.1021/LA701522P/SUPPL_FILE/LA701522P-FILE002.PDF
150. Esaka, T.; Sakaguchi, H.; Kobayashi, S. Hydrogen Storage in Proton-Conductive Perovskite-Type Oxides and their Application to Nickel-Hydrogen Batteries. *Solid State Ion.* **2004,** *166* (3-4), 351–357. DOI: 10.1016/J.SSI.2003.11.023
151. Kojima, Y. Hydrogen Storage Materials for Hydrogen and Energy Carriers. *Int. J. Hydrog. Energy* **2019,** *44* (33), 18179–18192. DOI: 10.1016/j.ijhydene.2019.05.119

CHAPTER 17

Removal of Heavy Metal Ions and Magnetic Materials from Water

ARUNA JOSEPH, MARIYAM THOMAS, and ANJU K. NAIR

Department of Physics and Centre for Research, St. Teresa's College (Autonomous) Ernakulam, Kerala, India

ABSTRACT

Heavy metals are presently found in the majority of industrial effluents and are one of the highly serious ecological hazards. As a result, removing contaminants from water or wastes is crucial for human health as well as for protection of our living environment. The adsorption approach is increasingly employed by the researchers due to its good results and necessitating the development of low-cost adsorbents to remove these metal pollutants. Magnetic materials have recently become popular due to their magnetic behavior. These are being used as effective as well as cost-effective adsorbent materials for eliminating heavy metal pollutants owing to their high absorption efficiencies.

17.1 INTRODUCTION

Water plays one of the most important and vital roles in the smooth functioning of the earth's ecosystem.[1] Though industrialization contributes to the major part of the development and prosperity of a country, unscientific anthropogenic activities like deforestation, increased consumption of natural resources, unscientific waste disposal, etc., to a large scale have disrupted the natural equilibrium and ecological balance of the nature. Global warming, extinction of wild life, and environmental pollution are only some of them.

Environmental Technology and Sustainability: Physical, Chemical, and Biological Technologies for Environmental Protection. Tamara Tatrishvili, Ann Rose Abraham, & A. K. Haghi (Eds.)

Water pollution caused by the introduction of inadequately treated industrial effluents into water bodies is a very serious and the most widespread kind of water pollution. Industrial effluents from processes like tanning, metal processing, agricultural activities (such as fertilizers and pesticides), electroplating, textile, metal finishing, chemical manufacturing, glass and ceramic industries, medicinal activities, and storage batteries have produced a rapid increase in the discharge of wastewater contaminated with heavy metal ions.[2–4]

17.2 HEAVY METALS AND THEIR SOURCES

Heavy metals are the elements which are natural components of the earth crust having atomic weights between 63.5 and 180.6; having a relatively high density ranging from 3.5 to 7 g $cm^{-3}$and specific gravity greater than 5.[5,6] They can exist in surface waters in colloidal or particulate form as hydroxides, oxides, silicates, sulfides or adsorbed to clay, silica, and organic matter, displaying a high affinity for humic acids, organo-clays, and oxides coated with organic matter or in dissolved form generally as ions or unionized organometallic chelates or complexes. Heavy metal ions are characterized by their high solubility and persistence. Owing to their nonbiodegradable property, unlike organic contaminants, the heavy metal ions dissolved in aquatic environment can be absorbed by aquatic organisms and get accumulated in them. Some of the heavy metals like selenium, cobalt, copper, iron, manganese, molybdenum, vanadium, strontium, and zinc are necessary for human body to maintain the metabolic activities. But at higher concentrations, these metal ions can cause serious, adverse, and detrimental effects to environment and health due to its toxicity, which has affected millions of people across the globe so far. Also sometimes, nonessential heavy metals such as lead and cadmium interfere with essential nutrients of similar appearance, such as calcium and zinc.[2,6]

But now, anthropogenic input has exceeded the natural input. Generally, the sources of heavy metal ion into nature can be classified as point and non-point sources. Natural input falls in the second category. Human input includes both point and non-point sources. Examples of some of the non-point sources are the surface runoff from mining operations and the urban stormwater runoff containing metals from roadways and atmospheric fallout, etc. Point sources mainly consist of industrial effluents, waste sludge, and domestic wastewater containing metals from consumer products.[4]

Many heavy metals like lead, cadmium, mercury, and arsenic and their detrimental effect on environment and human health have been extensively and regularly reviewed by many international bodies like World Health Organization. Acute heavy metal exposure has been implicated in several degenerative diseases and may damage central nervous function, the cardiovascular and gastrointestinal systems, lungs, kidneys, liver, endocrine glands, and bones and even can cause cancer.[7]

Even at low concentrations, certain heavy metal ions have reported to show toxic effects on organisms. At the same time the quality of drinking water sources is being depleted day by day due to the direct and indirect discharge of wastewater contaminated with heavy metal into environment. Today, heavy metal ions are considered one of the most priority pollutants. As a consequence, many developed countries like Europe (European Drinking Water Directive by European Environmental Agency, EEA) and USA (United States Environmental Protection Agency, USEPA) have specified standards and guidance values in drinking water to minimize human and environmental exposure to heavy metals. For countries that do not have such standards, the World Health Organization publishes guidelines on the standards that should be achieved (WHO, 189) forming an authoritative basis for the setting of national regulations and standards for water safety in support of public health.[8] BIS, Indian Standards (IS 8500:1791) has specified the desirable limit of chromium (Cr^{6+}) to be 0.05 mg/L. The World Health Organization Guideline for maximum allowable concentration for Cr^{6+} is also 0.05 mg/L.[9]

Some of the common heavy metals, its sources, health risks and The Maximum Contaminated Level (MCL) standards established by different health organizations for those heavy metals in drinking water are summarized in Table 17.1.

17.3 REMOVAL OF HEAVY METALS FROM WATER

Bioaccumulation of heavy metals in food chain can cause severe toxicity to biological systems. This threat has posed a severe pressure in the separation and purification of heavy metals from wastewater. Heavy metals can enter into household uses and from commercial applications. Several technologies like precipitation, coagulation, floatation, ion exchange, membrane separation, electrochemical treatments, photocatalytic process, and adsorption have been developed so far for the removal of a large range of heavy metals

TABLE 17.1 Common Heavy Metals, its Sources, Health Risks, and The Maximum Contaminated Level (MCL) Standards.

Heavy metal	Sources	Health hazards	MCL(mg/L)
Arsenic	• Mining and smelting processes, pesticide use, and coal combustion	• Skin manifestations, visceral cancers, vascular disease • Causes toxicological and carcinogenic effects • Immunotoxic • Genotoxicity through generation of reactive oxygen species and lipid peroxidation • Causes melanosis, keratosis, and hyperpigmentation in humans • Modulation of co-receptor expression	• WHO: 8 • EEA: 0.01 • Regulation of water quality (India): 0.05
Cadmium	• Metal-ore refining • Phosphate fertilizers and waste dumping	• Found to cause of itai-itai disease in Japan (bone degeneration), osteomalacia • Lung cancer, Bronchitis, emphysema • Liver and blood damage, anemia • Serious kidney damage, renal disorder	• USEPA: 0.005 • EEA: 0.2 • Regulation of water quality (India): 0.001
Chromium (total)	• Burning of fossil fuels, production of chromates, plastic manufacturing, electroplating of metals, and extensive use in the leather and tannery industries • Volcanic eruptions, geological weathering of rocks, soils, and sediments	• Necrosis, nephritis and death in man (8 mg $kg^{-1}$of body weight as Cr^{6+}) • Carcinogen • Headache, diarrhea, nausea, vomiting • High levels of exposure cause liver and kidney damage, skin ulceration, and also affect the gastrointestinal mucosa • Reduces the rate of photosynthesis in plant species. Toxic effects on hematological problems and immune response in freshwater fish	• USEPA: 0.1 (total hexavalent and trivalent) • EEA: 0.5 • Regulation of water quality (India): 0.1
Copper	• Mining, metallurgy, and industrial applications	• Wilson disease • Mucosal irritation and corrosion • Insomnia, diarrhea, vomiting, loss of strength • Continuous exposure may lead to kidney damage and even death • Toxic to a variety of aquatic organisms even at very low concentrations • Liver and kidney damage • Phytotoxic	• USEPA: 1.0 • EEA: 3 • Regulation of water quality (India): 0.01

TABLE 17.1 *(Continued)*

Heavy metal	Sources	Health hazards	MCL(mg/L)
Lead		• Damage the fetal brain, diseases of the kidneys, gastro intestinal tract, circulatory system, reproductive system, and central nervous system • Irreversible brain damage and encephalopathic symptoms • Affects hemoglobin synthesis, tiredness, anemia	• USEPA: 0.1 • EEA: 0.5 • Regulation of water quality (India): 0.1
Mercury (inorganic)	• Volcanic eruption, weathering of rocks and soils • Extensive use of the metal in industrial applications, its mining and processing • Applications in batteries and mercury vapor lamps	• Cause of Minamata disease in Japan. Mentally disturbed and physically deformed babies were born to mothers who were exposed to toxic mercury due to consumption of contaminated fish • Blindness and double vision • Physiological stress, abortion, and tremors • Disturbs the cholesterol • Rheumatoid arthritis • Diseases of the kidneys, digestive, circulatory system, and nervous system • Highly toxic methyl mercury can cause toxic effects on the CNS	• USEPA: 0.002 • EEA: 0.001 • Regulation of water quality (India): 0.004
Nickel	• Weathering of rocks and soils the leaching of the minerals • Paint formulation and enameling industries, industrial applications such as in electroplating, automobile, and aircraft parts, batteries, coins, spark plugs, cosmetics, and stainless steel	• Decreases body weight and damage the liver and heart • Dermatitis, nausea, chronic asthma, coughing, human carcinogen • Reduction in cell growth, cancer, and nervous system damage • High phytotoxicity	• USEPA: 0.1 • EEA: 0.1 • Regulation of water quality (India): 0.1

TABLE 17.1 *(Continued)*

Heavy metal	Sources	Health hazards	MCL(mg/L)
Zinc	• Burning of coal • Mining and metallurgical processing of zinc ores and its industrial application	• Depression, lethargy • neurological signs and nervous system, lack of muscular coordination • Nausea and vomiting in children • Higher concentration of zinc may cause anemia and cholesterol problems in human beings • Phytotoxic	• USEPA: 5 • EEA: 5 • Regulation of water quality (India) 0.1

Source: Adapted from Ref. [10]

from drinking water and wastewater effluents. These remediation technologies may be classified as conventional and unconventional remediation technologies.[11–14]

17.3.1 CONVENTIONAL PROCESSES

17.3.1.1 CHEMICAL PRECIPITATION

Precipitation is one of the most applicable, widely used, and economical approaches. The mechanism of this conventional chemical precipitation process involves the addition of chemical reagents like alum, lime, limestone, iron salts, etc., to produce insoluble precipitates of heavy metals by the reaction of dissolved metals and precipitant as hydroxides, sulphides, carbonates, and phosphates, followed by the separation of the precipitated solids from the cleaned water. Removal percentage of metal ions from the solution can be improved to optimum by changing parameters like pH, temperature, initial concentration, charge of the ions, etc. It can be employed to effectively treat high metal concentration of more than 800 mg/L. The main advantages of using chemical precipitation technique are simplicity of the process, inexpensive equipment requirement, and convenient and safe operations. The most widely used precipitation technique is the hydroxide treatment obtained with lime by adjusting pH to basic conditions. However, the chemical precipitation technique requires a long time for the settling of suspended solids. Also, the process requires a large amount of chemicals and produces an excessive amount of precipitate sludge which requires additional treatment.[15,16]

17.3.1.2 COAGULATION AND FLOCCULATION

Coagulation and flocculation mechanism is a commonly used remediation method in the wastewater remediation. A chemical that acts as a coagulant is added to wastewater. After coagulation, the water is slowly and gently mixed and the flocs grow to a size which will easily settle down which is known as flocculation. After flocculation, it can be separated and removed by filtration, straining, or floatation. The coagulation-flocculation mechanism is primarily based on the measurement of the zeta potential (ζ) which serves as the criteria to define the electrostatic interaction between coagulant-flocculent agents and pollutants. Process of coagulation reduces the net surface charge of the colloidal particles and thereby stabilizes by electrostatic repulsion, whereas the process of flocculation continually increases the particle size to discrete particles through collisions and interactions with inorganic polymers formed by the organic polymers added. The main drawbacks of this process are the production of sludge and transfer of toxic compounds into solid phase.[17,18]

17.3.1.3 FLOATATION

Floatation is a mechanism originated from mineral processing which has got potential industrial applications. It is a physical separation process, where separation of heavy metals from liquid phase is achieved by bubble attachment. The main flotation processes for the removal of metal ions from solution are dissolved air flotation (DAF), ion flotation, and precipitation flotation. It is mainly employed to treat effluents with heavy metal concentration lesser than 50 ppm or greater than 130 ppm. Better removal of small particles, shorter hydraulic retention times, and low cost are the advantages which make floatation process attractive. However, low removal efficiency demands subsequent treatments to improve the removal efficiency of heavy metal. Nowadays, a combination of floatation and other physico-chemical treatments are employed for the removal of metal ions from solution.[19]

17.3.1.3.1 Ion Exchange

Ion exchange process normally involves low-cost materials and convenient operation. Metal ions in solutions are removed by cations or anions containing a water insoluble special ion exchange, like synthetic organic ion exchange resin where soluble ions from the liquid phase or electrolyte are attracted to the solid phase and other ions with the same charges are released into the solution in an equivalent amount. The cations in cationic resins such as hydrogen and sodium ions are exchanged with cations like Ni^{2+}, Cu^{2+}, and Zn^{2+} in the solutions and negative ions in the resins. The main advantages of this method are the low sludge generation and less time consumption. Ion exchange is nonselective and is highly sensitive to the pH of the solution.[20,21]

17.3.1.3.2 Membrane Filtration

Recently, in the treatment of inorganic effluents, membrane filtration has received considerable attention. Suspended solids such as heavy metals can be treated by this method. Depending on the size of the particle that can be retained, different types of membrane filtration can be employed viz., ultrafiltration, nanofiltration, and reverse osmosis, to the nature and concentration of materials present in the wastewater, pH, and temperature.[20–22]

17.3.1.4 ULTRAFILTRATION (UF)

Ultrafiltration is a pressure-driven purification process where permeable membranes with pore sizes in the range of 0.1–0.001 μm are used to water and low molecular weight substances permeate a membrane while heavy metals, macromolecules and suspended solids, organic and inorganic polymeric molecules are left behind. Ultra filtration has a smaller space requirement due to its high packing density.[23]

17.3.1.5 REVERSE OSMOSIS (RO)

In the process of reverse osmosis, pressure is exerted to the concentrated side of cellophane like semi-permeable membrane to force the solution through the membrane to separate the pure solvent to other side (dilute side) and retain the solute or rejected impurities on the concentrated side itself. The separation efficiency in reverse osmosis is dependent on solute concentration, pressure, and water flux rate as it involves a diffusive mechanism. Though RO can effectively reduce metal ions, its applications are limited by disadvantages like high operational cost and limited pH range.[24]

17.3.1.6 NANOFILTRATION (NF)

Nanofiltration is a recently developed membrane process for liquid-phase separations which is pressure-driven.

This method is efficient in the removal of heavy metals from solution at moderate to high concentrations. When applied to dilute metal waste it is either not cost-effective or ineffective. Thus it is not applied by many end-users.[9]

A number of approaches like electrochemical treatments, photocatalytic processes, new adsorbent/reactive media of mineral, organic or biological origin, etc., are being studied recently in the physico-chemical water treatment field for the development of cheaper and more effective technologies, to decrease the amount of wastewater produced as well as to improve the quality of the treated effluent.[16]

17.3.2 UNCONVENTIONAL WATER TREATMENT PROCESSES

17.3.2.1 ELECTRODIALYSIS

Electrodialysis is a membrane process in which, by applying an electric potential, ionized species in the solution are passed through an ion exchange membrane. These membranes are generally thin sheets of plastic material which are cation or anion selective, which basically means that either positive ions or negative ions will flow through. For example, cation-selective membranes are polyelectrolytes with negatively charged matter, which repels anions and permits cations to flow through. Thus, when a solution containing ionic species passes through the cell compartments, the negatively charged ions migrate toward the anode and the positively charged ions toward the cathode, crossing the anion exchange and cation exchange membrane. Studies showed that membranes with higher ion exchange capacity increase in voltage and temperature improved the cell performance whereas an increase in the flow rate decreased the separation percentage. The main advantages of this process are the ability to produce a highly concentrated stream for recovery and high separation selectivity. However, as it is a membrane process, this technique requires clean feed, careful operation, and periodic maintenance to prevent stack damages thereby increasing process costs. The main limiting factor of this method is the corrosion process and membrane replacement.[25]

17.3.2.2 ELECTROCOAGULATION

It is a technique which is governed by many physicochemical processes such as oxidation, reduction, coagulation, and adsorption. On an industrial scale, electrocoagulation is a cost-effective and easy-handling technique. Though conventionally this technique has been used for the treatment of dyes, nitrates, fluorides, and phenolic compounds from wastewater, recently, various workers have investigated electrocoagulation for the removal of heavy metals from wastewater. Electrocoagulation consists of electrodes that act as the anode and the cathode, where oxidation and reduction takes place. This technique involves the plating-out of metal ions on a cathode surface and thus metals can be recovered in the elemental metal state. However, the main limiting factors of this technique are high energy cost and corrosion of the electrodes which demands its frequent replacement.[26]

17.3.2.3 PHOTOCATALYTIC PROCESS

This is an innovative and promising technique for efficient treatment of complex wastewater systems containing both heavy metals and organic pollutants. This method explores the utilization of novel photocatalysts which consume cheap photons from the UV-near visible region to relay electrons from the organic contaminants to heavy metal ions, thus inducing both degradation of organic pollutants and recovery of metals in one-pot systems. This technique is operable at concentrations less than ppm of the target compounds.[27]

17.3.3 ADSORPTION

Nowadays, extensive research on adsorbents with preferred properties such as low cost and high metal-binding capacity is being carried out and adsorption has emerged as an effective alternative treatment. Adsorption is a well-known, highly efficient, and low-cost method for water purification with the help of suitable adsorbents. Adsorption using low-cost adsorbents is an effective and affordable method for water purification. Adsorption has many benefits over other techniques because of its sludge free clean operation and the total removal of contaminants even from the diluted solution. Adsorption process offers operational flexibility and simple design. Also adsorption is inexpensive, simple in operation, and applicable at different scales ranging from household users to large industrial plants. Adsorbents may be of various origins: biological, mineral, or organic. For the removal of heavy metals from wastewater, a number of adsorbents are being studied worldwide such as carbon nanotubes, synthetic polymers, redox reactive materials, low-cost adsorbents, biopolymers, etc. Anyway technical applications and cost-effectiveness are the major considerations in the selection of the adsorbent to treat inorganic effluent.[28–30]

"Sorption" implies the transfer of heavy metals from liquid phase to the surface of a solid and becomes bound through chemical or physical interactions. Sorption actually implies a group of processes among which include adsorption and precipitation reactions. Adsorption has recently emerged as an effective and alternate technique for the treatment of wastewater-carrying metal ions. Adsorption is a process in which a gas or liquid solute gets attached on the surface of a solid or a liquid (adsorbent) to form an atomic or molecular layer (adsorbate). The principle of adsorption is adhesion.[7]

The process of adsorption is operative in most chemical, biological, and systems. In adsorption, there are three main steps in general:[7]

1. Transport of pollutant from wastewater (bulk solution) to adsorbent surface,
2. Adsorption on the particle surface,

Transport of adsorbate within the adsorbent particle.[16]

Another important method for wastewater treatment is biosorption. It is an effective, user-friendly technique for the removal of heavy metals from industrial wastewater with special advantages like low cost, simple design, specific affinity, high availability, etc. This technique has been adopted extensively for wastewater treatment due to their low-cost characteristics. Adsorbents for biosorption are usually made out of agricultural remnants. By chemically modifying or by burning to activated carbon, biological remnants such as rice husk, pecan shells, rice straw, coconut shell, hazelnut shell, jack-fruit, maze husk, etc. can be made adsorbents after chemical modification. A study found out that maximum metal removal by these biomasses was due to the presence of cellulose, silica, lignin, carbohydrates, etc., in the adsorbents.[31]

Activated carbon is one of the most widely used candidates as adsorbent in wastewater treatment. It is usually prepared in small pellets or a powder with a highly porous and amorphous structure consisting of micro crystal-lites with a graphite lattice. It can remove a wide variety of toxic metals and contaminants. However, the use of activated carbon in wastewater treat-ment has been oriented mainly toward the removal of organic contaminants. Research efforts on the removal of inorganic contaminants, specifically metal ions by activated carbon, have been markedly limited. It is very expensive, because the higher the quality, the greater is the cost of activated carbon. Moreover, activated carbon needs complexing agents to improve its perfor-mance for the removal of inorganic pollutants. Also, activated carbon fails in reducing the concentration of pollutants at ppb levels.[28–30]

But the major problems in using these materials as adsorbents are their separation, poor adsorption efficiency, and low recyclability. But recent advancements in nanotechnology have offered better alternatives for water purification.

Magnetic nanoparticles are one of the most attractive materials in the material science field and are used in a wide range of areas such as bio medicine, MRI, data storage, magnetic fluids, etc. They are recently studied to be effective in removing pollutants from solutions. Magnetic nanomaterials are less toxic, highly stable, easy to synthesize, and can be recycled easily. Different approaches are used for the synthesis of magnetic nanoparticles. Studies show that magnetic nanoparticles perform best when size of the particle is around 8–18 nm. This also depends on source material and area of application. Bing et al. have demonstrated that the biochar is derived from anaerobically digested sludge and explored the competitive adsorption behavior of commonly coexisted Pb(II) and Cd(II) ions (Fig. 17.1).[32]

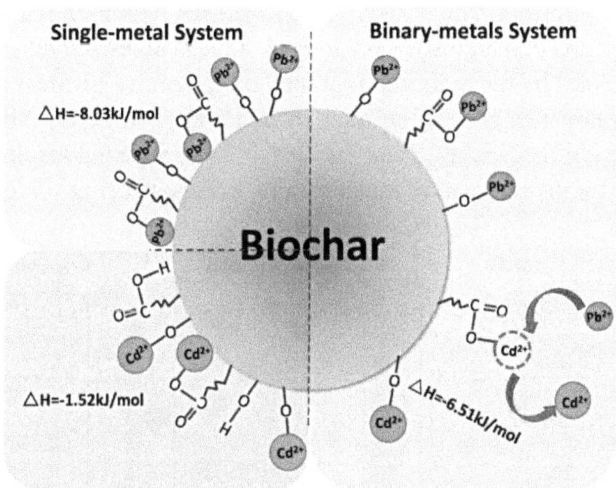

FIGURE 17.1 Adsorption capacity of single metal and bimetal system.

Source: Reproduced with permission from Ref. [32]. Copyright reserved to Elsevier.

In biologically and environmentally relevant targets, magnetic nanomaterials are used for separation and purification owing to high precision. Heavy metals like As(III) and As(V), Cr(VI), and Pd(II) are removed by using iron minerals in an effective media. Iron and iron oxide nanostructures are found highly efficient for the removal of heavy metals by adsorption or reduction.[9] Magnetic nanoparticles like iron oxide based nanoparticles as adsorbents are suitable candidates for the adsorption of heavy metal ions. The main attraction of this candidate is that the heavy metal together with magnetic nanoparticles can be easily separated from the water by an external magnetic field. [33] The desorption process of magnetic nanoparticles can also be easily achieved by pH adjustment which enables the nanoparticles to be recycled several times. Application of iron oxide-based nanomaterial is more attractive for removal of heavy metals from water because of their important features like small size, high surface area, and magnetic properties.[33] Experiment in ferrite nanoparticles with the general formula MFe_2O_4 (M = Fe, Co, Ni, Zn, Cu, Mn, etc.) as an adsorbent has been conducted in several studies.[34,35] The advantages of these particles like large surface area, low toxicity, ease of preparation, etc. make them very popular as adsorbents. These particles are capable of removing pollutants even at low concentration under varied conditions of pH and temperature. The dose of nanoparticles required is very low, making their application economical. Properties of ferrite materials strongly depend on the synthesis method. Metal ferrites are synthesized by several methods. Among these the sol-gel technique gives compositionally stoichiometric ferrites with high purity, chemical homogeneity, and crystallinity.[35] Zhao et al. have synthesized magnetic Fe_3O_4-MnO_2 nanoplates by the hydrothermal method to remove bivalent heavy metal ions from water. The adsorption rates of metal ions on Fe_3O_4–MnO_2 have in the order as: $Pb^{2+} > Cu^{2+} > Cd^{2+} > Zn^{2+} > Ni^{2+}$. The reported results have great applications for the removal of metal cations from water (Fig. 17.2).[36]

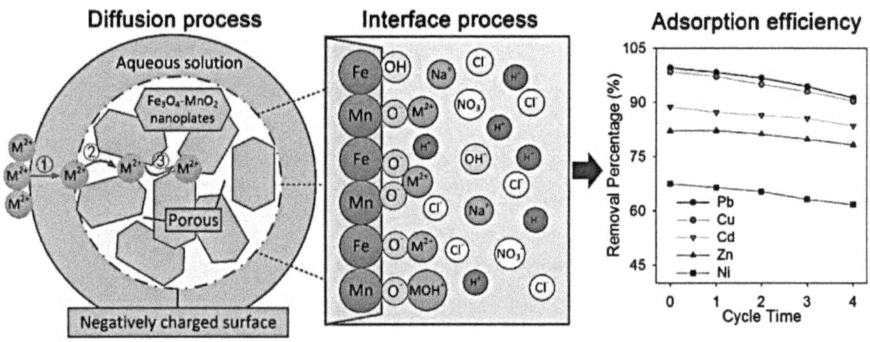

FIGURE 17.2 The adsorption capacity of Fe_2O_4–MnO_2 nanoparticles over different heavy metal ions.[36] Copyright reserved to Elsevier.

Source: Reproduced with permission from Ref. [36]. Copyright Elsevier

17.4 CONCLUSION

Science and technology post industrialization have tremendously changed our world. Scientific advancements made the impossible possible for humans. From household chores to revealing galactic secrets, scientific research and inventions helped a lot. But there is another side for these developments. A profit-driven economic and social system came into being which considered nature taken for granted. Natural resources now face a multidimensional threat. On one hand, these resources are being used up extensively for various profit-driven activities and on the other hand, the remaining reserves are polluted by the same activities. It has been found that ground water resources across the globe are getting exhausted whereas the demand for clean water to support human population, industries, and other organisms is rising. Along with that the quality of our water resources is declining steeply. Ground water depletion, pollution of fresh water reserves (such as rivers, lakes, wells, and ponds), and an increase in demand for fresh water are the main water-related problems. Governments levy taxes on water and people pay for bottled water nowadays. Organisms are adversely affected by the shortage of clean water which in turn disturbs the entire ecosystem. It is in this time we have to consider and tackle this problem. Water pollution is a serious problem of our times affecting numerous living organisms including humans. Various organic, inorganic, and biological pollutants are identified as water contaminants. Urban and industrial wastewater containing deadly chemicals such as heavy metal ions, dyes, hospital/pharmaceutical discharge, etc., which contaminate sources of drinking water results in serious health problems for humankind and living organisms.

Wastewater treatment is thus imperative for clean water recovery and its reuse. Decontamination or disinfection of wastewater is a matter of public health. Wastewater treatment and distribution of purified water will be costly and many developing countries cannot afford it because of other immediate priorities. Water purification methods should be of low cost and thus be affordable to developing nations. Thus, it is the responsibility of concerned researchers to develop an efficient and affordable purification technique. This is where adsorption technique using proper solid adsorbents comes to help. Adsorbents offer high performance; they are effective and simple to use, cost effective, and eco-friendly in application.

KEYWORDS

- **heavy metals**
- **ecological hazards**
- **adsorption**
- **cost effective**
- **magnetic materials**
- **external magnetic field**

REFERENCES

1. Agrawal, S.; Singh, N. B. Removal of Arsenic from Aqueous Solution by an Adsorbent Nickel Ferrite-Polyaniline Nanocomposite. *Indian J. Chem. Technol.* **2016,** *23* (5), 374–383.
2. Balasubramanian, N.; Kojima, T.; Basha, C. A.; Srinivasakannan, C. Removal of Arsenic from Aqueous Solution using Electrocoagulation. *J. Hazard. Mater.* **2009,** *167* (1-3), 966–969. DOI: 10.1016/j.jhazmat.2009.01.081.
3. Jian, M.; Liu, B.; Zhang, G.; Liu, R.; Zhang, X. Adsorptive Removal of Arsenic from Aqueous Solution by Zeolitic Imidazolate Framework-8 (ZIF-8) Nanoparticles. *Colloids Surf. A Physicochem. Eng. Asp.* **2015,** *465*, 67–76. DOI: 10.1016/j.colsurfa.2014.10.023.
4. Chammui, Y.; Sooksamiti, P.; Naksata, W.; Thiansem, S.; Arqueropanyo, O. anong. Removal of Arsenic from Aqueous Solution by Adsorption on Leonardite. *Chem. Eng. J.* **2014,** *240*, 202–210. DOI: 10.1016/j.cej.2013.11.083.
5. Ezeonuegbu, B. A.; et al. Agricultural Waste of Sugarcane Bagasse as Efficient Adsorbent for Lead and Nickel Removal from Untreated Wastewater: Biosorption, Equilibrium Isotherms, Kinetics and Desorption Studies. *Biotechnol. Rep.* **2021,** *30*, e00614. DOI: 10.1016/j.btre.2021.e00614.
6. Chang, Y.; et al. Electrochemical Heavy Metal Removal from Water using PVC Waste-Derived N, S Co-doped Carbon Materials. *RSC Adv.* **2020,** *7*. DOI: 10.1039/c9ra09237d.
7. Arora, R. Adsorption of Heavy Metals-A Review. *Mater. Today Proc.* **2019,** *18*, 4745–4750. DOI: 10.1016/j.matpr.2019.07.462.
8. Joseph, L.; Jun, B. M.; Flora, J. R. V.; Park, C. M.; Yoon, Y. Removal of Heavy Metals from Water Sources in the Developing World using Low-Cost Materials: A Review. *Chemosphere* **2019,** *229*, 142–159. DOI: 10.1016/j.chemosphere.2019.04.198.
9. Zhu, J.; et al. Magnetic Nanocomposites for Environmental Remediation. *Adv. Powder Technol.* **2013,** *24* (2), 459–467. DOI: 10.1016/j.apt.2012.10.012.
10. Gautam, R. K.; Sharma, S. K.; Mahiya, S.; Chattopadhyaya, M. C. CHAPTER 1: Contamination of Heavy Metals in Aquatic Media: Transport, Toxicity and Technologies for Remediation. In *Heavy Metals In Water*, 2014. DOI: 10.1039/9781782620174-00001.
11. Esfahani, A. R.; Zhang, Z.; Sip, Y. Y. L.; Zhai, L.; Sadmani, A. H. M. A. Removal of Heavy Metals from Water using Electrospun Polyelectrolyte Complex Fiber Mats. *J. Water Process Eng.* **2020,** *37*, 101438. DOI: 10.1016/j.jwpe.2020.101438.
12. Wołowiec, M.; Komorowska-Kaufman, M.; Pruss, A.; Rzepa, G.; Bajda, T. Removal of Heavy Metals and Metalloids from Water using Drinking Water Treatment Residuals as Adsorbents: A Review. *Minerals* **2019,** *9* (8), 487. DOI: 10.3390/min9080487.
13. Tang, X.; et al. Chemical Coagulation Process for the Removal of Heavy Metals from Water: A Review. *Desalin. Water Treat.* **2016,** *57* (4), 1–16. DOI: 10.1080/19443994.2014.977959.
14. Borji, H.; Ayoub, G. M.; Bilbeisi, R.; Nassar, N.; Malaeb, L. How Effective Are Nanomaterials for the Removal of Heavy Metals from Water and Wastewater? *Water Air Soil Poll.* **2020,** *231*. DOI: 10.1007/s11270-020-04681-0.
15. Singh, K.; Renu, N. A.; Agarwal, M. Methodologies for Removal of Heavy Metal Ions from Wastewater: An Overview. *Interdiscip. Environ. Rev.* **2017,** *18* (2), 124–142. DOI: 10.1504/ier.2017.10008828.
16. Barakat, M. A. New Trends in Removing Heavy Metals from Industrial Wastewater. *Arabian J. Chem.* **2011,** *4* (4), 361–377. DOI: 10.1016/j.arabjc.2010.07.019.

17. Iwuozor, K. O. Prospects and Challenges of Using Coagulation-Flocculation Method in the Treatment of Effluents. *Adv. J. Chem. A* **2019,** *2* (2), 105–127. DOI: 10.29088/sami/ajca.2019.2.105127.
18. Matilainen, A.; Vepsäläinen, M.; Sillanpää, M. Natural Organic Matter Removal by Coagulation During Drinking Water Treatment: A Review. *Adv. Colloid Interface Sci.* **2010,** *159* (2), 189–197. DOI: 10.1016/j.cis.2010.06.007.
19. Dong, J.; Xu, M. In *Application of Natural Zeolite for the Removal of Diethylenetriamine (deta) and Deta-Metal Complexes from Flotation Process Water*, XXV International Mineral Processing Congress 2010, IMPC 2010, 2010.
20. Hasanpour, M.; Hatami, M. Application of Three Dimensional Porous Aerogels as Adsorbent for Removal of Heavy Metal Ions from Water/Wastewater: A Review Study. *Adv. Colloid Interface Sci.* **2020,** *284*, 102247. DOI: 10.1016/j.cis.2020.102247.
21. Amin, M. T.; Alazba, A. A.; Manzoor, U. A Review of Removal of Pollutants from Water/Wastewater using Different Types of Nanomaterials. *Adv. Mater. Sci. Eng.* **2014,** *2014*, 825910. DOI: 10.1155/2014/825910.
22. Maity, S.; Dubey, A.; Chakraborty, S. A Review on Polypyrrole-Coated Bio-composites for the Removal of Heavy Metal Traces from Waste Water. *J. Ind. Text.* **2021**. DOI: 10.1177/1528083719871272.
23. Li, X.; Jiang, L.; Li, H. Application of Ultrafiltration Technology in Water Treatment. *IOP Conf. Ser.: Earth Environ. Sci.* **2018,** *1876*, 012009. DOI: 10.1088/1755-1315/186/3/012009.
24. Yang, Z.; et al. A Review on Reverse Osmosis and Nanofiltration Membranes for Water Purification. *Polymers* **2019,** *11* (8), 1252. DOI: 10.3390/polym11081252.
25. Gurreri, L.; Tamburini, A.; Cipollina, A.; Micale, G. Electrodialysis Applications in Wastewater Treatment for Environmental Protection and Resources Recovery: A Systematic Review on Progress and Perspectives. *Membranes* 2020, 10 (7), 146. DOI: 10.3390/membranes10070146.
26. Moussa, D. T.; El-Naas, M. H.; Nasser, M.; Al-Marri, M. J. A Comprehensive Review of Electrocoagulation for Water Treatment: Potentials and Challenges. *J. Environ. Manag.* **2017,** *186* (Pt 1), 24–41. DOI: 10.1016/j.jenvman.2016.10.032.
27. Loeb, S. K.; et al. The Technology Horizon for Photocatalytic Water Treatment: Sunrise or Sunset? *Environ. Sci. Technol.* **2019,** *53* (6), 2937–2947. DOI: 10.1021/acs.est.8b05041.
28. Li, S.; et al. Water Purification: Adsorption over Metal-Organic Frameworks. *Chinese J. Chem.* 2016, 34 (2), 175–185. DOI: 10.1002/cjoc.201500761.
29. Khaliha, S.; et al. Defective Graphene Nanosheets for Drinking Water Purification: Adsorption Mechanism, Performance, and Recovery. *FlatChem* **2021,** *29*, 100283. DOI: 10.1016/j.flatc.2021.100283.
30. Jiuhui, Q. Research Progress of Novel Adsorption Processes in Water Purification: A Review. *J. Environ. Sci.* **2008,** *20* (1), 1–13.
31. Bhattacharjee, C.; Dutta, S.; Saxena, V. K. A Review on Biosorptive Removal of Dyes and Heavy Metals from Wastewater using Watermelon Rind as Biosorbent. *Environ. Adv.* **2020,** *2*, 100007. DOI: 10.1016/j.envadv.2020.100007.
32. Ni, B. J.; et al. Competitive Adsorption of Heavy Metals in Aqueous Solution onto Biochar Derived from Anaerobically Digested Sludge. *Chemosphere* **2019,** *219*, 351–357. DOI: 10.1016/j.chemosphere.2018.12.053.

33. Aji, W. W. & Suharyadi, E. Study of Heavy Metal Ions Mn(II), Zn(II), Fe(II), Ni(II), Cu(II), and Co(II) Adsorption Using MFe2O4 (M=CO2+, Mg2+, Zn2+, Fe2+, Mn2+, and Ni2+) Magnetic Nanoparticles as Adsorbent. *Mater. Sci. Forum* **2017,** *901*, 142–148. DOI: 10.4028/www.scientific.net/MSF.901.142.
34. Kirankumar, V. S., Lakshmi Priya, V., Kavi Priya, A. & Sumathi, S. Adsorption of chromium (VI) on bismuth incorporated cobalt ferrite nanoparticles. *IOP Conf. Ser.: Mater. Sci. Eng.*, **2017,** *263* (2), 022022. DOI: 10.1088/1757-899X/263/2/022022.
35. Vazquez-Olmos, A. R.; et al. Mechanosynthesis of MFe2O4 (M = Co, Ni, and Zn) Magnetic Nanoparticles for Pb Removal from Aqueous Solution. *J. Nanomater.* **2016,** *2016*, 1–9. DOI: 10.1155/2016/9182024.
36. Zhao, J.; et al. Highly Efficient Removal of Bivalent Heavy Metals from Aqueous Systems by Magnetic Porous Fe3O4-MnO2: Adsorption Behavior and Process Study. *Chem. Eng. J.* **2016,** *304*, 737–746. DOI: 10.1016/j.cej.2016.07.003.

Index

H

V

W

X

Z

www.ingramcontent.com/pod-product-compliance
Ingram Content Group UK Ltd.
Pitfield, Milton Keynes, MK11 3LW, UK
UKHW022214190926
475702UK00003B/358